Transportation Network Modeling 2018

Transportation Research Record: Journal of the Transportation Research Board Volume 2672, Issue 48

ISBN: 978-1-5108-8254-6

Printed from e-media with permission by:

Curran Associates, Inc.
57 Morehouse Lane
Red Hook, NY 12571

Some format issues inherent in the e-media version may also appear in this print version.

Published by Sage Publications USA

ISBN (Print) 978-1-5108-8254-6
ISBN 2018 Printed Set (All Issues) 978-1-5108-7735-1

Printed by Curran Associates, Inc. (2019)

Additional copies of this publication are available from:

Curran Associates, Inc.
57 Morehouse Lane
Red Hook, NY 12571 USA
Phone: 845-758-0400
Fax: 845-758-2633
Email: curran@proceedings.com
Web: www.proceedings.com

Volume 2672 Number 48 December 2018

TRR
JOURNAL OF THE TRANSPORTATION RESEARCH BOARD

Contents

Transportation Network Modeling

Articles

* Membership as of February 2019.

TRANSPORTATION RESEARCH BOARD

CORE SPONSORS

Transportation Departments of the 50 States and the District of Columbia

Federal Government

U.S. Department of Transportation
- Federal Aviation Administration (FAA)
- Federal Highway Administration
- Federal Motor Carrier Safety Administration (FMCSA)
- Federal Railroad Administration (FRA)
- Federal Transit Administration
- National Highway Traffic Safety Administration (NHTSA)
- Office of the Assistant Secretary of Transportation for Research and Technology (OST-R)

U.S. Air Force Civil Engineer Center (AFCEC)
U.S. Coast Guard (USCG)
U.S. Department of Energy
U.S. Department of the Interior (DOI)
U.S. Environmental Protection Agency (EPA)

Nongovernmental Organizations

American Public Transportation Association (APTA)
Association of American Railroads (AAR)
California Air Resources Board

The National Academies of
SCIENCES · ENGINEERING · MEDICINE

The **National Academy of Sciences** was established in 1863 by an Act of Congress, signed by President Lincoln, as a private, nongovernmental institution to advise the nation on issues related to science and technology. Members are elected by their peers for outstanding contributions to research. Dr. Marcia McNutt is president.

The **National Academy of Engineering** was established in 1964 under the charter of the National Academy of Sciences to bring the practices of engineering to advising the nation. Members are elected by their peers for extraordinary contributions to engineering. Dr. C. D. Mote, Jr., is president.

The **National Academy of Medicine** (formerly the Institute of Medicine) was established in 1970 under the charter of the National Academy of Sciences to advise the nation on medical and health issues. Members are elected by their peers for distinguished contributions to medicine and health. Dr. Victor J. Dzau is president.

The three Academies work together as the National Academies of Sciences, Engineering, and Medicine to provide independent, objective analysis and advice to the nation and conduct other activities to solve complex problems and inform public policy decisions. The Academies also encourage education and research, recognize outstanding contributions to knowledge, and increase public understanding in matters of science, engineering, and medicine.

Learn more about the National Academies of Sciences, Engineering, and Medicine at **www.national-academies.org.**

The **Transportation Research Board** is one of seven major programs of the National Academies of Sciences, Engineering, and Medicine. The mission of the Transportation Research Board is to increase the benefits that transportation contributes to society by providing leadership in transportation innovation and progress through research and information exchange, conducted within a setting that is objective, interdisciplinary, and multimodal. The Board's varied activities annually engage about 7,000 engineers, scientists, and other transportation researchers and practitioners from the public and private sectors and academia, all of whom contribute their expertise in the public interest. The program is supported by state transportation departments, federal agencies including the component administrations of the U.S. Department of Transportation, and other organizations and individuals interested in the development of transportation.

Learn more about the Transportation Research Board at **www.TRB.org**.

Article

TRR
JOURNAL OF THE TRANSPORTATION RESEARCH BOARD

Transportation Research Record
2018, Vol. 2672(48) 1–11

DOI: 10.1177/0361198118787803
journals.sagepub.com/home/trr
SAGE

Balanced Scheduling of School Bus Trips using a Perfect Matching Heuristic

Ali Shafahi[1], Sanaz Aliari[2], and Ali Haghani[2]

Abstract

In the school bus scheduling problem, the main contributing factor to the cost is the number of buses needed for the operations. However, when subcontracting the pupils' transportation, unbalanced tours can increase the costs significantly as the lengths of some tours can exceed the daily fixed driving goal and will result in over-hour charges. This paper proposes a mixed integer programming (MIP) model and a matching-based heuristic algorithm to solve the "balanced" school bus scheduling problem with fixed start times in a multi-school setting. The heuristic solution always has the minimum number of buses as it starts with a minimal number of tours and does not alter the number of tours during its balancing stage. The effectiveness of the heuristic is tested by comparing its solutions with results from solving the MIP using commercial solvers whenever solvers could find a good solution. To illustrate the performance of the MIP and the heuristic, 11 problems were examined with different numbers of trips which are all based on two real-world problems: a California case study with 54 trips and the Howard County Public School System with 994 trips. Our numerical results indicate the proposed heuristic algorithm can find reasonable solutions in a significantly shorter time. The balanced solutions of our algorithm can save up to 16% of school bus operation costs compared with the best solution found by solvers from optimizing the MIP model after 40 hours. The balancing stage of the heuristic decreases the standard deviation of the tour durations by up to 47%.

Optimal scheduling of bus services can lead to significant savings in operating costs. An example of such a service exists in the school bus service of Howard County Public School System (HCPSS) in the state of Maryland, which serves 78 schools with a total of 994 afternoon trips. Each afternoon trip starts from a school and finishes at the last passenger's delivery point. Planning and implementation of cost-efficient school bus transportation systems are indeed a challenging task for such a large demand network. It is noteworthy that the bell-times of these schools are different and hence, it is possible to form links of sets of trips into blocks/tours that each can be served by a single bus. The problem of chaining trips, each being served by a single bus is called the school bus blocking/scheduling problem. Optimal bus blocking in a way that the number of blocks (and hence the number of required buses) is minimized can significantly reduce the fixed operating costs of a school district. Considering the very high annual fixed cost for each bus (between \$50,000 and \$100,000 in the state of Maryland), the minimization of the total number of buses is a major contributor to the cost-efficiency of such school bus operations. The variable component of cost, which is proportional to the traveled time/distance of buses, is the second largest contributor to costs and should also be minimized. Given that in blocking/scheduling problems, the trip durations are fixed, the latter can be achieved by minimizing the aggregate portions of operation times of each bus, during which they have no passengers (deadhead). Figure 1 visualizes an example of a small school bus system and one of the corresponding possible blockings.

In this paper, a variation of the scheduling/blocking problem for school buses is introduced that can further improve the cost efficiency of the service. More precisely, the considered problem is the balanced school bus scheduling problem (SBSP) in which the objective is not only to seek minimization of the total number of buses but also to provide comparable (balanced) block/tour durations for all the buses. The balanced blocking is beneficial because more savings can be gained by reducing the over-hour operating costs and avoiding extra charges.

[1]Department of Computer Science, University of Maryland, College Park, MD
[2]Department of Civil and Environmental Engineering, University of Maryland, College Park, MD

Corresponding Author:
Address correspondence to Ali Shafahi: ashafahi@umd.edu

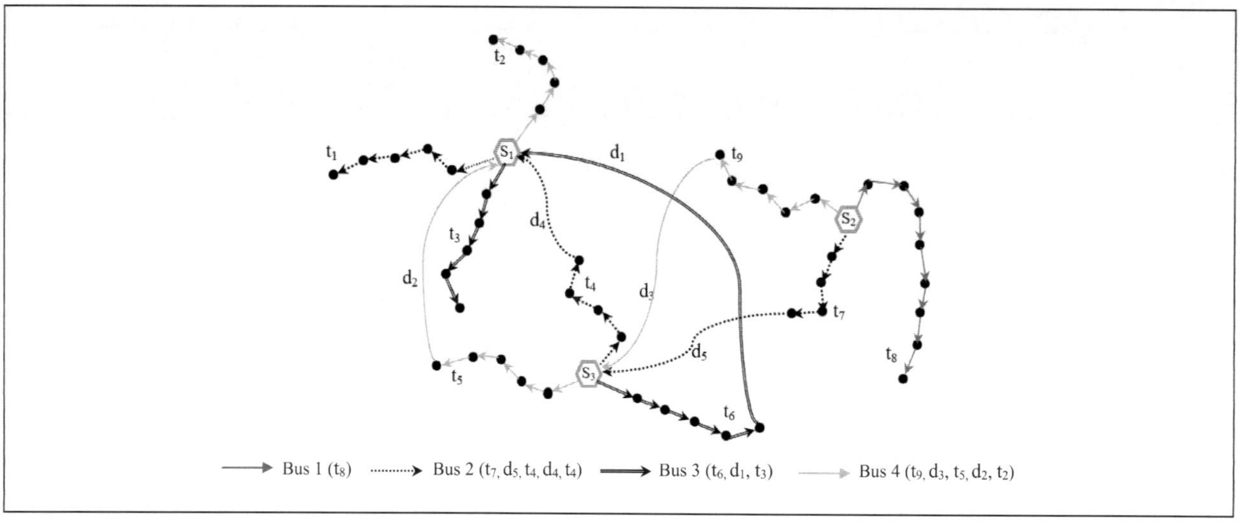

Figure 1. Illustration of an unbalanced blocking for an afternoon nine-trip, three-school scenario using four buses.

Moreover, such a balanced blocking would be more equitable for the drivers. This problem is formulated as a mixed integer programming (MIP) model. However, since finding the exact solution of such a model is an NP-hard problem, a two-stage heuristic approach is proposed to find a balanced blocking solution for large-size problems. First, the generic SBSP with fixed trip start times (without the balancing constraints) is solved to find an initial bus blocking schedule that simultaneously minimizes the number of blocks and the aggregate deadhead. Then, the schedules are made more balanced by building a bipartite graph and using a perfect matching approach.

The remainder of the paper is organized as follows. The next section presents a brief literature review of the school bus scheduling problem. Then a detailed description of the problem is provided. Subsequently, the MIP and the matching-based heuristic are introduced. Afterwards, the results from the mathematical model and the heuristic algorithm are presented and compared and the effects of the balancing stage of the heuristic on the set of tours are illustrated. Finally, the conclusions and the possible future extension of the studied problem are presented.

Literature Review

The SBSP belongs to the general class of open vehicle routing problems (OVRP) (*1*). The goal of the OVRP in general is to design a set of Hamiltonian paths (open routes) to serve customer demand. In the case of school bus scheduling, the problem consists of chaining a set of scheduled trips to an optimal number of blocks that are bound by the number of available vehicles. Moreover, there are more characteristics to the SBSP that differentiate it from the generalized version of OVRP such as consideration of heterogeneous fleet, mixed student load, multiple depots, and work-load balancing for the school bus trips. The OVRP has been extensively studied in the literature. A survey on the algorithms for solving OVRP is provided in Li et al. (*2*). These studies differ in the characteristics of the problem, the objective function and constraints used to develop the mathematical model, and the strategy used to find the solution for the model.

As for the characteristics of the SBSP, many different aspects of a practical transportation system are considered. For instance, some of the earliest research considers single depot for buses (*3*, *4*), while some others take the possibility of having multiple depots into account (*5–7*). Some of the more practical aspects of the bus scheduling problem such as fuel consumption (*5*), fixed/flexible time windows (*6*, *7*) and multiple vehicle types (*8*, *9*) have also been taken into account.

The SBSP might then be approached through different practical objectives; one might be solely interested in maximizing profit (minimizing costs), while others take measures of the equity of service (fairness) and level of service into account. The most common objective functions are the minimization of total travel distance or time, the total number of trips, and the total cost that is the combination of the aggregate travel distance and number of trips (*10–14*). Setting aside the fixed costs for each trip, the bus operational/variable cost is directly related to the total travel distance. Multi-objective cost functions, composed of conflicting objectives, are also widely used. In Corberan et al. (*15*) and Kim et al. (*16*), the two conflicting objectives of minimizing the number of buses while minimizing the maximum ride time for riders are considered. These are conflicting objectives since fewer buses would require the routes to be longer. A similar objective function is used in Li and Fu (*17*) where the authors formulate the blocking problem as an OVRP with fixed time windows (by considering each trip as a virtual stop) and present two exact solution algorithms based on the assignment problem (AP). One for the special case where the start and end service time of each virtual stop are fixed, and the buses are homogenous, and another AP-based solution algorithm for the more general case. The second approach is based on an iterative heuristic algorithm, composed of an AP-based construction algorithm and an improvement step.

Balancing of school transportation systems has been studied in the past. However, these studies have all focused on balancing the routes/trips and not the tours which are a collection of the trips. Balanced route/trip length is sometimes used in multi-objective cost functions and is motivated by the need for equity and fairness of the service (*10*). In Bowerman et al. (*10*), some equity criteria including balanced route/trip length for different buses are considered in the form of a multi-objective, nonlinear integer program. The author presents a heuristic method based on an initial clustering of students into fixed clusters of routes. Although the heuristic did not explicitly enforce the balancing restrictions, the author found that the durations of the routes are close to each other. Load balancing is another example of balancing in the school transportation operations. The goal of load balancing is to have routes which have a similar number of students. A case study of the load balancing of school bus routing in Hong Kong is presented in Li and Fu (*17*). They use a combinatorial multi-objective optimization approach and address the travel length by a modification stage to the initial solution, during which the spare capacity of buses with empty spaces are filled with some of the load of fuller buses that serve similar routes.

Matching-like algorithms have been previously used in the school bus transportation context. Perfect matching is used in Forbes et al. (*18*) to assign a morning route to an afternoon route and construct work shifts for bus drivers. In Wark and Holt (*19*) a repeated matching heuristic is introduced to solve the bus routing problem, to find optimal clustering of customers that can be served by each bus. For balancing purposes, matching has also been used. However, to the best of the authors' knowledge, it has not been used for balancing of school bus tours. In Harvey et al. (*20*), a relaxation of the matching problem, called semi-matching, is described and applied in the context of a job assignment system, aiming at balancing the load assigned to each machine. This approach cannot be directly applied to our problem since semi-matching cannot assure a compatible blocking when deadheads are considered.

Problem Statement

This study presents a mathematical model and a solution heuristic for the balanced school bus scheduling (tour generation) task for a school district with multiple schools. The inputs to this problem are the set of afternoon trips (starting from school) and their information, and a goal hour which is used for balancing and as a soft cut-off for the maximum duration/mileage of any single tour (block). The input trips (their sequence of stops, duration, and start-time) are fixed and will not be altered. The goal of the balanced scheduling problem is to assign the trips to tours such that primarily the number of tours is minimized, and secondarily the tours are mostly shorter than a given goal and hopefully balanced. The main objective is minimizing the number of tours since as previously mentioned the number of tours is the main contributor to the transportation costs of school districts as each additional tour requires the deployment of an additional bus for a school year.

The secondary objective is to ensure that the aggregate duration that tours exceed the given goal is minimized. This objective is particularly useful for the cases of transportation services of a school district that are outsourced. Usually, each contracted bus has a fixed mileage/time for its service and any additional service beyond that is charged extra at a high over-time rate. In this study, the tour duration is defined to be the student ride time (trip duration) plus the deadhead (a portion of operational times with no students on board).

It is worth noting that fixing the school trips are justified since many of the large school districts such as the Philadelphia school district continue to manually build the school trips themselves via help from their schedulers because of many local and social constraints that need to be taken into account when building every single trip. Examples of these constraints are parents' specific requests, safety considerations based on the crime level of the school district, possibilities for students to cross streets for different types of streets, etc.

An assumption implied throughout this study is that all trips can be assigned to all tours (homogenous fleet). This assumption can be relaxed by solving different problems for a different set of trips all having similar loads. Although this approach may not yield the optimal solution, it can lead to finding an approximate solution. However, it is worth noting that the homogeneous fleet assumption is not very limiting as it is valid for many school districts which outsource their logistics. Our real-world instance case, HCPSS, is an example which outsources all their regular education buses and uses a homogeneous fleet.

Methodology

In this section, a mathematical model that is used for finding the optimal solution of the balanced school bus blocking problem is presented. Then, a solution heuristic algorithm that can be used for solving large instances of the problem is introduced.

Mixed Integer Programming Model

The set $T = \{1, 2, \ldots, N\}$ of trips are given to serve the demand within the school district. It is desirable to link the trips into the minimum K tours to hopefully reduce the number of vehicles needed for operations. Since each tour is operated by one bus, the set $V = \{1, 2, \ldots, K\}$ denotes both the set of vehicles and tours which have more than one trip assigned to them. Note that since there are N trips, K would be at most $N/2$. Consequently, $K = N/2$ is set. This way of defining tours reduces the number of variables which are related to both trips and tours since the number of potential tours are cut into half by only considering tours that have more than one trip assigned to them. Due to the homogeneous-fleet assignment, the trips which are done just by themselves would not need to be assigned to a particular tour. The other inputs to the problem include the trips' travel times, the fixed start time and the pairwise deadhead matrix between all pairs of the compatible trips. A pair of trips are compatible if it is possible to chain them in a tour without violating the time window constraint. The following notations are defined to model the balanced school bus blocking problem.

x_{ij}^k: 1 if trips i and j are assigned to tour k and trip i is served right before trip j; 0, otherwise.
m_i^k: 1 if trip i has at least one trip preceding and at least one trip succeeding it in tour k;
0, otherwise.
n_i^k: 1 if trip i is served by tour k; 0, otherwise.
a_i: 1 if trip i belongs to a single-trip tour (no trips succeeding or preceding it)
b^k: 1 if tour k is being used to serve at least two trips; 0, otherwise
p^k: The period of time that tour k exceeds the goal

The notations for the inputs of our problem are listed as

T_i : Travel time duration for trip i. This is equal to the student ride time for trip i.
D_{ij} : Deadhead travel time between trips i and j.
G : Maximum goal duration
B : Maximum number of trips that can be assigned to a single tour
M_{bus}: Penalty term for dispatching each additional bus
M_G : Penalty term for exceeding the set goal duration G.

The objective function (1) aims at minimizing the total penalty costs associated with dispatching additional vehicles to satisfy the demand (M_{bus})and exceeding the goal (M_G) for a scheduling plan.

$$\text{Minimize } z = M_{\text{bus}} \cdot \text{Nbuses} + \sum_k M_G.p^k$$
$$\text{Nbuses}: = \sum_k \sum_i \sum_j x_{ij}^k - \sum_k \sum_i m_i^k + \sum_i a_i \tag{1}$$

This objective function is subject to a number of constraints. To find the penalty associated with the overtime operations, the duration of time that exceeds the set goal must be obtained. Constraints (2) calculate the amount of time that each generated tour exceeds the set goal. Which means a value is assigned to p^k which acts as a slack variable during the calculations.

$$\sum_i \sum_j D_{ij} x_{ij}^k + \sum_i \sum_j (T_i + T_j) x_{ij}^k - \sum_i T_i m_i^k - p^k \le G \quad \forall k \in V \tag{2}$$

Constraints (3) enforce that all trips must be assigned to exactly one tour. Either a tour that is only composed of one trip (when $a_i = 1$) or a tour which has more than one trip assigned to it.

$$\sum_j \sum_k x_{ij}^k + \sum_j \sum_k x_{ji}^k + a_i - \sum_k m_i^k = 1 \quad \forall i \in T \tag{3}$$

Constraints (4) and (5) find non-zero m_i^k which indicates the middle trips.

$$\sum_j x_{ij}^k + \sum_j x_{ji}^k \ge 1 + m_i^k - (1 - n_i^k) \quad \forall i \in T, \forall k \in V \tag{4}$$

$$\sum_j x_{ij}^k + \sum_j x_{ij}^k \le 2 \times n_i^k \quad \forall i \in T, \forall k \in V \tag{5}$$

Each generated tour must have exactly one starting trip and exactly one ending trip that are not middle trips.

$$\sum_i n_i^k - \sum_i m_i^k = 2 \times b^k \quad \forall k \in V \tag{6}$$

Constraints (7) find non-zero a_i which indicates that trip i belongs to a tour with a single trip.

$$\sum_j \sum_k x_{ij}^k + \sum_j \sum_k x_{ji}^k \le 2(1 - a_i) \quad \forall i \in T \tag{7}$$

Constraints (8) and (9) find non-zero b^k which indicates the non-single tours.

$$\sum_i \sum_j x_{ij}^k \le (B - 1) \cdot b^k \quad \forall k \in V \tag{8}$$

$$\sum_i \sum_j x_{ij}^k \geq b^k \quad \forall k \in V \tag{9}$$

Each trip can appear at most once as a preceding trip and once as a succeeding trip.

$$\sum_j \sum_k x_{ij}^k \leq 1 \quad \forall i \in T \tag{10}$$

$$\sum_j \sum_k x_{ji}^k \leq 1 \quad \forall i \in T \tag{11}$$

Constraints (12) reduce the symmetry of the problem resulting from homogenous fleet.

$$b^k \leq b^{k-1} \quad \forall k \in \{2, \ldots, N\} \tag{12}$$

And finally binary and non-negativity constraints of

$$a_i, m_i^k, n_i^k, b^k, x_{ij}^k \in \{0, 1\} \quad \forall i,j \in T, \quad \forall k \in V \tag{13}$$

$$p^k \geq 0 \quad \forall k \in V \tag{14}$$

This formulation is based on the work in Shafahi et al. (*21*) with some modifications to address the balancing characteristics. It is important to note that since this formulation is based on preprocessing the data, the time window and sub-tour elimination constraints are not needed. Indeed, time window constraints are implicitly considered by excluding incompatible trips from the input data. For the same reason, not many sub-tours are possible: for instance, if trip i is compatible with trip j, trip j cannot be compatible with trip i since one of them starts later than the other.

Perfect Matching Heuristic Approach

As mentioned before, OVRP are categorized as NP-hard problems, and it is impossible to obtain optimal solutions for large networks. Indeed, in our case, the commercial solver used (Xpress) for solving the proposed mathematical model yielding a suboptimal solution with a 15% optimality gap after 42 hours for a problem with 250 trips. Therefore, a heuristic approach to tackle the large size problems is proposed. The main idea of this heuristic is based on breaking the problem into two stages: first, in the blocking stage, an optimal solution will be obtained for the blocking problem using the assignment-based approach discussed in Kim et al. (*16*) which applies to the blocking problems with fixed start times. The first stage generates solutions that have the minimum number of tours, and the aggregate deadhead is also minimized. Second, in the balancing stage, the solution will be modified to satisfy the duration goal as much as possible. The modification is done through finding a minimum weighted perfect matching in a bipartite graph. The minimum weight perfect matching problem is also called the AP and can be solved in polynomial time using different algorithms such as Hungarian algorithm, modified primal simplex, and the auction algorithm. Since the balancing step is based on a perfect matching of tours to the trips, the number of tours that are minimal from the blocking step will not be altered. The perfect matching step ensures that the heuristic always gives the minimal number of tours and the main fixed-cost contributor (number of buses) is minimal. The steps of this heuristic are described below, and a summary of the steps is depicted in Figure 2.

- **Stage 1:** ***Optimal blocking***: Relax the time balancing constraint and obtain an optimal schedule for the input trips using the assignment-based approach described in Kim et al. (*16*).
- **Stage 2:** ***Balancing modification via perfect matching***:
  - **Step 2-1.** ***Bipartite graph vertex construction***: Build the bipartite graph as follows: Strip the last trip from each tour, and associate a vertex with each stripped tour, and build set A.
  - Similarly, associate a vertex with each stripped trip and build set B. Two sets of vertices of equal size are obtained ready for the bipartite graph.
  - **Step 2-2.** ***Compatible edge construction***: Add an edge between set A, the stripped tours, and set B, stripped trips, in the bipartite graph, if the linked pair has a compatible* time window. Afterward, calculate the corresponding weights**.
  - * Trips a and b are compatible if the end time of current trip (trip a) plus the deadhead to the next trip (D_{ab}) will not exceed its (trip b) start time.
  - ** The weight for a newly constructed tour is the amount of time the tour operates beyond the goal (p^{knew}).
  - **Step 2-3.** ***Minimum weight perfect matching***: Perform the Hungarian algorithm (Kuhn–Munkres algorithm) to find the optimal matching.

Granted that the problem is broken into two APs, a drastic decrease in the computation time is expected.

Computational Results

In this section, the results of both the mathematical model and the heuristic algorithm are discussed for 11 different scenarios. All scenarios are built based on two real-world problems. Nine scenarios are built by sampling from the trips of a real-world case study. For these

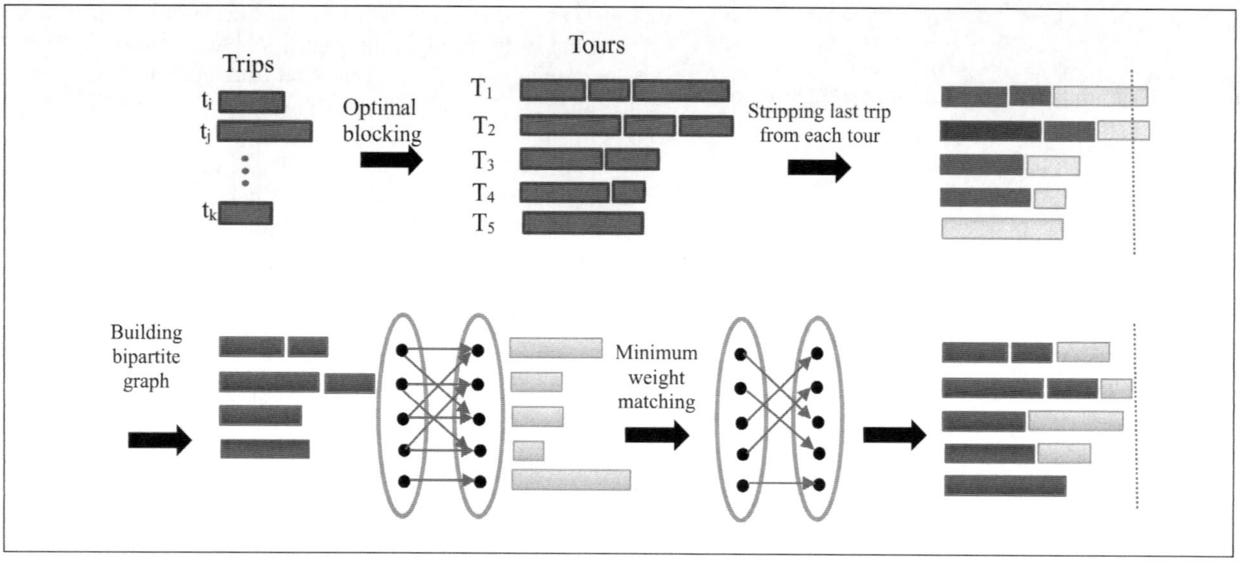

Figure 2. Schematic representation of the proposed matching-based heuristic.

nine scenarios, a scenario with n trips is built by taking n (n = 10, 20, 30, 40, 50, 100, 200, 250, 300, 500) random trips out of the maximum 994 trips of the real-world data of HCPSS. The scenario names are set such that they reflect the number of trips associated with them. The average duration of the trips is around 25 min with the standard deviation (SD) of 10 min. The minimum and the maximum duration of the HCPSS trips are 7 and 84 min, respectively. A smaller-size real world problem is also considered for a school district in California that has 54 trips (average duration: 40 min, minimum duration: 22 min, maximum duration: 66 min, and the SD: 10 min). Given the distribution of the duration for trips and deadheads, it was noticed that a 75-min goal would illustrate the balancing effects best thus, the goal is set to be equal to 75 min in all problems.

The MIP problem is solved by Xpress on a dual processor Intel Xeon computer with a total 192 GB of memory and 40 cores each 2.20 GHz. Xpress uses a combination of branch and bound and heuristic searches at the nodes to search for the optimal solution. Xpress reported the optimality of the solution for all scenarios with up to 50 trips. However, for larger problems, it either was not able to load the problem on memory or was not able to report optimality even after 40 hours.

To see the effect of the coefficients of the objective function on the performance, the model is run using two values for the penalty of additional buses, M_{bus}. A larger M_{bus} (60,000) and a smaller M_{bus} (600). The larger value is the average cost for each additional bus in Maryland. The smaller value is set such that the cost of adding a new bus would never exceed the total cost of exceeding the goal. Figure 3 illustrates various performance indicators for mainly the scenarios in which Xpress was able to find a solution. As it can be seen in the results (Figure 3), the performance of the solutions found by solving the MIP using Xpress is sensitive to the coefficient of buses. This behavior originates from the rounding errors and the precision of the machine. When the lower bound of the objective function is a large number, say in the order of 1 million, then an objective function value of 1,000,001 has a gap of %1e-6 and is practically optimal and is less than the default tolerance of Xpress (%1e-4).

The maximum running time for the scenarios with less than 200 trips is set to 10 hours and the scenarios exceeding 250 trips to 40 hours. For 300 trips, no valuable solution was found even after 40 hours as the problem size is so big that loading the problem is almost impossible. The number of variables, constraints, and nonzero elements for the 300-trip scenario is roughly 3.68 million, 77.21 thousand, and 29.87 million respectively. Each simplex iteration takes 11,743 s. Letting Xpress run for 150,000 s will only give a solution that requires 300 buses (each trip assigned to its exclusive tour). The problems with 500 trips and 994 trips were not loaded on the computer after 150,000 s.

Evaluation of the Heuristic by Comparing with Good Solution of the MIP

To compare the solution of the proposed matching heuristic, the outputs of the heuristic are compared with the solutions found by solving the MIP using Xpress. Xpress uses a combination of branch and cut and local search heuristics for solving an MIP. For smaller sized problems, Xpress can find the optimal solution. One

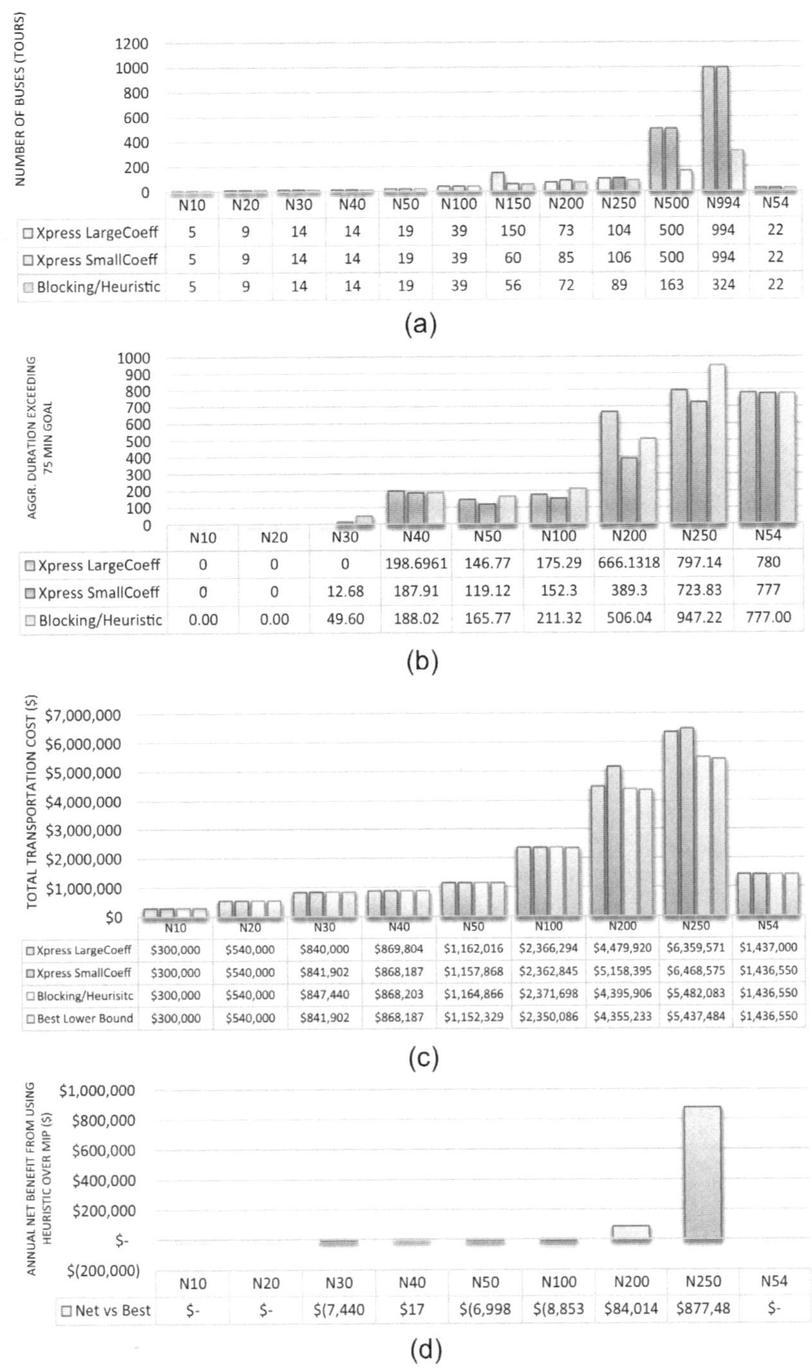

	N10	N20	N30	N40	N50	N100	N150	N200	N250	N500	N994	N54
Xpress LargeCoeff	5	9	14	14	19	39	150	73	104	500	994	22
Xpress SmallCoeff	5	9	14	14	19	39	60	85	106	500	994	22
Blocking/Heuristic	5	9	14	14	19	39	56	72	89	163	324	22

(a)

	N10	N20	N30	N40	N50	N100	N200	N250	N54
Xpress LargeCoeff	0	0	0	198.6961	146.77	175.29	666.1318	797.14	780
Xpress SmallCoeff	0	0	12.68	187.91	119.12	152.3	389.3	723.83	777
Blocking/Heuristic	0.00	0.00	49.60	188.02	165.77	211.32	506.04	947.22	777.00

(b)

	N10	N20	N30	N40	N50	N100	N200	N250	N54
Xpress LargeCoeff	$300,000	$540,000	$840,000	$869,804	$1,162,016	$2,366,294	$4,479,920	$6,359,571	$1,437,000
Xpress SmallCoeff	$300,000	$540,000	$841,902	$868,187	$1,157,868	$2,362,845	$5,158,395	$6,468,575	$1,436,550
Blocking/Heurisitc	$300,000	$540,000	$847,440	$868,203	$1,164,866	$2,371,698	$4,395,906	$5,482,083	$1,436,550
Best Lower Bound	$300,000	$540,000	$841,902	$868,187	$1,152,329	$2,350,086	$4,355,233	$5,437,484	$1,436,550

(c)

	N10	N20	N30	N40	N50	N100	N200	N250	N54
Net vs Best	$-	$-	$(7,440	$17	$(6,998	$(8,853	$84,014	$877,48	$-

(d)

Figure 3. Comparison of (*a*) number of buses needed, (*b*) aggregate minutes exceeding the 75-min goal, (*c*) annual transportation costs, and (*d*) net annual benefit/cost of using the heuristic compared with the best solution found by solving the MIP using Xpress for different scenarios.

important measure and contributor to the overall transportation cost is the number of buses/tours. The number of buses for the different scenarios are depicted in Figure 3*a*. We can see that for cases in which the problem size is small/medium (i.e., the number of trips is less than 100), the Xpress solver can find the minimum number of buses. However, as the problem size exceeds 100 trips, the MIP solver is no longer able to find the minimum number of buses during its set maximum runtime. For the scenarios with 500 trips or more in which the MIP solver was not loaded within a 150,000 second time limit, the number of tours was set equal to the number of trips. Based on these results, it is observable that the MIP solver loses its value very soon and is only practical for small to medium size scenarios. Thus, it cannot be used for the HCPSS problem as it will not give any "good" solutions within any foreseeable time. However, it can find the least number of buses for the small size California problem.

The second contributor to cost is the excess minutes (Figure 3*b*). The transportation authorities have to pay extra hour fees for those. Comparing the aggregate over time minutes of the heuristic solutions with the MIP solutions for all scenarios is not fair since in many of the scenarios the Xpress was not able to find a solution with minimum number of tours within the set time limit. At the extreme case, if each bus just makes one trip then there will be no excess minutes, but obviously, the overall cost is a lot more. To make the comparison fair, it is only focused on the scenarios in which the MIP solver (Xpress) was able to find the minimum number of tours (N10, N20, N30, N40, N50, N100, N54) or its number of tours was very close to the minimum number (N200). As it can be seen in Figure 3*b*, the heuristics solutions are comparable with the solutions from the MIP solution. A better measure to compare the solutions of the heuristic with those of the MIP might be the annual cost of each solution that combines the number of buses and the over-hour costs.

Assuming an extra-hour rate of $50/hr and 180 school days per year, the cost for each additional minute per year is $150. Also, assuming $60,000, the overall cost of the solutions gained from the heuristic and MIPs as well as the best lower bound found by Xpress are summarized in Figure 3*c*. Comparison of the best lower bounds of the MIP found by Xpress and the heuristic solutions indicates a maximum gap of 1.1% and an average gap of 0.5% which accentuates the effectiveness of the proposed heuristic. The net annual benefit/cost of using the heuristic over the best result from the MIP is depicted in Figure 3*d*. As it can be seen, for those scenarios which are small enough that the MIP solver can find a good solution, the additional annual cost of the heuristic's solution is always less than $10,000. In the California example (N54), the heuristic finds the optimal solution. However, as the problem size increases, the savings from the heuristic increase. The heuristic can find a solution that has about 16% less cost for the scenario with 250 trips. For larger scenarios, the MIP solver cannot find any reasonable solution at all even when it is given a 150,000 s time limit.

Another measure that shows how well the heuristic performs is the number of tours exceeding the goal. This measurement is interesting as over-hour costs have to be paid for each of the tours exceeding the goal. If there is a fixed cost component to over-hour costs (such as having a one-hour minimum duration for over-hours), this measurement turns into a significant one. For example, if two trips exceed the goal by 5 min, the transportation district might need to purchase an additional hour for each of the buses due to contractual obligations. While this measurement is not optimized, it is still exciting to see that the heuristic performs okay on this measurement. As illustrated in Figure 4, the heuristic's performance is close to the MIP solutions in all scenarios except for scenario N250 for a 75 minute goal.

How the Heuristic Affects the Schedule's Key Performance Indicators (KPI)

The number of buses is a critical KPI. Due to the optimality of the solution during the blocking stage, it is known that the heuristic provides the minimum number

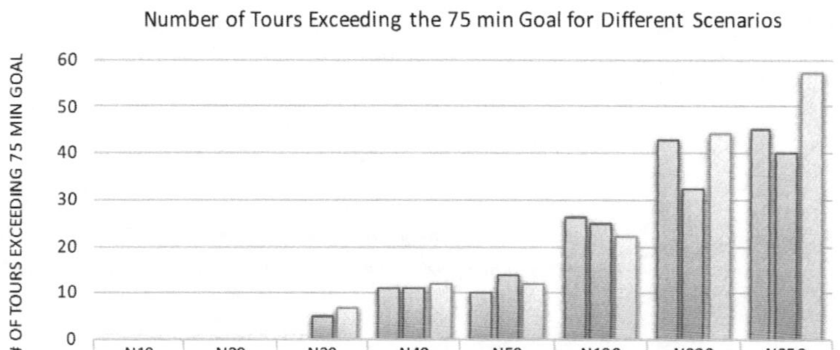

	N10	N20	N30	N40	N50	N100	N200	N250
Xpress LargeCoeff	0	0	0	11	10	26	43	45
Xpress SmallCoeff	0	0	5	11	14	25	32	40
Blocking/Heurisitc	0	0	7	12	12	22	44	57

Figure 4. Comparison of different methods regarding number of tours exceeding the goal.

Table 1. How the Heuristic Changes the Tours

	Perfect matching heuristic						
	Statistics						
Scenario	Min(old)	Min(new)	Max(old)	Max(new)	Avg(old)	Avg(new)	Heuristic running time(s)
N10	37.32	37.32	71.00	71.00	53.26	53.26	0
N20	45.72	45.72	72.14	72.14	60.04	60.04	<0.01 s
N30	36.46	46.98	109.63	91.86	68.89	72.47	<0.01 s
N40	62.15	69.70	112.54	112.54	87.76	87.91	<0.01 s
N50	41.10	47.34	112.23	112.23	78.37	80.73	0.01
N100	42.31	51.72	124.02	113.74	75.58	78.54	0.12
N200	33.15	44.82	110.89	120.89	77.59	79.70	0.5
N250	37.78	51.19	170.02	128.04	82.48	83.49	0.64
N500	22.64	70.77	145.58	150.31	86.89	87.20	2.94
N994	47.70	61.56	153.68	148.73	87.87	88.20	13.99
N54	69.00	76.00	145.00	145.00	110.32	110.32	0.01

Table 2. Key Performance Indicators of the Scenarios Before and After Applying the Heuristic

	Perfect matching heuristic					
	KPI					
Scenario	SD(old)	SD(new)	Rat. exceed. tours(old)	Rat. exceed. tours(new)	Exceeding mins(old)	Exceeding mins(new)
N10	13.18	13.18	0/5	0/5	0.00	0.00
N20	9.11	9.11	0/9	0/9	0.00	0.00
N30	20.83	12.25	6/14	7/14	88.86	49.60
N40	14.44	13.15	11/14	12/14	198.70	188.02
N50	18.97	15.16	12/19	12/19	182.59	165.77
N100	19.71	10.45	19/39	22/39	321.08	211.32
N200	18.79	12.52	43/72	44/72	660.66	506.04
N250	21.17	15.69	51/89	57/89	1077.75	947.22
N500	18.08	13.40	126/163	184/163	2319.05	2008.42
N994	16.94	13.43	242/324	296/324	4811.17	4342.37
N54	23.46	23.83	21/22	22/22	783.00	777.00

of buses. To illustrate the effect of the balancing step, the optimal solutions of the blocking problem are compared with the new solution from the matching heuristic. Recall that the optimal solutions to the blocking problem primarily minimize the number of tours (buses) and secondarily minimize the aggregate deadhead time. The heuristic's performance can be measured based on statistics regarding the tour durations before and after applying the second stage (balancing-stage). In particular, the focus is on three KPIs: a) SD of the tours, b) aggregate minutes exceeding the set goal, and c) percentage of tours exceeding the set goal before (old) and after (new) the balancing stage. Also, three statistics about the tours are reported, namely, their average, minimum tour duration, and maximum tour duration. These results are summarized in Tables 1 and 2. Table 1 shows that the heuristic acts by increasing the minimum tour duration and usually decreasing the maximum tour duration. This result is making the tours more balanced as can be seen in Table 2. The only case in which the SD increases after applying the heuristic is the California case study (N54). This increase is only a marginal increase. Therefore, it is observable that the heuristic's performance is very satisfactory regarding making more balanced tours. The first two scenarios with 10 and 20 trips remain unchanged after the balancing step because the duration of their tours is not over the 75-min goal and hence all the weights of the matching are equal to zero, and the balancing step does nothing.

From Table 1 it also can be seen that the heuristic increases the aggregate duration of the tours by increasing the average durations. This behavior is expected because the initial set of tours are optimized to have the shortest durations using the blocking problem. Any changes to those set of tours would cause the overall duration to increase. It is worth noting that this increase is very small. The other key performance indicators summarized in Table 2 (i.e. the aggregate minutes exceeding the 75-min goal and the ratio of the number of tours exceeding the goal over the number of tours) also show that the heuristic performs well. The heuristic only negatively affects the percentage of the tours exceeding goal KPI in a meaningful way. However, by making modifications to the weight calculation procedure in step 2-2 of the algorithm, we can improve this KPI at the cost of worsening some other KPIs. Overall the algorithm does well in balancing as it reduces the Standard Deviation (SD) by up to 47%. It also reduces the minutes exceeding the goal by up to 44%.

Conclusion

In this paper, a MIP formulation of the optimal balanced SBSP is proposed. Solving this formulation yields a balanced schedule which corresponds to the optimum fleet size. Due to the complexity of the problem, for larger instances, a matching-based heuristic algorithm is proposed. The proposed method can obtain leading solutions in significantly shorter time for any size problem. However, using the Xpress MIP solver for solving the proposed MIP is recommended whenever the problem size is tiny so that the solver can find the optimal solution.

To illustrate the performance of the balancing step of the heuristic, it is depicted how the heuristic affects some statistics and KPIs of the tours such as the SD of the tours and aggregate minutes exceeding the set goal. Statistic measures of the 11 test cases show that the balancing step of the heuristic (the matching phase) always decreases the aggregate minutes exceeding the goal. In fact, it reduces those by up to 44% in some scenarios. It is also indicated that in all scenarios related to HCPSS, the SD of the tours decreases. This reduction in the SD is up to an astonishing 47%.

This study opens the door to research related to balancing of school bus tours. Some possible future research directions could be: performing the balancing using more general methods such as min-cost flow; considering more specific versions of the SBSP such as heterogeneous fleet, mixed load, and multiple depots; and incorporation of uncertainties in balanced scheduling.

Author Contributions

The authors confirm contribution to the paper as follows: study conception and design: A. Shafahi, S. Aliari, A. Haghani; data collection: A. Shafahi, S. Aliari, A. Haghani; analysis and interpretation of results: A. Shafahi, S. Aliari; draft manuscript preparation: A. Shafahi, S. Aliari. All authors reviewed the results and approved the final version of the manuscript.

References

1. Fu, Z., R. Eglese, and L. Y. Li. A New Tabu Search Heuristic for the Open Vehicle Routing Problem. *Journal of the Operational Research Society*, Vol. 56, No. 3, 2005, pp. 267–274.
2. Li, F., B. Golden, and E. Wasil. The Open Vehicle Routing Problem: Algorithms, Large-Scale Test Problems, and Computational Results. *Computers & Operations Research*, Vol. 34, No. 10, 2007, pp. 2918–2930.
3. Orloff, C. S. Route Constrained Fleet Scheduling. *Transportation Science*, Vol. 10, No. 2, 1976, pp. 149–168.
4. Gavish, B., and E. Shlifer. An Approach for Solving a Class of Transportation Scheduling Problems. *European Journal of Operational Research*, Vol. 3, No. 2, 1979, pp. 122–134.
5. Haghani, A., and M. Banihashemi. Heuristic Approaches for Solving Large-Scale Bus Transit Vehicle Scheduling Problem with Route Time Constraints. *Transportation Research Part A: Policy and Practice*, Vol. 36, No. 4, 2002, pp. 309–333.
6. Desaulniers, G., J. Lavigne, and F. Soumis. Multi-Depot Vehicle Scheduling Problems with Time Windows and Waiting Costs. *European Journal of Operational Research*, Vol. 111, No. 3, 1998, pp. 479–494.
7. Kliewer, N., B. Amberg, and B. Amberg. Multiple Depot Vehicle and Crew Scheduling with Time Windows for Scheduled Trips. *Public Transport*, Vol. 3, No. 3, 2012, pp. 213–244.
8. Kliewer, N., T. Mellouli, and L. Suhl. A Time–Space Network Based Exact Optimization Model for Multi-Depot Bus Scheduling. *European Journal of Operational Research*, Vol. 175, No. 3, 2006, pp. 1616–1627.
9. Ceder, A. A. Public-Transport Vehicle Scheduling with Multi Vehicle Type. *Transportation Research Part C: Emerging Technologies*, Vol. 19, No. 3, 2011, pp. 485–497.
10. Bowerman, R., B. Hall, and P. Calamai. A Multi-Objective Optimization Approach to Urban School Bus Routing: Formulation and Solution Method. *Transportation Research Part A: Policy and Practice*, Vol. 29, No. 2, 1995, pp. 107–123.
11. Ripplinger, D. Rural School Vehicle Routing Problem. *Transportation Research Record: Journal of the Transportation Research Board*, 2005. 1922: 105–110.
12. Kamali, B., S. J. Mason, and E. A. Pohl. An Analysis of Special Needs Student Busing. *Journal of Public Transportation*, Vol. 16, No. 1, 2013, p. 2.
13. Pacheco, J., A. Alvarez, S. Casado, and J. L. González-Velarde. A Tabu Search Approach to an Urban Transport Problem in Northern Spain. *Computers & Operations Research*, Vol. 36, No. 3, 2009, pp. 967–979.
14. Desrosiers, J., M. Sauvé, and F. Soumis. Lagrangian Relaxation Methods for Solving the Minimum Fleet Size Multiple Traveling Salesman Problem with Time Windows. *Management Science*, Vol. 34, No. 8, 1988, pp. 1005–1022.
15. Corberán, A., E. Fernández, M. Laguna, and R. A. F. A. E. L. Marti. Heuristic Solutions to the Problem of Routing School Buses with Multiple Objectives. *Journal of the*

Operational Research Society, Vol. 53, No. 4, 2002, pp. 427–435.
16. Kim, B. I., S. Kim, and J. Park. A School Bus Scheduling Problem. *European Journal of Operational Research*, Vol. 218, No. 2, 2012, pp. 577–585.
17. Li, L. Y. O., and Z. Fu. The School Bus Routing Problem: A Case Study. *Journal of the Operational Research Society*, Vol. 53, No. 5, 2002, pp. 552–558.
18. Forbes, M., J. Holt, P. Kilby, and A. Watts. A Matching Algorithm with Application to Bus Operations. *Australian Journal of Combinatorics*, Vol. 4, 1991, pp. 71–85.
19. Wark, P., and J. Holt. A Repeated Matching Heuristic for the Vehicle Routeing Problem. *Journal of the Operational Research Society*, Vol. 45, No. 10, 1994, pp. 1156–1167.
20. Harvey, N. J., R. E. Ladner, L. Lovász, and T. Tamir. Semi-Matchings for Bipartite Graphs and Load Balancing. *Journal of Algorithms*, Vol. 59, No. 1, 2006, pp. 53–78.
21. Shafahi, A., Z. Wang, and A. Haghani. Solving the School Bus Routing Problem by Maximizing Trip Compatibility. *Transportation Research Record: Journal of the Transportation Research Board*, 2017. 2667: 17–27.

The Standing Committee on Transportation Network Modeling (ADB30) peer-reviewed this paper (18-01108).

Article

TRR
JOURNAL OF THE TRANSPORTATION RESEARCH BOARD

Transportation Research Record
2018, Vol. 2672(48) 12–23

DOI: 10.1177/0361198118773565
journals.sagepub.com/home/trr
SAGE

Modeling Day-to-Day Dynamics of Travelers' Risky Route Choices under the Influence of Predictive Traffic Information

Xiaoming Lou[1], Lin Cheng[2], Dawei Li[3], Shibei Zhu[1], and Jing Zhou[4]

Abstract

In stochastic transport networks, travelers tend to reserve a "travel time budget" (TTB) for every alternative route to assure their punctual arrival with a predefined confidence level. This paper incorporates the concept of TTB into a deterministic nonlinear dynamic system to model the day-to-day evolution process of travelers' risk-taking route choices in stochastic degradable road networks. This paper also introduces consistent traffic information prediction, which can be provided by advanced traveler information systems (ATIS), into the dynamic system to investigate its influence on travelers' route choice behaviors and the corresponding day-to-day network flow evolutions. In the proposed dynamic model, ATIS-equipped travelers are distinguished from those not so equipped to reflect traveler heterogeneity in access to traffic information. Some theoretical analyses are further conducted to investigate the existence and uniqueness of the fixed point, and the stability of the day-to-day dynamic model. The analysis results show that consistent predictive traffic information can help to stabilize day-to-day network flow fluctuations and thus can enhance fixed-point stability of the dynamic model. In addition, this paper investigates the effects of travelers' attitudes to risk on the network flow evolution process. An illustrative example is presented to demonstrate the characteristics of the dynamic model as well as to verify the theoretical analysis results.

Uncertainty in travel time, which may be caused by the stochastic degradations of road capacity, is an adverse element always present in urban traffic networks. It reduces the reliability of link and route travel time and thus brings risks of commuting delays to travelers. Several empirical surveys (e.g. *1–3*) have verified that travel time variability is one of the most important factors influencing travelers' route choices. A concept closely related to the reliability of travel time is travelers' attitudes to risk, which measures their aversion or preference toward uncertainty.

Travelers' risk-taking in their choice of route under uncertain travel time has been studied extensively (*4–12*). The core idea behind this research is based on the concept of "travel time budget" (TTB) (*13*), which reveals a behavioral mechanism whereby travelers add a margin to the expected duration of a journey so as to maintain a preferred probability of arriving within budgeted time, referred to as the within-budget time reliability (WBTR). However, TTB is unable to cope with the case of extremely long travel time in route choice equilibrium. To address this issue, Chen and Zhou (*14*) proposed a mean-excess travel time (METT) model, to explicitly consider both reliability and unreliability aspects of travel time variability in travelers' route choice process.

Besides the uncertainty of travel time, traffic information is considered to be an important factor influencing travelers' route choices. Advanced traveler information systems (ATIS) can potentially improve travelers' access to real-time traffic information through a range of communication technologies. With this traffic information, travelers can conduct both pre-trip path selection and en-route path adjustment to avoid unexpected traffic congestion.

Travelers' route choices are usually described by the static traffic equilibrium assignment approach, which

[1]Department of Urban and Rural Construction, Zhejiang Development & Planning Institute, Hangzhou, China
[2]School of Transportation, Southeast University, Nanjing, China
[3]Jiangsu Key Laboratory of Urban ITS, Jiangsu Province Collaborative Innovation Center of Modern Urban Traffic, School of Transportation, Southeast University, Nanjing, China
[4]Zhejiang Institute of Communications, Hangzhou, China

Corresponding Author:
Address correspondence to Lin Cheng: gist@seu.edu.cn

mainly focuses on the steady traffic flow equilibrium distribution and does not account for the disequilibrium of dynamic evolution before reaching the ultimate steady state. The day-to-day (DTD) dynamic assignment model can be seen as a generalization of the static equilibrium paradigm to reproduce the network flow evolution process. It can better reflect travelers' memory and learning mechanisms, with route choices on a particular day conditional on various influence parameters including uncertain travel time and traffic information.

Since the 1990s, research into DTD dynamic traffic systems has grown to encompass a wide range of approaches. Readers may refer to Cantarella and Cascetta (*15*), Watling and Hazelton (*16*), Watling and Cantarella (*17*) and Cantarella and Watling (*18*, *19*) for both synthesis and development of the dynamic evolution process of traffic flow. Some authors have advanced the DTD systems to forecast how the traffic information provided by ATIS affects network flow evolution process (e.g. *20–24*). Under the framework of DTD dynamics, the effects of traffic information largely depend on the actual information strategy implemented by the ATIS device. An ideal strategy for ATIS is to provide consistent traffic information prediction, which means the predicted network flow on which the traffic information is based is the same as the actual travelers' route choice results affected by the traffic information (*25–28*). Consistent traffic information provides the potential capacity to optimize travelers' routing strategies; its effects on the DTD network flow dynamics still need further research.

This paper incorporates the concept of TTB into a deterministic nonlinear dynamic system to model the day-to-day evolution process of travelers' risk-taking route choices, which is a relatively unexplored problem in the existing literature. In stochastic degradable networks, travelers do not know their exact a priori travel times. Risk-averse individuals tend to depart earlier to allow for additional time or add a travel time margin to the expected trip time to avoid arriving late. From this point of view, TTB conforms to the fact of travelers' departure time choices, which further requires us to investigate the sophisticated within-day traffic network flow dynamics.

However, the doubly dynamic network flow model, which combines both within-day realization process and DTD evolution process, is beyond the scope of this paper and remains a research challenge. This study involves a simplification by only focusing on travelers' daily route-switching behaviors, and does not consider travelers' departure time choices, to bypass the complicated within-day dynamics problem. In consideration of travel time uncertainty, the concept of TTB is adopted in the proposed DTD model to reflect the route disutility evaluated by travelers. Subsequently, travelers consider the perceived route disutility as the basis on which to make their route choices and adjustments.

The main advantage of TTB expression is its simplicity and tractability. In particular, travelers' risk aversion attitudes can be reflected by the single parameter of WBTR in TTB, without resorting to any more complex behavior paradigm. This simplicity of TTB, on the one hand, will facilitate the formulation of risk-taking behaviors in our DTD dynamic model, and on the other hand, will facilitate the theoretical analysis about the effects of travelers' risk attitudes on network flow dynamics.

This paper also incorporates an ATIS consistent travel time prediction mechanism into the DTD dynamic model to capture the effects of predictive traffic information. Two traveler types are assumed to form the entire road-user population, according to their access to ATIS information. For ease of presentation, we named these two traveler groups, respectively, as informed and uninformed travelers. By this means, our DTD model reflects traveler heterogeneity in terms of their source of travel time information.

The remainder of this paper is organized as follows. The second section describes the degradable transport network in question and provides definitions of relevant notions used in this paper. The third section proposes the analytical DTD network flow evolution, to describe the interactive choice behaviors of informed and uninformed travelers. The fourth section presents some theoretical analyses to investigate the existence and uniqueness of the fixed point, and the asymptotic stability of the proposed DTD dynamic system. The fifth section applies the proposed DTD model to a network example to verify its theoretical properties. The conclusion and future work are summarized in the final section.

Travel Time Budget with Corresponding Equilibrium Model

Degradable Transport Network and Definitions of Relevant Notions

A transport network is a directed graph (N, L) where N represents the node set and L corresponds to the link or road set. The notations that will be used in this paper are listed as follows:

$l \in L$ link index;

W set of OD pairs, $W = \{(o, d)\}$;

w OD pair index, $w \in W$;

R_w set of routes on OD pair w;

R set of all routes, $R = \bigcup_w R_w$;

$r \in R$ route index;

Λ_{lrw} link-route index: if link l belongs to route r, $\Lambda_{lrw} = 1$, otherwise, $\Lambda_{lrw} = 0$;

$\mathbf{\Lambda}_w$ $|L| \times |R_w|$ link-route incidence matrix for OD pair w, $\mathbf{\Lambda}_w = (\Lambda_{lrw})$;

q_w travel demand on OD pair w;

$\mathbf{q}$ $|W| \times 1$ vector of OD demands, $\mathbf{q} = \left(q_1, q_2, \cdots, q_{|W|}\right)^T$;

c_l^{free} free travel time of link l;

x_l traffic flow on link l;

$\mathbf{x}$ $|L| \times 1$ vector of link traffic flows, $\mathbf{x} = \left(x_1, x_2, \cdots, x_{|L|}\right)^T$;

$y_{w,r}$ traffic flow on route r between OD pair w;

$\mathbf{y}_w$ $|R_w| \times 1$vector of $y_{w,r}$, $\mathbf{y}_w = (y_{w,1}, y_{w,2}, \cdots, y_{w,|R_w|})^T$;

$\bar{u}_l$ design capacity of link l;

$\underline{u}_l$ worst-degraded capacity of link l, defined as a fraction $\varepsilon_l (0 < \varepsilon_l < 1)$ of design capacity, namely $\underline{u}_l = \varepsilon_l \cdot \bar{u}_l$;

u_l stochastic capacity of degradable link l, its value varies from lower bound $\underline{u}_l$ to upper bound $\bar{u}_l$;

$\mathbf{u}$ $|L| \times 1$ vector of stochastic link capacities, $\mathbf{u} = \left(u_1, u_2, \cdots, u_{|L|}\right)^T$;

$c_l(x_l)$ link travel time function;

$\mathbf{c}(\mathbf{x})$ $|L| \times 1$ vector of $c_l(x_l)$, $\mathbf{c}(\mathbf{x}) = (c_1(x_1), c_2(x_2), \cdots, c_{|L|}(x_{|L|}))^T$;

$S_{w,r}$ stochastic travel time of route r between OD pair w;

$\mathbf{S}_w$ $|R_w| \times 1$ vector of $S_{w,r}$, $\mathbf{S}_w = \left(S_{w,1}, S_{w,2}, \cdots, S_{w,|R_w|}\right)^T$;

$f_{w,r}(S_{w,r})$ probability density function of $S_{w,r}$;

$F_{w,r}(S)$ cumulative distribution function of $S_{w,r}$, $F_{w,r}(S) = \int_0^S f_{w,r}(S_{w,r}) dS_{w,r}$;

ρ probability required for within-budget time arrivals, namely the WBTR for travelers;

$b_{w,r}(\mathbf{x}, \rho)$ route TTB value reserved by travelers according to their WBTR level and their observed link flows $\mathbf{x}$;

$\mathbf{b}_w(\mathbf{x}, \rho)$ $|R_w| \times 1$ vector of $b_{w,r}(\mathbf{x}, \rho)$, $\mathbf{b}_w(\mathbf{x}, \rho) = (b_{w,1}(\mathbf{x}, \rho), b_{w,2}(\mathbf{x}, \rho), \cdots, b_{w,|R_w|}(\mathbf{x}, \rho))^T$;

θ dispersion parameter (θ>0) in probabilistic route choice model;

$p_{w,r}(\mathbf{b}_w(\mathbf{x}, \rho), \theta)$ the probability travelers select route r between OD pair w;

$\mathbf{p}_w(\mathbf{b}_w(\mathbf{x}, \rho), \theta)$ $|R_w| \times 1$ vector of $p_{w,r}(\mathbf{b}_w(\mathbf{x}, \rho), \theta)$, $\mathbf{p}_w(\mathbf{b}_w(\mathbf{x}, \rho), \theta) = (p_{w,1}(\mathbf{b}_w(\mathbf{x}, \rho), \theta), p_{w,2}(\mathbf{b}_w(\mathbf{x}, \rho), \theta), \cdots, p_{w,|R_w|}(\mathbf{b}_w(\mathbf{x}, \rho), \theta))^T$.

For a degradable transportation network, the following relationships are valid: $\mathbf{S}_w = \mathbf{\Lambda}_w^T \cdot \mathbf{c}(\mathbf{x})$, $\forall w \in W$, which represents the relation between actual route travel time and link travel time, the superscript T denotes matrix transpose; $\mathbf{x} = \sum_w \mathbf{\Lambda}_w \cdot \mathbf{y}_w$, which represents the relation between link and route flows; and $q_w = \sum_{r \in R_w} y_{w,r} = 1^T \cdot \mathbf{y}_w, \forall w \in W$, which expresses the total flow conservation between each OD pair, the notation 1 denotes unit vector.

Travel Time Budget

Suppose that all travelers adopt a uniform WBTR to reserve their TTB for the alternative routes. Mathematically, the route TTB $b_{w,r}(\mathbf{x}, \rho)$ associated with WBTR $\rho (0 < \rho < 1)$ should meet the following condition:

$$\text{Pro}\{S_{w,r} \leq b_{w,r}(\mathbf{x}, \rho)\} = \rho, \forall r \in R_w, w \in W, \quad (1)$$

where Pro{E} represents the probability of occurrence of event E. Without loss of generality, we assume the stochastic link and route travel times are continuously distributed over their respective value ranges. Then, Equation 1 can be equivalently reformulated as

$$F_{w,r}(b_{w,r}(\mathbf{x}, \rho)) = \int_0^{b_{w,r}(\mathbf{x}, \rho)} f_{w,r}(S_{w,r}) dS_{w,r} = \rho, \forall r \in R_w, w \in W. \quad (2)$$

Notice that the specific route travel time distributions are implicitly determined by the stochastic link capacities $\mathbf{u}$.Taking the inverse of Equation 2,

$$b_{w,r}(\mathbf{x}, \rho) = F_{w,r}{}^{-1}(\rho), \forall r \in R_w, w \in W, \quad (3)$$

where $F_{w,r}{}^{-1}(\cdot)$ represents the inverse function of the cumulative distribution function $F_{w,r}(\cdot)$.

Figure 1 shows the relationship between route travel time distribution, WBTR ρ and TTB $b_{w,r}(\mathbf{x}, \rho)$.

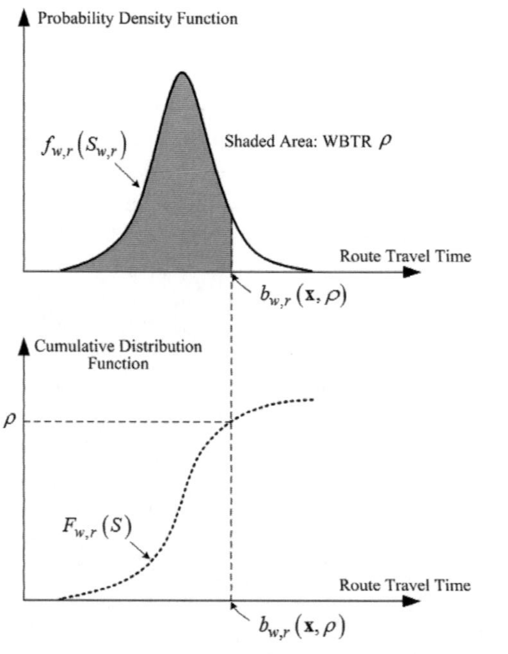

Figure 1. Relationship between route travel time distribution, WBTR, and TTB.

Obviously, a larger ρ-value corresponds to more extreme risk aversion, which means travelers will hold a longer TTB for their trip.

Stochastic Travel Time Budget Equilibrium Model

In their daily commuting activities, travelers tend to select the superior routes with shortest TTB. At the ultimate steady state, travelers would distribute themselves in the network such that no single individual could reduce his or her perceived TTB by unilaterally changing routes. In reality, travelers' perceived TTB is usually not equal to the actual TTB due to the inclusion of perception errors. This equilibrium pattern formed under the notion of TTB is usually called the stochastic travel time budget equilibrium (STTBE) model (9).

The route flow distribution can then be estimated according to travelers' route choice behavior, which is usually modeled through a probabilistic route choice function $P(\cdot)$ expressed as

$$\mathbf{p}_w(\mathbf{b}_w(\mathbf{x}, \rho), \theta) = P(\mathbf{b}_w(\mathbf{x}, \rho), \theta), \forall w \in W, \quad (4)$$

The functional form of $P(\cdot)$ depends on the specific distribution of travelers' perception error terms. If the error terms are assumed to be identically and independently Gumbel-distributed random variables, then the route choice model (Equation 4) can be specified as the following Logit-based formula,

$$p_{w,r}(\mathbf{b}_w(\mathbf{x}, \rho), \theta) = \frac{\exp\{-\theta \cdot b_{w,r}(\mathbf{x}, \rho)\}}{\sum_{k \in R_w} \exp\{-\theta \cdot b_{w,k}(\mathbf{x}, \rho)\}}, \quad \forall r \in R_w, w \in W. \quad (5)$$

In Equation 5, the positive dispersion parameter θ is inversely proportional to the standard error of the distribution of the perceived TTB.

The route flow distribution of travelers can now be formulated based on their route choice probabilities:

$$\mathbf{y}_w = q_w \cdot \mathbf{p}_w(\mathbf{b}_w(\mathbf{x}, \rho), \theta), \forall w \in W, \quad (6)$$

and link flow can also be easily achieved by applying the link-route topological relation:

$$\mathbf{x} = \sum_w q_w \cdot \mathbf{\Lambda}_w \cdot \mathbf{p}_w(\mathbf{b}_w(\mathbf{x}, \rho), \theta). \quad (7)$$

Equation 7 is actually a fixed-point problem (FPP) which defines a map of $\mathbf{x}$ to itself. The proposed FPP (Equation 7) is exactly equivalent to the STTBE model.

Description of the Day-to-Day Model

Experience Learning of Uninformed Travelers

The exponential-smoothing filter can approximately simulate travelers' experience learning behaviors and thus is widely used in the existing literature to capture travelers' day-to-day updating about the mean route travel time. This filter is also applied here to reflect travelers' dynamic updating processes of their reserved TTB values:

$$B_{w,r}^{\text{un},t+1} = \alpha \cdot b_{w,r}^t(\mathbf{x}^t, \rho) + (1 - \alpha) \cdot B_{w,r}^{\text{un},t}, \quad \forall r \in R_w, w \in W. \quad (8)$$

Note that in the context of the DTD dynamic system, a superscript t is added to the used notions to reflect the time unit 'day'. Therefore, in Equation 8, $\mathbf{x}^t$ represents the transient link flows occurring on day t. represents the route TTB renewed by uninformed travelers on day t. And $\alpha (0 < \alpha \leq 1)$ denotes a constant weight (independent of t) parameter, which reflects the preference between new knowledge $b_{w,r}^t(\mathbf{x}^t, \rho)$ and old experience $B_{w,r}^{\text{un},t}$ by uninformed travelers.

Define $\mathbf{B}_w^{\text{un},t}$ and $\mathbf{b}_w^t(\mathbf{x}^t, \rho)$, respectively, to represent the vectors of $B_{w,r}^{\text{un},t}$ and $b_{w,r}^t(\mathbf{x}^t, \rho)$. Equation 8 can then be reformulated in a vector form,

$$\mathbf{B}_w^{\text{un},t+1} = \alpha \cdot \mathbf{b}_w^t(\mathbf{x}^t, \rho) + (1 - \alpha) \cdot \mathbf{B}_w^{\text{un},t}, \forall w \in W, \quad (9)$$

or equivalently in a more compact form,

$$\mathbf{B}^{\text{un},t+1} = \alpha \cdot \mathbf{b}^t(\mathbf{x}^t, \rho) + (1 - \alpha) \cdot \mathbf{B}^{\text{un},t}, \quad (10)$$

where $\mathbf{B}^{\text{un},t}$ and $\mathbf{b}^t(\mathbf{x}^t, \rho)$, respectively, represent the collection of $\mathbf{B}_w^{\text{un},t}$ and $\mathbf{b}_w^t(\mathbf{x}^t, \rho)$ on all OD pairs.

ATIS Consistent Travel Time Prediction

Define $B_{w,r}^{\text{in},t+1}$ to denote the predictive route TTB information provided by ATIS on day $t+1$. Under the situation of consistent prediction, the value of $B_{w,r}^{\text{in},t+1}$ is exactly equal to the actual TTB which is associated with the future network traffic conditions appearing on the subsequent day $t+1$. That is,

$$B_{w,r}^{\text{in},t+1} = b_{w,r}^{t+1}(\mathbf{x}^{t+1}, \rho), \forall r \in R_w, w \in W. \quad (11)$$

Equation 11 can also be reformulated equivalently as a vector form,

$$\mathbf{B}_w^{\text{in},t+1} = \mathbf{b}_w^{t+1}(\mathbf{x}^{t+1}, \rho), \forall w \in W, \quad (12)$$

where $\mathbf{B}_w^{\text{in},t+1}$ represents the vector of $B_{w,r}^{\text{in},t+1}$. The more compact form is given as

$$\mathbf{B}^{\text{in},t+1} = \mathbf{b}^{t+1}(\mathbf{x}^{t+1}, \rho), \quad (13)$$

where $\mathbf{B}^{\text{in},t+1}$ denotes the collection of all $\mathbf{B}_w^{\text{in},t+1}$.

Dynamic Traffic Flow Evolution

It is not possible for either informed or uninformed travelers to obtain identical TTB predictions or perceptions for the same route and, as a result, their travel choices are not identical but dispersed across multiple alternative routes. Therefore, the probabilistic route choice function, as introduced earlier in this section, is also adopted here to model the route choice results of two traveler groups.

Specifically, for the uninformed travelers, their route choice probabilities can be estimated according to the renewed TTB perceptions:

$$\mathbf{p}_w^{\text{un},t}\left(\mathbf{B}_w^{\text{un},t},\theta^{\text{un}}\right) = P^{\text{un}}\left(\mathbf{B}_w^{\text{un},t},\theta^{\text{un}}\right), \forall w \in W, \tag{14}$$

where $P^{\text{un}}(\cdot)$ denotes the probabilistic route choice function for the uninformed travelers. $\mathbf{p}_w^{\text{un},t}\left(\mathbf{B}_w^{\text{un},t},\theta^{\text{un}}\right)$ represents the $|R_w| \times 1$ vector with element $p_{w,r}^{\text{un},t}\left(\mathbf{B}_w^{\text{un},t},\theta^{\text{un}}\right)$, which represents the probability for an uninformed traveler to select route $r \in R_w$ on day t. The positive dispersion parameter θ^{un} is related to the variation of TTB perceptions of uninformed travelers.

For the informed travelers, their route choice probabilities can be formulated by another probabilistic route choice model accounting for the effect of random errors:

$$\mathbf{p}_w^{\text{in},t}\left(\mathbf{b}_w^t(\mathbf{x}^t,\rho),\theta^{\text{in}}\right) = P^{\text{in}}\left(\mathbf{b}_w^t(\mathbf{x}^t,\rho),\theta^{\text{in}}\right), \forall w \in W, \tag{15}$$

where $P^{\text{in}}(\cdot)$ denotes the probabilistic route choice function for the informed travelers. $\mathbf{p}_w^{\text{in},t}\left(\mathbf{b}_w^t(\mathbf{x}^t,\rho),\theta^{\text{in}}\right)$ represents the $|R_w| \times 1$ vector with element $p_{w,r}^{\text{in},t}\left(\mathbf{b}_w^t(\mathbf{x}^t,\rho),\theta^{\text{in}}\right)$, which represents the probability for an informed traveler to select route $r \in R_w$ on day t. The positive dispersion parameter θ^{in} is adopted to reflect the predictive information quality of ATIS.

Assume that all travelers will reconsider yesterday's decisions, and make current route choices which are consistent with the probabilistic route choice functions defined in Equations 14 and 15, then the route flows driven by the two traveler groups can be calculated respectively as follows:

$$\mathbf{y}_w^{\text{un},t} = (1-\eta)\cdot q_w \cdot \mathbf{p}_w^{\text{un},t}\left(\mathbf{B}_w^{\text{un},t},\theta^{\text{un}}\right), \forall w \in W, \tag{16}$$

$$\mathbf{y}_w^{\text{in},t} = \eta \cdot q_w \cdot \mathbf{p}_w^{\text{in},t}\left(\mathbf{b}_w^t(\mathbf{x}^t,\rho),\theta^{\text{in}}\right), \forall w \in W, \tag{17}$$

where the constant parameter $\eta(0<\eta<1)$ represents ATIS market penetration level, or equivalently the proportion of informed travelers over all network users for every OD pair. $\mathbf{y}_w^{\text{un},t}$ denotes the $|R_w| \times 1$ vector with element $y_{w,r}^{\text{un},t}$, which represents the number of uninformed travelers selecting route $r \in R_w$ on day t. And $\mathbf{y}_w^{\text{in},t}$ denotes the $|R_w| \times 1$ vector with element $y_{w,r}^{\text{in},t}$, which represents the number of informed travelers selecting route $r \in R_w$ on day t.

Now the mixed link traffic flows can be achieved by applying the link-route topological relation:

$$\begin{aligned}\mathbf{x}^t = {} & (1-\eta)\sum_w q_w \boldsymbol{\Lambda}_w \mathbf{p}_w^{\text{un},t}\left(\mathbf{B}_w^{\text{un},t},\theta^{\text{un}}\right) \\ & + \eta \sum_w q_w \boldsymbol{\Lambda}_w \mathbf{p}_w^{\text{in},t}\left(\mathbf{b}_w^t(\mathbf{x}^t,\rho),\theta^{\text{in}}\right).\end{aligned} \tag{18}$$

Equation 18 presents the dynamic link flow evolution process in our DTD model.

Integrated Day-to-Day Dynamic Model

Having modeled the dynamic learning and route adjustment behaviors of two traveler groups as well as the consistent travel time prediction of ATIS, we can now integrate them together to formulate the DTD dynamic evolution model as below,

$$\begin{cases}\mathbf{B}^{\text{un},t+1} = \alpha \cdot \mathbf{b}^t(\mathbf{x}^t,\rho) + (1-\alpha)\cdot \mathbf{B}^{\text{un},t}, \\ \mathbf{x}^t = (1-\eta)\cdot \sum_w q_w \boldsymbol{\Lambda}_w \mathbf{p}_w^{\text{un},t}\left(\mathbf{B}_w^{\text{un},t},\theta^{\text{un}}\right) + \eta \cdot \sum_w q_w \boldsymbol{\Lambda}_w \mathbf{p}_w^{\text{in},t}\left(\mathbf{b}_w^t(\mathbf{x}^t,\rho),\theta^{\text{in}}\right).\end{cases} \tag{19}$$

Note that the second equation of system (19) is equivalent to a FPP which defines a map of $\mathbf{x}^t$ to itself. The self-mapping property of this FPP is due to the consistent traffic information prediction of ATIS.

The above-mentioned FPP implies that $\mathbf{x}^t$ can be regarded as an implicit function of $\mathbf{q}$, $\mathbf{B}^{\text{un},t}$, η, ρ, θ^{un} and θ^{in}. For ease of presentation, we define $\text{H}(\cdot)$ to represent this functional relation, namely,

$$\mathbf{x}^t = \text{H}\left(\mathbf{q},\mathbf{B}^{\text{un},t},\eta,\rho,\theta^{\text{un}},\theta^{\text{in}}\right). \tag{20}$$

Obviously, the specific form of $\text{H}(\cdot)$ depends on the FPP defined by the second equation of the dynamic system (Equation 19).

Now we can substitute Equation 20 into the first equation of the system (19), to get a more concise DTD dynamic evolution process, as follows:

$$\begin{aligned}\mathbf{B}^{\text{un},t+1} = {} & \alpha \cdot \mathbf{b}^t\left(\text{H}\left(\mathbf{q},\mathbf{B}^{\text{un},t},\eta,\rho,\theta^{\text{un}},\theta^{\text{in}}\right),\rho\right) \\ & + (1-\alpha)\cdot \mathbf{B}^{\text{un},t}.\end{aligned} \tag{21}$$

This means the proposed DTD system (19) can be reformulated equivalently as a simpler evolution process about the TTB perceptions $\mathbf{B}^{\text{un},t}$, as long as we can derive the function relation $\text{H}(\cdot)$ emerging in Equation 20.

Theoretical Properties of the Day-to-Day Model

Fixed Point and Equilibrium State

Fixed point is a particular point of a dynamic system; it plays an important role in revealing the laws and structures of the stable state of the system. The following theorem establishes the relation between fixed point and equilibrium state.

- **Theorem 1.** *If the fixed point of the DTD dynamic system (Equation 19) exists, then it is equivalent to the extended STTBE state containing two traveler groups.*

Proof. Fixed point of the dynamic system (Equation 19) can be obtained from the stable condition of perceived TTB values, namely $\mathbf{B}^{\mathrm{un},t+1} = \mathbf{B}^{\mathrm{un},t} = \mathbf{B}^{\mathrm{un},*}$. For every variable in the dynamic system, the superscript * is marked to represent its stationary value at the ultimate fixed point. Thus,

$$\mathbf{B}^{\mathrm{un},*} = \alpha \cdot \mathbf{b}^*(\mathbf{x}^*, \rho) + (1-\alpha)\cdot \mathbf{B}^{\mathrm{un},*}, \tag{22}$$

therefore,

$$\mathbf{B}^{\mathrm{un},*} = \mathbf{b}^*(\mathbf{x}^*, \rho). \tag{23}$$

Substitute the result (23) into Equations 16 and 17, we can get

$$\mathbf{y}_w^{\mathrm{un},*} = (1-\eta)\cdot q_w \cdot \mathbf{p}_w^{\mathrm{un},*}\left(\mathbf{b}_w^*(\mathbf{x}^*, \rho), \theta^{\mathrm{un}}\right), \forall w \in W, \tag{24}$$

$$\mathbf{y}_w^{\mathrm{in},*} = \eta \cdot q_w \cdot \mathbf{p}_w^{\mathrm{in},*}\left(\mathbf{b}_w^*(\mathbf{x}^*, \rho), \theta^{\mathrm{in}}\right), \forall w \in W, \tag{25}$$

and hence,

$$\mathbf{x}^* = (1-\eta)\cdot \sum_w q_w \mathbf{\Lambda}_w \mathbf{p}_w^{\mathrm{un},*}\left(\mathbf{b}_w^*(\mathbf{x}^*, \rho), \theta^{\mathrm{un}}\right) + \eta \cdot \sum_w q_w \mathbf{\Lambda}_w \mathbf{p}_w^{\mathrm{in},*}\left(\mathbf{b}_w^*(\mathbf{x}^*, \rho), \theta^{\mathrm{in}}\right). \tag{26}$$

Equation 26 defines a self-mapping of the stationary link flow distribution $\mathbf{x}^*$. This coincides with conventional stochastic equilibria. Actually, Equation 26 is an extended version of the original STTBE model (7) by considering two traveler groups explicitly.

Fixed-Point Existence and Uniqueness

We can now discuss the fixed-point existence and uniqueness properties of the proposed DTD system. We begin with four basic assumptions.

Assumption 1. *The link travel time function $\mathbf{c}(\mathbf{x}^t)$ is continuous over the feasible link flow set.*
Assumption 2. *The link travel time function $\mathbf{c}(\mathbf{x}^t)$ is strictly monotonic increasing with respect to the link flows, i.e.,*

$$(\mathbf{x}^t - \tilde{\mathbf{x}}^t)^T(\mathbf{c}(\mathbf{x}^t) - \mathbf{c}(\tilde{\mathbf{x}}^t)) > 0, \mathbf{x}^t \neq \tilde{\mathbf{x}}^t. \tag{27}$$

Assumption 3. *The probabilistic route choice functions $P^{\mathrm{un}}(\cdot)$ and $P^{\mathrm{in}}(\cdot)$ are both continuous with respect to their respectively arguments $\mathbf{B}_w^{\mathrm{un},t}$ and $\mathbf{b}_w^t(\mathbf{x}^t, \rho)$.*
Assumption 4. *The probabilistic route choice functions $P^{\mathrm{un}}(\cdot)$ and $P^{\mathrm{in}}(\cdot)$ are both non-increasing with respect to their respectively arguments $\mathbf{B}_w^{\mathrm{un},t}$ and $\mathbf{b}_w^t(\mathbf{x}^t, \rho)$, i.e.,*

$$\left(\mathbf{B}_w^{\mathrm{un},t} - \tilde{\mathbf{B}}_w^{\mathrm{un},t}\right)^T \left(P^{\mathrm{un}}\left(\mathbf{B}_w^{\mathrm{un},t}, \theta^{\mathrm{un}}\right) - P^{\mathrm{un}}\left(\tilde{\mathbf{B}}_w^{\mathrm{un},t}, \theta^{\mathrm{un}}\right)\right) \leq 0, \quad \mathbf{B}_w^{\mathrm{un},t} \neq \tilde{\mathbf{B}}_w^{\mathrm{un},t}, \tag{28}$$

$$\left(\mathbf{b}_w^t(\mathbf{x}^t, \rho) - \tilde{\mathbf{b}}_w^t(\tilde{\mathbf{x}}^t, \rho)\right)^T \left(P^{\mathrm{in}}\left(\mathbf{b}_w^t(\mathbf{x}^t, \rho), \theta^{\mathrm{in}}\right) - P^{\mathrm{in}}\left(\tilde{\mathbf{b}}_w^t(\tilde{\mathbf{x}}^t, \rho), \theta^{\mathrm{in}}\right)\right) \leq 0, \; \mathbf{b}_w^t(\mathbf{x}^t, \rho) \neq \tilde{\mathbf{b}}_w^t(\tilde{\mathbf{x}}^t, \rho). \tag{29}$$

Assumptions 1 and 2 are reasonable in the context of transportation planning and can be satisfied by the widely-used Bureau of Public Roads (BPR) function. Assumptions 3 and 4 are also quite mild conditions and can be easily ensured by an invariant random utility model such as the Logit-based formula. As shown in the following theorem, fixed-point existence conditions mainly require continuity of all involved functions.

Theorem 2. (fixed-point existence) *Suppose that each OD pair is connected at least one route. Then Assumptions 1 and 3 together ensure fixed-point existence of the DTD dynamic system (Equation 19) (or equivalently (Equation 21)).*

To meet the fixed-point uniqueness condition, the monotonicity of the involved functions is further introduced in the following theorem.

Theorem 3. (fixed-point uniqueness) *Suppose the fixed point of the DTD dynamic system (Equation 19) exists (namely Assumptions 1 and 3 are satisfied). Assumptions 2 and 4 further ensure the uniqueness of the fixed point.*

The proofs of Theorems 2 and 3 are similar to that of Cantarella and Cascetta (*15*), and thus they are omitted here for reasons of simplicity.

Fixed-Point Asymptotic Stability

Stability is an important property of a dynamic model for its applicability in practice. A dynamic model is characterized as stable or asymptotically stable system with respect to its fixed point. A DTD model is said to be stable if all solutions starting at points near to the fixed point stay nearby. Furthermore, a DTD model is said to be asymptotically stable if all the solutions starting at points near the fixed point tend toward the fixed point with time. To investigate the fixed-point stability of a deterministic, discrete-time dynamic model, a common method is to conduct a spectral analysis of the Jacobian matrix of the transition process contained in the underlying DTD evolution system.

As stated earlier, the proposed DTD system is equivalent to an evolution process regarding the TTB perceptions $\mathbf{B}_w^{\mathrm{un},t}$. It can be reformulated in an abstract form as follows:

$$\mathbf{B}_w^{\mathrm{un},t+1} = \Phi\left(\mathbf{B}_w^{\mathrm{un},t}\right), \tag{30}$$

where $\Phi(\cdot)$ represents the transition function implied in evolution process (21).

Let the $|R| \times |R|$ matrix $\mathbf{M}_{\mathrm{B}}(\mathbf{B}^{\mathrm{un},t})$ denote the Jacobian matrix of $\Phi(\mathbf{B}^{\mathrm{un},t})$ with respect to $\mathbf{B}^{\mathrm{un},t}$, namely, $\mathbf{M}_{\mathrm{B}}(\mathbf{B}^{\mathrm{un},t}) = \mathbf{Jac}[\Phi(\mathbf{B}^{\mathrm{un},t})]$, and let $\mathbf{M}_{\mathrm{B}}^*$ denote the result of $\mathbf{M}_{\mathrm{B}}(\mathbf{B}^{\mathrm{un},t})$ computed at the fixed point $\mathbf{B}^{\mathrm{un},t} = \mathbf{B}^{\mathrm{un},*}$. According to nonlinear dynamic system theory, the fixed-point asymptotic stability is guaranteed if and only if the eigenvalues of $\mathbf{M}_{\mathrm{B}}^*$ are all within the unit circle.

Define $\mathbf{J}_{\mathrm{b}}(\mathbf{x}^t)$ to indicate the Jacobian matrix of $\mathbf{b}^t(\mathbf{x}^t, \rho)$ with respect to $\mathbf{x}^t$, $\mathbf{J}_{\mathrm{b}}(\mathbf{x}^t) = \mathbf{Jac}[\mathbf{b}^t(\mathbf{x}^t, \rho)]$. Meanwhile, using $\mathbf{J}_{\mathrm{L}}^{\mathrm{un}}(\mathbf{B}^{\mathrm{un},t})$ to indicate the Jacobian of the link flow loading for uninformed travelers with respect to the route TTB perceptions $\mathbf{B}^{\mathrm{un},t}$, that is, $\mathbf{J}_{\mathrm{L}}^{\mathrm{un}}(\mathbf{B}^{\mathrm{un},t}) = \mathbf{Jac}\left[\sum_w q_w \boldsymbol{\Lambda}_w \mathbf{p}_w^{\mathrm{un},t}(\mathbf{B}_w^{\mathrm{un},t}, \theta^{\mathrm{un}})\right]$. Similarly, using $\mathbf{J}_{\mathrm{L}}^{\mathrm{in}}(\mathbf{b}^t)$ to indicate the Jacobian of the link flow loading for informed travelers with respect to the actual route TTBs $\mathbf{b}^t$, $\mathbf{J}_{\mathrm{L}}^{\mathrm{in}}(\mathbf{b}^t) = \mathbf{Jac}\left[\sum_w q_w \boldsymbol{\Lambda}_w \mathbf{p}_w^{\mathrm{in},t}(\mathbf{b}_w^t, \theta^{\mathrm{in}})\right]$. In addition, let $\mathbf{J}_{\mathrm{b}}^*$, $\mathbf{J}_{\mathrm{L}}^{\mathrm{un},*}$ and $\mathbf{J}_{\mathrm{L}}^{\mathrm{in},*}$ denote, respectively, the results of the matrices $\mathbf{J}_{\mathrm{b}}(\mathbf{x}^t)$, $\mathbf{J}_{\mathrm{L}}^{\mathrm{un}}(\mathbf{B}^{\mathrm{un},t})$ and $\mathbf{J}_{\mathrm{L}}^{\mathrm{in}}(\mathbf{b}^t)$ evaluated at the fixed point. The following lemma establishes the expression of $\mathbf{M}_{\mathrm{B}}^*$ by using $\mathbf{J}_{\mathrm{b}}^*$, $\mathbf{J}_{\mathrm{L}}^{\mathrm{un},*}$ and $\mathbf{J}_{\mathrm{L}}^{\mathrm{in},*}$.

Lemma 1. *At the fixed point, the $|R| \times |R|$ Jacobian matrix of the DTD dynamic system (Equation 21), namely $\mathbf{M}_{\mathrm{B}}^*$, can be expressed as the combination of $\mathbf{J}_{\mathrm{b}}^*$, $\mathbf{J}_{\mathrm{L}}^{\mathrm{un},*}$ and $\mathbf{J}_{\mathrm{L}}^{\mathrm{in},*}$ by the following formula:*

$$\mathbf{M}_{\mathrm{B}}^* = (1-\alpha)\cdot \mathbf{I}' + \alpha \cdot (1-\eta)\cdot \mathbf{J}_{\mathrm{b}}^* \left[\mathbf{I} - \eta \cdot \mathbf{J}_{\mathrm{L}}^{\mathrm{in},*}\mathbf{J}_{\mathrm{b}}^*\right]^{-1} \mathbf{J}_{\mathrm{L}}^{\mathrm{un},*}, \tag{31}$$

where $\mathbf{I}$ and $\mathbf{I}'$ represent two unit matrices with respective dimensions $|L| \times |L|$ and $|R| \times |R|$.

Proof. Note that the DTD dynamic system (21) contains an implicit function $\mathrm{H}(\cdot)$, which defines the relation between the link flows $\mathbf{x}^t$ and the route TTB perceptions $\mathbf{B}^{\mathrm{un},t}$. Define $\mathbf{J}_x(\mathbf{B}^{\mathrm{un},t})$ to denote the Jacobian matrix of $\mathbf{x}^t$ with respect to $\mathbf{B}^{\mathrm{un},t}$, which is implicit in Equation 18. By computing the partial deviations of both sides of Equation 18 with respect to $\mathbf{B}^{\mathrm{un},t}$, we obtain an equation about $\mathbf{J}_x(\mathbf{B}^{\mathrm{un},t})$ as below:

$$\mathbf{J}_x(\mathbf{B}^{\mathrm{un},t}) = (1-\eta)\cdot \mathbf{J}_{\mathrm{L}}^{\mathrm{un}}(\mathbf{B}^{\mathrm{un},t}) + \eta \cdot \mathbf{J}_{\mathrm{L}}^{\mathrm{in}}(\mathbf{b}^t)\mathbf{J}_{\mathrm{b}}(\mathbf{x}^t)\mathbf{J}_x(\mathbf{B}^{\mathrm{un},t}), \tag{32}$$

and thus,

$$\mathbf{J}_x(\mathbf{B}^{\mathrm{un},t}) = (1-\eta)\cdot \left[\mathbf{I} - \eta \cdot \mathbf{J}_{\mathrm{L}}^{\mathrm{in}}(\mathbf{b}^t)\mathbf{J}_{\mathrm{b}}(\mathbf{x}^t)\right]^{-1} \mathbf{J}_{\mathrm{L}}^{\mathrm{un}}(\mathbf{B}^{\mathrm{un},t}), \tag{33}$$

provided that the matrix $\left[\mathbf{I} - \eta \cdot \mathbf{J}_{\mathrm{L}}^{\mathrm{in}}(\mathbf{b}^t)\mathbf{J}_{\mathrm{b}}(\mathbf{x}^t)\right]$ is invertible. Now, we can calculate the partial deviations of Equation 21 to derive out the expression of $\mathbf{M}_{\mathrm{B}}(\mathbf{B}^{\mathrm{un},t})$ as below,

$$\begin{aligned}\mathbf{M}_{\mathrm{B}}(\mathbf{B}^{\mathrm{un},t}) &= (1-\alpha)\cdot \mathbf{I}' + \alpha \cdot \mathbf{J}_{\mathrm{b}}(\mathbf{x}^t)\mathbf{J}_x(\mathbf{B}^{\mathrm{un},t}) \\ &= (1-\alpha)\cdot \mathbf{I}' + \alpha \cdot (1-\eta)\cdot \mathbf{J}_{\mathrm{b}}(\mathbf{x}^t) \\ &\quad \left[\mathbf{I} - \eta \cdot \mathbf{J}_{\mathrm{L}}^{\mathrm{in}}(\mathbf{b}^t)\mathbf{J}_{\mathrm{b}}(\mathbf{x}^t)\right]^{-1} \mathbf{J}_{\mathrm{L}}^{\mathrm{un}}(\mathbf{B}^{\mathrm{un},t}).\end{aligned} \tag{34}$$

By substituting the fixed point into the above formula, we can obtain the result (Equation 31).

In the following, we will introduce some additional assumptions into the dynamic system, to simplify the theoretical analysis of fixed-point stability.

Assumption 5. *The probabilistic route choice model for the uninformed travelers $P^{\mathrm{un}}(\cdot)$ follows the same functional form and dispersion parameter as that for the informed travelers $P^{\mathrm{in}}(\cdot)$.*

Mathematically, Assumption 5 contributes to the result: $\mathbf{J}_{\mathrm{L}}^{\mathrm{un},*} = \mathbf{J}_{\mathrm{L}}^{\mathrm{in},*} = \mathbf{J}_{\mathrm{L}}^*$. Actually, this assumption implies that the random errors produce identical effects on the route choice results of two traveler groups. By adopting this assumption, the following lemma presents the specific eigenvalues of $\mathbf{M}_{\mathrm{B}}^*$, which can be further used to deduce the fixed-point asymptotic stability condition later.

Lemma 2. *Suppose that Assumption 5 holds. Then the eigenvalues of matrix* $\mathbf{M}_B{}^*$*, denoted as* λ*, can be related to the eigenvalues of* $\mathbf{J}_b{}^*\mathbf{J}_L{}^*$ *as follows*:

$$\lambda = 1 - \alpha + \alpha \cdot (1 - \eta) \cdot \frac{\omega}{1 - \eta \cdot \omega}, \tag{35}$$

where $\omega(\omega \neq 1/\eta)$ *denotes one of the eigenvalues of matrix* $\mathbf{J}_b{}^*\mathbf{J}_L{}^*$.

Proof. Under the situation of Assumption 5, the Jacobian matrix $\mathbf{M}_B{}^*$ presented in lemma 1 can be reformulated as below:

$$\mathbf{M}_B{}^* = (1 - \alpha) \cdot \mathbf{I}' + \alpha \cdot (1 - \eta) \cdot \mathbf{J}_b{}^*[\mathbf{I} - \eta \cdot \mathbf{J}_L{}^*\mathbf{J}_b{}^*]^{-1}\mathbf{J}_L{}^*. \tag{36}$$

Define λ' to indicate one of $|R|$ eigenvalues of the matrix $\mathbf{J}_b{}^*[\mathbf{I} - \eta \cdot \mathbf{J}_L{}^*\mathbf{J}_b{}^*]^{-1}\mathbf{J}_L{}^*$. Then, through matrix algebra, λ' is corresponding to an eigenvalue of $\mathbf{J}_b{}^*\mathbf{J}_L{}^*$ and can be expressed as below,

$$\lambda' = \frac{\omega}{1 - \eta \cdot \omega}, \tag{37}$$

provided that ω is not equal to $1/\eta$. Therefore,

$$\begin{aligned}\lambda &= 1 - \alpha + \alpha \cdot (1 - \eta) \cdot \lambda' \\ &= 1 - \alpha + \alpha \cdot (1 - \eta) \cdot \frac{\omega}{1 - \eta \cdot \omega}.\end{aligned}$$

We can now reach the following theorem about the specific fixed-point asymptotic stability condition for the proposed DTD dynamic system.

Theorem 4. (fixed-point asymptotic stability) *Suppose that Assumption 5 holds. If the Jacobian matrix of the link travel time function is diagonal and positive definite (as assumed for the separable BPR formula), and the Jacobian matrix of the probabilistic route choice function, namely* $\mathbf{J}_L{}^*$*, is symmetric and negative semidefinite (as it occurs for the regular Logit choice model), then all the eigenvalues of matrix* $\mathbf{J}_b{}^*\mathbf{J}_L{}^*$ *are real and non-positive. Under this situation, the fixed-point asymptotic stability condition is given as below*:

$$\alpha < \frac{2 \cdot (1 - \eta \cdot \omega)}{1 - \omega}. \tag{38}$$

Proof. Given that all the eigenvalues of matrix $\mathbf{J}_b{}^*\mathbf{J}_L{}^*$ are non-positive real numbers, $\omega \leq 0$, then the eigenvalues of $\mathbf{M}_B{}^*$ are also real and less than one. That is,

$$\begin{aligned}\lambda &= 1 - \alpha + \alpha \cdot (1 - \eta) \cdot \frac{\omega}{1 - \eta \cdot \omega} \\ &= 1 + \frac{\omega - 1}{1 - \eta \cdot \omega} \cdot \alpha < 1.\end{aligned} \tag{39}$$

To guarantee that the fixed point is asymptotically stable, the absolute value of λ should be less than one, $-1 < \lambda < 1$, and therefore,

$$1 + \frac{\omega - 1}{1 - \eta \cdot \omega} \cdot \alpha > -1. \tag{40}$$

By putting the parameter α alone on the left side, we can reformulate the above inequality equivalently as the stability condition (Equation 38).

Theorem 4 shows that, to assure the fixed-point asymptotic stability condition of the proposed DTD dynamic system, the adopted preference parameter α should not exceed an upper bound. Define $\Gamma(\eta, \omega)$ to represent the stability bound of α,

$$\Gamma(\eta, \omega) = \frac{2 \cdot (1 - \eta \cdot \omega)}{1 - \omega}. \tag{41}$$

then the asymptotic stability condition (Equation 38) can be expressed as below,

$$\alpha < \Gamma(\eta, \omega). \tag{42}$$

In the next subsection, we will conduct some analyses of $\Gamma(\eta, \omega)$ to investigate the relations between fixed-point stability, ATIS market penetration level, and travelers' attitudes to risk.

Factors Influencing Fixed-Point Asymptotic Stability

We can first compute the partial derivative of $\Gamma(\eta, \omega)$ with respect to η, to check the effect of ATIS market penetration level on the fixed-point asymptotic stability. That is,

$$\frac{\partial \Gamma(\eta, \omega)}{\partial \eta} = \frac{-2\omega}{1 - \omega}. \tag{43}$$

If all the conditions in Theorem 4 are ensured, then we can know $\omega \leq 0$, and therefore $\partial \Gamma(\eta, \omega)/\partial \eta \geq 0$. This result shows that the stability bound of α will extend with the ATIS penetration level. In other words, the consistent predictive traffic information can enhance the fixed-point stability of the dynamic model. Actually, when $\eta \geq 1/2$, $\Gamma(\eta, \omega) > 1$, and then the stability condition (Equation 42) always holds.

Besides η, the matrix eigenvalue ω is also an important factor to influence the stability bound $\Gamma(\eta, \omega)$. The partial derivation of $\Gamma(\eta, \omega)$ with respect to ω can be calculated as follows:

$$\frac{\partial \Gamma(\eta, \omega)}{\partial \omega} = \frac{2 \cdot (1 - \eta)}{(1 - \omega)^2}. \tag{44}$$

We can also easily conclude that $\partial \Gamma(\eta, \omega)/\partial \omega > 0$. Note that the value of eigenvalue ω is determined by the matrix $\mathbf{J}_b{}^*\mathbf{J}_L{}^*$, which further depends on the degree of congestion in the transport network flow, travelers' perception errors about route TTB values, and their risk attitudes toward travel time uncertainty.

In general, an increase of network flow congestion (appearing as an increase of OD demand or a decrease of link capacity), or a decrease in travelers' perception errors (reflected by the increase of the dispersion parameter θ), may cause lower ω, and further lead to the decrease of $\Gamma(\eta, \omega)$. Therefore, more serious traffic congestion in the transport network, or more accurate perceptions of travelers, may be more likely to lead the fixed point toward instability.

In the proposed model, a higher value of ρ leads to a longer route TTB value $b_{w,r}(\mathbf{x}, \rho)$, which corresponds to a greater degree of degradation of link capacities. Take the risk-averse travelers, for instance, they may pay more attention on the serious capacity degradation situation since it will result in extremely long travel times, and thus they would mainly focus on extreme traffic congestion rather than the other mild situations. That is, as travelers' attitudes change from risk-prone to risk-averse, their concerns will shift from free flow condition to congestion condition. From the above analysis, we can infer that when travelers apply more extreme risk-averse attitudes to making route choices, the aggregated network flow will become more unstable.

Numerical Examples

The proposed DTD model is applied to the Nguyen Dupuis network, whose topology is illustrated in Figure 2*a*. There are four OD pairs in this network, 1–2, 1–3, 4–2, and 4–3 and the numbers of feasible routes between these OD pairs are 8, 6, 5, and 6 respectively. The OD demands are given as $q_{12} = 600$ flow units/hour, $q_{13} = 1000$ flow units/hour, $q_{42} = 800$ flow units/hour, $q_{43} = 400$ flow units/hour.

The link travel time function is of the BPR type:

$$c_l(x_l^t) = c_l^{\text{free}} \cdot \left[1 + 0.15 \cdot \left(\frac{x_l^t}{u_l}\right)^4\right], l = 1, 2, u_l \in [\varepsilon_l \bar{u}_l, \bar{u}_l]. \tag{45}$$

The free flow travel time c_l^{free} and the link design capacities $\bar{u}_l$ are given in Figure 2*b*. The worst-degraded capacity fractions for these links are set as $\varepsilon_l = 0.6, \forall l \in L$. The Logit model as formulated by Equation 5 is applied to describe travelers' stochastic route choices.

Suppose that, for each link l, its stochastic link capacity u_l is uniformly distributed in the variation range $[\varepsilon_l \bar{u}_l, \bar{u}_l]$. The mean and variance of link travel time can be derived as follows:

$$E(c_l(x_l^t)) = c_l^{\text{free}} - 0.05 c_l^{\text{free}} (x_l^t)^4 \frac{(1 - \varepsilon_l^{-3})}{\bar{u}_l^4 (1 - \varepsilon_l)}, \forall l \in L, \tag{46}$$

$$var(c_l(x_l^t)) = 0.0225 (c_l^{\text{free}})^2 (x_l^t)^8 \left\{ \frac{(1 - \varepsilon_l^{-7})}{-7 \bar{u}_l^8 (1 - \varepsilon_l)} - \frac{1}{9} \left[\frac{(1 - \varepsilon_l^{-3})}{\bar{u}_l^4 (1 - \varepsilon_l)} \right]^2 \right\}, \forall l \in L. \tag{47}$$

The derivation of Equations 46 and 47 can be found in Lo and Tung (5). This study assumes that link capacity (and hence link travel time) distributions are independent. Then the stochastic route travel time variable is

$$S_{w,r}^t = \sum_{l \in L} \Lambda_{lrw} \cdot c_l(x_l^t), \forall r \in R_w, w \in W. \tag{48}$$

Regardless of the perceived link travel time distribution, as long as the Lindeberg's condition is satisfied, the stochastic route travel time follows a normal distribution according to the central limit theorem. That is,

$$S_{w,r}^t \tilde{N}\left(E\left(S_{w,r}^t\right), var\left(S_{w,r}^t\right)\right), \forall r \in R_w, w \in W, \tag{49}$$

$$E\left(S_{w,r}^t\right) = \sum_{l \in L} \Lambda_{lrw} \cdot E(c_l(x_l^t)), \forall r \in R_w, w \in W, \tag{50}$$

$$var\left(S_{w,r}^t\right) = \sum_{l \in L} \Lambda_{lrw} \cdot var(c_l(x_l^t)), \forall r \in R_w, w \in W. \tag{51}$$

On this basis, the route TTB value $\mathbf{b}^t(\mathbf{x}^t, \rho)$ can be calculated as follows:

$$b_{w,r}^t(\mathbf{x}^t, \rho) = E\left(S_{w,r}^t\right) + \Psi^{-1}(\rho) \cdot \sqrt{var\left(S_{w,r}^t\right)}, \quad \forall r \in R_w, w \in W, \tag{52}$$

where $\Psi(\cdot)$ represents the cumulative distribution function of the standard normal distribution, and $\Psi^{-1}(\cdot)$ represents the inverse function of $\Psi(\cdot)$.

For this example, we first execute the DTD dynamic process (Equation 19) according to different combinations of parameters α and η, and meanwhile keep the WBTR level ρ and the dispersion parameters θ^{un}, θ^{in} unchanged as $\rho = 0.8$, $\theta^{\text{un}} = \theta^{\text{in}} = 0.3$. Here we introduce an index G^t to reflect the dynamic network flow evolution process:

$$G^t = \left(\mathbf{x}^t - (1 - \eta) \cdot \sum_w q_w \mathbf{\Lambda}_w \mathbf{p}_w^{\text{un},t}(\mathbf{b}_w^t(\mathbf{x}^t, \rho), \theta^{\text{un}}) - \eta \cdot \sum_w q_w \mathbf{\Lambda}_w \mathbf{p}_w^{\text{in},t}(\mathbf{b}_w^t(\mathbf{x}^t, \rho), \theta^{\text{in}})\right)^T \cdot \left(\mathbf{x}^t - (1 - \eta) \cdot \sum_w q_w \mathbf{\Lambda}_w \mathbf{p}_w^{\text{un},t}(\mathbf{b}_w^t(\mathbf{x}^t, \rho), \theta^{\text{un}}) - \eta \cdot \sum_w q_w \mathbf{\Lambda}_w \mathbf{p}_w^{\text{in},t}(\mathbf{b}_w^t(\mathbf{x}^t, \rho), \theta^{\text{in}})\right). \tag{53}$$

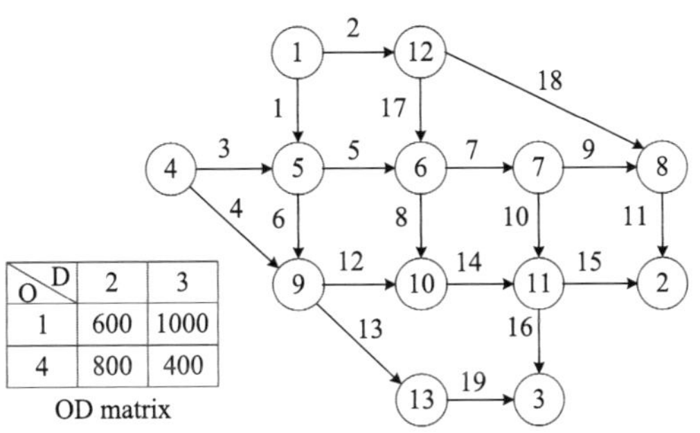

(a)

l	c_l^{free} (min)	$\bar{u}_l$ (flow units/hour)	l	c_l^{free} (min)	$\bar{u}_l$ (flow units/hour)
1	5	900	11	9	500
2	7	700	12	6	800
3	7	700	13	12	500
4	12	500	14	6	800
5	6	800	15	5	900
6	5	800	16	6	800
7	7	900	17	7	600
8	5	800	18	20	300
9	7	900	19	10	600
10	6	900			

(b)

Figure 2. Test network topology and link parameters.

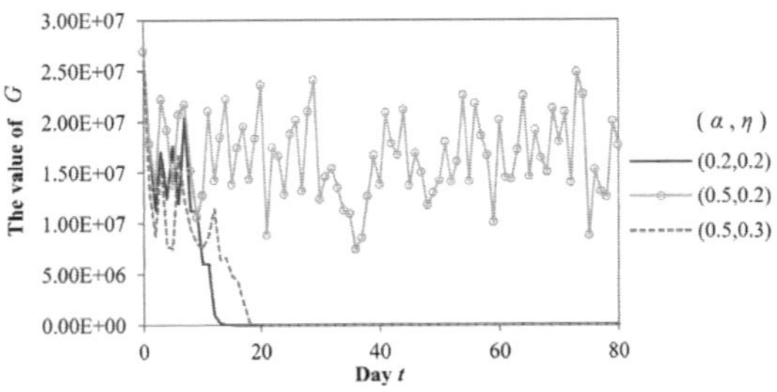

Figure 3. Evolutions of G^t under three different combinations of parameters α and η ($\rho = 0.8$, $\theta^{un} = \theta^{in} = 0.3$).

Obviously, $G^t = 0$ if and only if

$$\mathbf{x}^t = (1-\eta)\cdot\sum_w q_w \mathbf{\Lambda}_w \mathbf{p}_w^{\text{un},t}\left(\mathbf{b}_w^t(\mathbf{x}^t,\rho),\theta^{\text{un}}\right) + \eta\cdot\sum_w q_w \mathbf{\Lambda}_w \mathbf{p}_w^{\text{in},t}\left(\mathbf{b}_w^t(\mathbf{x}^t,\rho),\theta^{\text{in}}\right), \tag{54}$$

which exactly conforms to the extended STTBE state (Equation 26).

Figure 3 compares the dynamic evolution trajectories of G^t under three different combinations of α and η. When $(\alpha,\eta) = (0.2, 0.2)$, the DTD dynamic system can converge to the unique fixed point (namely STTBE state) after performing a period of flow fluctuation process. When the preference parameter α increases to 0.5, however, the value of G^t will always fluctuate and can never drop to zero. This actually corresponds to an unstable network flow situation. Under the case of $\alpha = 0.5$, we increase the value of η from 0.2 to 0.3, to examine its effect on the network flow evolution process. We can find the increase of η can mitigate network flow fluctuations caused by large preference parameter α. The above comparison results verify the positive influence of consistent traffic information prediction on the stability of DTD network flow evolution.

We can further investigate the effect of travelers' attitudes to risk by designing an additional sensitivity experiment to measure the WBTR level ρ. Specifically, we assign two values to the parameter ρ to reflect different traveler risk attitudes. For each ρ we substitute different values of α and η into the DTD model, to test the stability of the network flow evolution. Then we can roughly depict the stability and instability regions on the $\alpha - \eta$ coordinate plane, according to the above test results. Figure 4 compares the stability (instability) region of the DTD model under two different values of parameter ρ.

As we have verified, larger α or smaller η is more likely to cause an unstable network flow evolution process. Therefore, the instability regions in Figure 4 are both located in the lower right corner of the $\alpha - \eta$ coordinate plane. We can find that the instability region enlarges with the increase of travelers' WBTR level ρ. This result verifies that travelers' extreme risk-averse route choice behaviors, which are associated with high ρ-value, will give rise to more unstable network flow evolution.

Conclusion and Future Work

Our study contributes to the literature by developing an extended DTD dynamic model to describe the interactive route choice behaviors of ATIS-equipped and unequipped travelers over a degradable transport network. Theoretical analysis and numerical experiment results both verify that ATIS consistent traffic information prediction can enhance the stability of DTD network flow evolution, while travelers' extreme risk-averse route choice behaviors, on the other hand, would cause greater network flow fluctuations. This study provides important understanding of travelers' risk-taking behaviors under the influence of

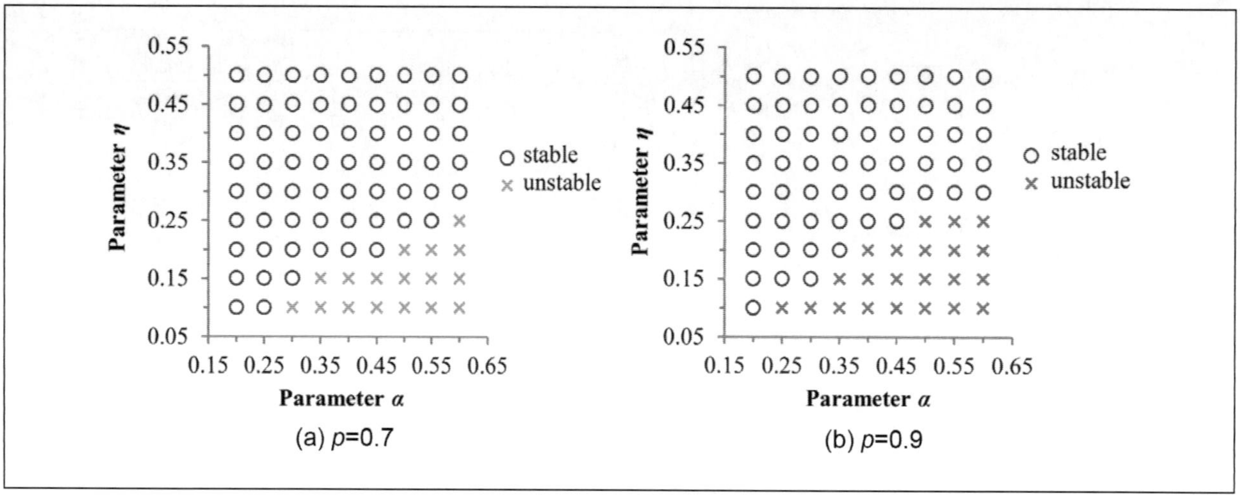

Figure 4. Comparison of stability of the DTD model under different traveler risk attitudes ($\theta^{un} = \theta^{in} = 0.3$).

traffic information, which will assist decision makers in designing effective traffic information release strategies that consider behavioral dynamics of individuals' travel choices on stochastic transport networks.

For the needs of practical application, further studies should be carried out with collection of empirical data for model calibration and validation. This study employs a consistent prediction mechanism to simulate the traffic information strategy of ATIS. However, the consistent condition is an idealized assumption which may not completely conform to the actual case. Therefore, it is meaningful to consider more realistic traffic information prediction processes in the future work. In addition, to simplify analysis of fixed-point stability of the proposed DTD dynamic system, this paper imposes several strict assumptions for the link cost function and the probabilistic route choice model. It would be interesting to relax these assumptions and to drive more universal stability conditions.

Acknowledgments

This work is supported by the Natural Science Foundation of China (grant nos. 51378119, 51578150, 51608115), the Natural Science Foundation of Jiangsu Province (BK20150613), and the Projects of International Cooperation and Exchange of the National Natural Science Foundation of China (No. 51561135003). Comments provided by three anonymous referees are much appreciated.

Author Contributions

Conceptualization: Lin Cheng, Xiaoming Lou; Methodology: Xiaoming Lou, Lin Cheng, Dawei Li; Numerical simulation: Xiaoming Lou, Jing Zhou; Analysis and interpretation of results: Xiaoming Lou, Dawei Li; Supervision: Lin Cheng; Validation: Lin Cheng, Dawei Li; Draft manuscript preparation: Xiaoming Lou, Shibei Zhu, Jing Zhou.

References

1. Abdel-Aty, M. A., R. Kitamura, and P. P. Jovanis. Exploring Route Choice Behavior Using Geographical Information System-based Alternative Paths and Hypothetical Travel Time Information Input. *Transportation Research Record: Journal of the Transportation Research Board*, 1995. 1493: 74–80.
2. Brownstone, D., A. Ghosh, T. F. Golob, C. Kazimi, and D.V. Amelsfort. Drivers' Willingness-to-pay to Reduce Time: Evidence from the San Diego I-15 Congestion Pricing Project. *Transportation Research Part A: Policy and Practice*, Vol. 37, 2003, pp. 373–387.
3. Asensio, J., and A. Matas. Commuters' Valuation of Travel Time Variability. *Transportation Research Part E: Logistics and Transportation Review*, Vol. 44, 2008, pp. 1074–1085.
4. Uchida, T., and Y. Iida. Risk Assignment: A New Traffic Assignment Model Considering the Risk Travel Time Variation. *Proc., 12th International Symposium on Transportation and Traffic Theory*, Berkeley, California, 1993, pp. 89–105.
5. Lo, H. K., and Y. K. Tung. Network with Degradable Links: Capacity Analysis and Design. *Transportation Research Part B: Methodological*, Vol. 37, 2003, pp. 345–363.
6. Shao, H., W. H. K. Lam, and M. L. Tam. A Reliability-based Stochastic Traffic Assignment Model for Network with Multiple User Classes under Uncertainty in Demand. *Network and Spatial Economics*, Vol. 6, 2006, pp. 173–204.
7. Siu, B. W. Y., and H. K. Lo. Doubly Uncertain Transport Network: Degradable Link Capacity and Perception Variations in Traffic Conditions. *Transportation Research Record: Journal of the Transportation Research Board*, 2006. 1964: 59–69.
8. Szeto, W. Y., L. O'Brien, and M. O'Mahony. Risk-Averse Traffic Assignment with Elastic Demands: NCP Formulation and Solution Method for Assessing Performance Reliability. *Networks and Spatial Economics*, Vol. 6, 2006, pp. 313–332.

9. Shao, H., W. H. K. Lam, M. L. Tam, and X. M. Yuan. Modeling Rain Effects on Risk-Taking Behaviors of Multi-User Classes in Road Network with Uncertainty. *Journal of Advanced Transportation*, Vol. 42, 2008, pp. 265–290.
10. Siu, B. W. Y., and H. K. Lo. Travel Time Budget in Doubly Uncertain Transport Network: Degradable Capacity and Stochastic Demand. *European Journal of Operational Research*, Vol. 191, 2008, pp. 166–181.
11. Lam, W. H. K., H. Shao, and A. Sumalee. Modeling Impacts of Adverse Weather Conditions on a Road Network with Uncertainties in Demand and Supply. *Transportation Research Part B: Methodological*, Vol. 42, 2008, pp. 890–910.
12. Sumalee, A., K. Uchida, and W. H. K. Lam. Stochastic Multi-Modal Transport Network under Demand Uncertainties and Adverse Weather Condition. *Transportation Research Part C: Emerging Technologies*, Vol. 19, 2011, pp. 338–350.
13. Lo, H. K., X. W. Luo, and B. W. Y. Siu. Degradable Transport Network: Travel Time Budget of Travelers with Heterogeneous Risk Aversion. *Transportation Research Part B: Methodological*, Vol. 40, 2006, pp. 792–806.
14. Chen, A., and Z. Zhou. The α-Reliable Mean-Excess Traffic Equilibrium Model with Stochastic Travel Times. *Transportation Research Part B: Methodological*, Vol. 44, 2010, pp. 493–513.
15. Cantarella, G. E., and E. Cascetta. Dynamic Processes and Equilibrium in Transportation Networks: Towards a Unifying Theory. *Transportation Science*, Vol. 29, 1995, pp. 305–329.
16. Watling, D. P., and M. L. Hazelton. The Dynamics and Equilibria of Day-to-Day Assignment Models. *Networks and Spatial Economics*, Vol. 3, 2003, pp. 349–370.
17. Watling, D. P., and G. E. Cantarella. Modelling Sources of Variation in Transportation Systems: Theoretical Foundations of Day-to-Day Dynamic Models. *Transportmetrica B: Transport Dynamics*, Vol. 1, 2013, pp. 3–32.
18. Cantarella, G. E., and D. P. Watling. A General Stochastic Process for Day-to-Day Dynamic Traffic Assignment: Formulation, Asymptotic Behavior, and Stability Analysis. *Transportation Research Part B: Methodological*, Vol. 92, 2016, pp. 3–21.
19. Cantarella, G. E., and D. P. Watling. Modelling Road Traffic Assignment as a Day-to-Day Dynamic, Deterministic Process: A Unified Approach to Discrete- and Continuous-Time Models. *EURO Journal on Transportation and Logistics*, Vol. 5, 2016, pp. 69–98.
20. Friesz, T. L., D. Bernstein, N. J. Mehta, R. L. Tobin, and S. Ganjalizadeh. Day-to-Day Dynamic Network Disequilibria and Idealized Traveler Information Systems. *Operations Research*, Vol. 42, 1994, pp. 1120–1136.
21. Huang, H. J., T. L. Liu, and H. Yang. Modeling the Evolutions of Day-to-Day Route Choice and Year-to-Year ATIS Adoption with Stochastic User Equilibrium. *Journal of Advanced Transportation*, Vol. 42, 2008, pp. 111–127.
22. Cantarella, G. E. Day-to-Day Dynamic Models for Intelligent Transportation Systems Design and Appraisal. *Transportation Research Part C: Emerging Technologies*, Vol. 29, 2013, pp. 117–130.
23. Bifulco, G. N., G. E. Cantarella, F. Simonelli, and P. Velona. Advanced Traveler Information Systems under Recurrent Traffic Conditions: Network Equilibrium and Stability. *Transportation Research Part B: Methodological*, Vol. 92, 2016, pp. 73–87.
24. Lou, X. M., L. Cheng, and Z. M. Chu. Modelling Travellers' En-Route Path Switching in a Day-to-Day Dynamical System. *Transportmetrica B: Transport Dynamics*, Vol. 5, 2017, pp. 17–41.
25. Herbert. W., F. Mili. Route Guidance: State of the Art vs. State of the Practice, In *Proceedings of the 2008 IEEE Intelligent Vehicles Symposium*, Eindhoven, Netherlands, 2008, pp. 1167–1174
26. Li, D., T. Miwa, T. Morikawa, and P. Liu. Incorporating Observed and Unobserved Heterogeneity in Route Choice Analysis with Sampled Choice Sets. *Transportation Research Part C: Emerging Technologies*, Vol. 67, 2016, pp. 31–46.
27. Li, D., T. Miwa, and T. Morikawa. Modeling Time-of-Day Car Use Behavior: A Bayesian Network Approach. *Transportation Research Part D: Transport and Environment*, Vol. 47, 2016, pp. 54–66.
28. Li, D., T. Miwa, and T. Morikawa. Considering En-Route Choices in Utility-Based Route Choice Modelling. *Networks and Spatial Economics*, Vol. 14, No. 3–4, 2014, pp. 581–604.

The Standing Committee on Transportation Network Modeling (ADB30) peer-reviewed this paper (18-01438).

Article

TRR
JOURNAL OF THE TRANSPORTATION RESEARCH BOARD

Transportation Research Record
2018, Vol. 2672(48) 24–35

DOI: 10.1177/0361198118780706
journals.sagepub.com/home/trr
SAGE

Level-Change Stackelberg Games Model for the Combined Traffic Assignment–Signal Control Equilibrium on Networks

Hang Yang[1], Zhongyu Wang[2], Yajie Zou[1], Bing Wu[1], and Xuesong Wang[1]

Abstract

Combined traffic assignment–signal control equilibrium is usually integrated into a non-cooperative games model between the network authority and road users. Unlike a pure Wardropian equilibrium, in reality there may be both competition and cooperation between authority and users. Authority has always been regarded as the upper level in classical bi-level formulations, but this placement may increase the difficulty of obtaining a global optimal solution between authority and users. This paper proposes a level-change Stackelberg (LC Stackelberg) model that embraces both authority–user and user–authority formulations. The model is calibrated by a model predictive control (MPC) controller. A route-choice probability model is used to estimate flow burden on two parallel routes. Meanwhile, the difference of route-choice probability between the two parallel paths is regarded as the level-change threshold. A generalized autoregressive conditional heteroscedasticity (GJR-GARCH) model is used as a triggering function in the MPC controller to fulfill the level-change procedure. A modified wavelet neural network algorithm is used to seek the global optimal solution. Cournot, Stackelberg, and Monopoly, combined with a fixed-time control policy based on the Webster method, were chosen as benchmarks in a numerical example to test model validity. The results show that the LC Stackelberg model obtains the minimum total travel time compared with other models. Furthermore, the level-change between authority and users could also decrease route choice probability on one specific path, indicating the model's potential application in urban networks.

The combination of dynamic traffic assignment (DTA) and dynamic traffic control is regarded as one of the most effective ways to deal with urban congestion, especially in metropolitan areas (*1*). The competition and cooperation between transportation management authorities and road users result in improved traffic control strategies and re-assignment of traffic flow in the network. For the fixed cycle length, users should fully obey the rules set by the authority. For adaptive signal control, the authority is required to adjust control strategies according to users' responses, and the authority may not always play the leading role in games.

This study introduces a level-change Stackelberg (LC Stackelberg) model, through which the upper level and lower level of a bi-level formulation can be changed dynamically. As a consequence, the relationship between authority and users is also changed according to traffic states. The changing threshold is the biggest difference between route-choice probability of users on those two paths. Meanwhile, a generalized autoregressive conditional heteroscedasticity model (GJR-GARCH) is used to fulfill the level-change when the threshold is met. The procedure of level-change is achieved in a centralized model predictive control (MPC) controller to obtain the global optimal solution, rather than the local optimal solution of the whole network.

The objective of this study includes two aspects: equity and efficiency. The equity objective seeks optimal equilibrium distribution of traffic flow in the entire network. For the efficiency aspect, the method obtains global

[1]Key Laboratory of Road and Traffic Engineering of Ministry of Education, Tongji University, Shanghai, China
[2]College of Transport and Communications, Shanghai Maritime University, Shanghai, China

Corresponding Author:
Address correspondence to Bing Wu: wubing@tongji.edu.cn

optimal total travel time by a modified wavelet neural network (WNN). The main contributions of the proposed method are as follows:

- Unlike most existing Stackelberg models, this study explores network performance under the level-change procedure between authority and users; therefore both authority–user's and user–authority's effects on network performance under different control strategies are presented.
- The study proposes a modified WNN algorithm to accelerate the convergence rate of optimization and avoid local optimal solutions as much as possible by improving the shift factor and the constriction factor of wavelet transform.
- The study addresses the combined traffic assignment–signal control problem in a centralized framework MPC, in which both user route-choice behavior and total travel time of network can be observed simultaneously after the level-change procedure.

Literature Review

Allsop first pointed out that the effects of signal settings on traffic flow pattern should be taken into account explicitly by combining traffic control and route choice (*2*). Following this research, a number of studies have sought the optimal solution of the combined traffic assignment–signal control problem. The existing approaches can be categorized into two major strands: (a) the iterative optimization and assignment procedure, and (b) global optimization models. The iterative optimization and assignment procedure focuses on the consistency of control policies. It mainly explores the long-term effects brought by signal control strategies on traffic flow distributions. Smith explored the traffic assignment–signal control problem under the junction interactions condition (*3*). The objective function that measures the extent to which a traffic distribution departs from equilibrium was set, and the optimal solution was obtained when the network cost–flow function is monotonic and continuously differentiable. Van Vliet proposed a simulation and assignment of traffic to urban road networks (SATURN) model to evaluate the performance of entire networks (*4*). An iterative optimization algorithm based on consistency analysis was deployed in this study. Al-Malik and Gartner investigated the deterministic route-choice equilibrium under a fixed-time control policy (*5*).

Global optimization models determine optimal control policies that account for travelers' route choices in response to changes in signal control strategies. These models were proposed to obtain equilibrium solutions of a network or comprehensive system. Friesz et al. formed the DTA problem into two continuous time formulations (*6*). Both system performance and dynamic traffic flow patterns were optimized with equilibrium. Gartner included traffic flow as a decision variable in the objective function of the usual traffic signal timing problem to obtain a control consistent equilibrium (*7*). Tian et al. integrated traffic assignment and signal control operations of surface street and freeway systems using the VISSIM software package and reduced delay in the mixed network (*8*). Mahmassani introduced a DTA system that considers evolution of traffic patterns for given traffic loading under particular control measures, as well as route guidance information to users (*9*). Papageorgiou took splitting rate as the key variable to obtain the optimal signal control approach under dynamic user optimal conditions (*10*). Yang and Yagar formulated this combined problem into a bi-level program in which the upper level is traffic control strategy and the lower level is static user equilibrium (*11*). Papageorgiou et al. presented greater detail about these studies (*12*). Combining traffic assignment and signal control means integrating optimization problems from both macroscopic and microscopic views, so a mature method for solving this type of combined problem is very much needed. Ban et al. used the differential complementarity systems to solve the continuous time dynamic user equilibrium (DUE) problem (*13*). Numerical results were presented to validate the instantaneous dynamic traffic equilibria on a small network and also on the Sioux Falls network. Ban et al. proposed a link-node complementarity model for the basic deterministic DUE problem with single-user-class and fixed demands (*14*). The proposed model captured the exact flow propagation constraints and the results showed that predictive DUE is preferable. Yu et al. proposed an optimal traffic signal control framework to minimize total travel time in a road network with signal control strategy (*15*). Numerical tests showed that the adaptive control taking signal timing into account on drivers' route travel time performs best. Liu and Smith presented a combined dynamical mathematical model consisting of day-to-day traffic re-routing and traffic signal control (*16*). Results showed that the dynamical system entered an approximate equilibria in a finite number of days or steps under certain conditions.

Game-theoretic formulation was one of the most widely used frameworks in solving traffic assignment–signal control problems (*17*), the four major game structures being Cournot (*18*), Monopoly (*19*), Stackelberg (*20*), and Cournot-Nash (*21*). Stackelberg formulations were more suitable for networks where cooperation and competition both exist (*12*). In many Stackelberg formulations, the upper level was for authority and the lower level for users, which meant that users would implement fully the optimal strategies set by authority. In reality, traffic flow patterns may not follow this plan exactly, especially when traffic conditions change. Some exceptions existed in Chiu and Mahmassani (*22*, *23*). In this study, the upper level was not only for authority, but also for users when authority was not able to minimize the system cost merely by declaring new strategies to users. The authority-leading and user-leading formulations were both used to evaluate system performance. These findings call for further insights into the game relationship between traffic management authorities and road users. The leader and follower in games may change when fluctuations exist in the system.

Methodology

The LC Stackelberg model can be divided into two modules: the route-choice module and the signal-control module. The route-choice module is operated by a simple route-choice model which can be seen in Haddad et al.'s study (*24*). Meanwhile, the signal-control model is calibrated in a MPC controller. The connection of these two modules is the level information. For simplicity, this study defines authority–user (authority for upper level) and user–authority (user for upper level). The model structure is presented in Figure 1; all parameters will be introduced in the experiments.

LC Stackelberg Formulation

In Stackelberg games, the relationship between the upper and lower levels of the bi-level formulation is leader–follower. The classical formulation assumes all users' (followers') information is accessible in the authority's (leader's) database. Authority is the holistic decision maker and users are required to obey all the rules set by authority. The format of the classical bi-level model for traffic assignment and control (authority–user) is presented in Equation 1:

$$\text{upper level : authority}$$
$$M \text{ in } Z^{\text{authority}}(\Lambda, x) = \sum_a x_a t_a(x_a, \lambda_a)$$
$$st \sum_{a \in L_j} \lambda_a = B_j, \forall j$$
$$\lambda_a^{\max} \geq \lambda_a \geq \lambda_a^{\min}, \forall a$$

$$\text{lower level : users} \tag{1}$$
$$M \text{ in } Z^{\text{user}}(x) = \sum_a \int_0^{x_a} t_a(w, \lambda_a) dw$$
$$st \sum_k f_k^{\text{rs}} = q^{rs}, \forall r, s$$
$$f_k^{\text{rs}} \geq 0, \forall r, s, k$$
$$x_a = \sum_{\text{rs}} \sum_k f_k^{\text{rs}} \delta_{a,k}^{\text{rs}}, \forall a \in A$$

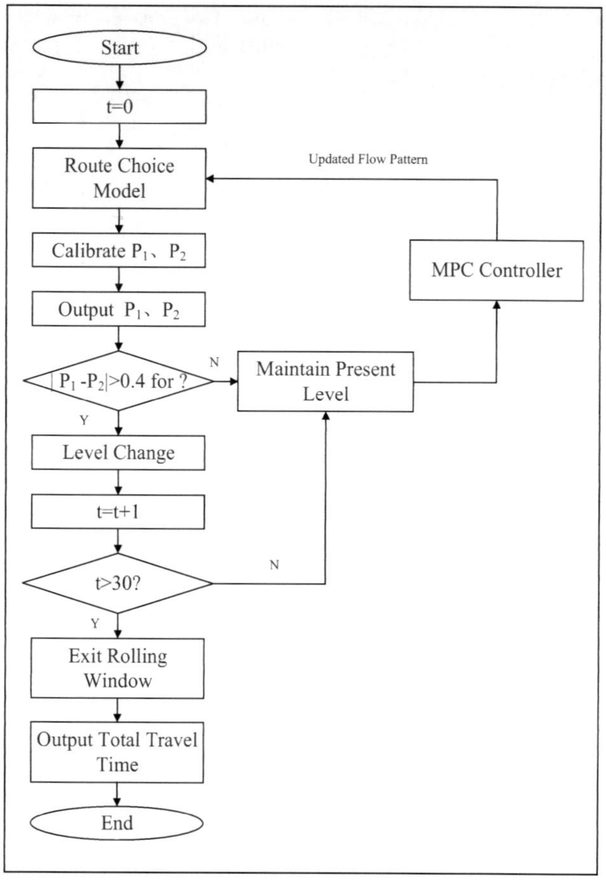

Figure 1. Calibration procedure for LC Stackelberg model.

In this equation, j is the set of intersections, Λ is the set of signal timing, x is the flow pattern, x_a is the flow on link a, t_a is the cost function on link a, q^{rs} is the demand of the (r, s) O-D pair, f_k^{rs} is the flow of the path k on (r, s), $\delta_{a,k}^{\text{rs}}$ is a 0-1 variable that decides whether link a is on path k, λ_a is the green time ratio of Approach a, $\lambda_a^{\min}$ and $\lambda_a^{\max}$ are the minimum and maximum values of green time ratio for each approach, L_j is the sum of approaches whose exit is in intersection j, B_j is the sum of green time ratio of all approaches of intersection j. In this format, the upper level is a signal optimization model that tries to minimize the total cost of the entire system. The lower level is a traffic distribution model that provides a signal optimization with link flow pattern that is decided by users' route choice. As the "sharing" variable, x is flow patterns information in the upper level and it is determined by users' optimal choice in the lower level.

However, in the combined traffic assignment–signal control problem, authority is unable to maintain its leading status in games. Authority tries to minimize the total cost for system optimization, while users hope to get the minimal travel cost, which is in accordance with user equilibrium condition. In such condition, no driver is able to minimize the travel cost just by changing route by himself. In reality, users' route-choice behavior is predicted by loop-detected data in adaptive control. However, the authority cannot always predict the actual route choice just by the loop-detected data in time. When such a situation occurs, users will act as the leading role on the basis of their initiative and practical experience. Then, users' route choices require the authority to follow their decisions and decide new control strategies, so user–authority is considered in this study as the extended form of authority–user:

$$
\begin{aligned}
&\text{upper level: user} \\
&Min\ Z^{user}(\Lambda, x) = \sum_a \int_0^{x_a} t_a(w, \lambda_a)dw \\
&st \sum_k f_k^{rs} = q^{rs}, \forall r, s \\
&f_k^{rs} \geq 0, \forall r, s, k \\
&x_a = \sum_{rs}\sum_k f_k^{rs}\delta_{a,k}^{rs}, \forall a \in A \\
&\text{lower level: authority} \\
&M\text{in } Z^{authority}(\Lambda) = \sum_a x_a t_a(x_a, \lambda_a) \\
&st \sum_{a \in L_j} \lambda_a = B_j, \forall j \\
&\lambda_a^{max} \geq \lambda_a \geq \lambda_a^{min}, \forall a
\end{aligned}
\tag{2}
$$

As the "sharing" variable, Λ is determined by the optimal signal control strategy set by the authority. It should be noted that level-change does not mean changing upper level and lower level only in single authority–user formulations or user–authority formulations. Both authority–user and user–authority formulations are put into the MPC controller to obtain the optimal total travel time. The controller will choose one of them according to the flow patterns on two parallel paths at each time t, and the exchange relies on the level-change function.

MPC Controller and Route-Choice Model

In this paper, the MPC controller simulates the level-change procedure of LC Stackelberg model, as well as obtaining the optimal total travel time for the entire network with the modified WNN, which will be discussed later. The application of MPC for traffic assignment and control was first introduced in Gartner (*25*). Other studies that deal with MPC approaches for traffic management are Peek Traffic (*26*), Kotsialos et al. (*27*), and Diakaki et al. (*28*). Bellemans et al. (*29*) and Hegyi (*30*, *31*) indicated that MPC can be extended to coordinated control of urban networks.

Because of the ability of handling multi-input and output systems, the MPC controller can be applied in solving the combined signal control-traffic assignment problem, in which the signal control and traffic assignment are input and output for each other during the rolling procedure. Furthermore, the fewer numbers of parameter-setting decrease the computation cost. Authority–user and user–authority are equivalent in the MPC controller, and optimal total travel time is calibrated in a centralized framework. The working principle of MPC in this study is shown in Figure 2*a*. Output of MPC is the updated saturation flow rate and green time ratio of each approach, and the controller brings these outputs back to the route-choice model.

The route-choice module is the other major part of the LC Stackelberg model. In this study, one existing simple route-choice model is used, which can be seen in Haddad et al.'s study (*24*). The probability distribution (P_1, P_2) at time t on two parallel paths is calculated on the basis of the travel time value in t-1. In real urban network applications, the information about level-change can be delivered to local signal control by the MPC controller, which is shown in Figure 2*b*. The detailed formulation of route-choice model will not be discussed in this paper. To accomplish the level-change, a GJR-GARCH model is used, which uses 0-1 variable to represent the upper level at each time t.

Level-Change Function

It is critical to determine the changing threshold of the LC Stackelberg model. As mentioned in the introduction, level-change is aimed at obtaining the global optimal equilibrium distribution of traffic flow. Smith (*3*, *32*) pointed out that if the difference of route-choice probability between two parallel paths exceeds 0.4,which means one probability is more than 70% and the other one is less than 30%, then the traffic flow distribution on these two links can be regarded as unbalanced. In this paper, the level-change begins when such an "unbalanced" situation is detected, and so 0.4 is chosen as the level-change threshold in LC Stackelberg model. When the probability difference exceeds 0.4, authority–user (user–authority) is replaced with user–authority (authority–user).

A modified GJR-GARCH model based on 0-1 variables is used to accomplish the level-change. The GJR-GARCH model is a modified version of the GARCH model, which has been used successfully in economics researches on both symmetric and asymmetric analyses of the game relationship between two stakeholders (*33*, *34*). The expression of conditional heteroscedastic models is:

$$e_t = z_t \times \sqrt{h_t} \tag{3}$$

where $h_t = E(e_t^2|F_{t-1})$ and $\{z_t\}$ make up a sequence of independent and identically distributed random variables

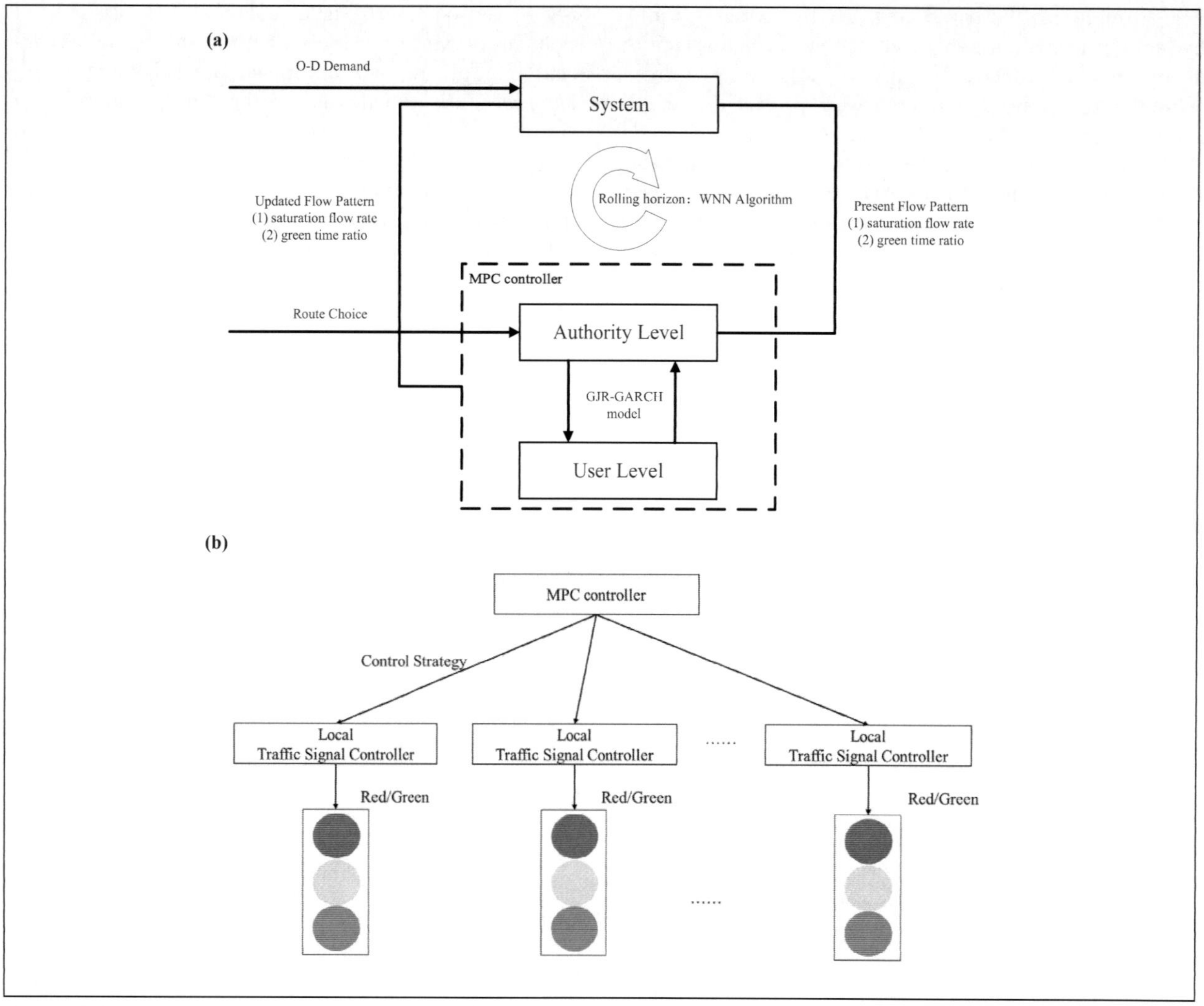

Figure 2. The MPC framework: (*a*) the working principle of MPC; (*b*) the potential application of the LC Stackelberg model in a real network.

with zero mean and unit variance, which can be assumed to follow a standard normal, standard student-t, or generalized error distribution. The difference of route-choice probability between two parallel paths is e_t. h_t is a transition variable in each time t, which acts as the connection between e_t and z_t. In this study, in order to make the level-change steadier, the change only happens when the defined "unbalanced" condition lasts more than three continuous time points. For example, if the difference is more than 0.4 at time t-2, t-1, and t, then level-change happens at time t+1.

Glosten et al. proposed the GJR-GARCH model with a conditional variance formulation, which successfully captured asymmetric responses of users and authorities in road networks (*34*). The GJR-GARCH approach is listed below:

$$h_t = a_0 + \sum_{i=1}^{r} \beta_i h_{t-i} + \sum_{j=1}^{u} (\alpha_j e_{t-j}^2 + \gamma_j e_{t-j}^2 I_{t-j}) \quad (4)$$

in which a_0 is a constant, and β_i, α_j, and γ_j are non-negative parameters. I_{t-j} is a binary variable for e_t:

$$I_{t-j} = \begin{cases} 1 \text{ if } e_{t-j} < 0 \\ 0 \text{ if } e_{t-j} \geq 0 \end{cases} \quad (5)$$

Equation 3 highlights the difference of e_{t-j} on the conditional variance according to whether the sign is positive or not. Variance I_{t-j} is an indicator as shown in Equation 5; whether it is 1 or 0 is set at first in the MPC controller by the state-setting module. The "sharing" variable is different in the authority–user format versus the user–authority format, so the GJR-GARCH model that has the ability to deal with asymmetric analysis (*35*, *36*) is used, rather than the conditional GARCH model. In the calibration procedure of the MPC controller, the value of variance I_{t-j} is recorded at each time point t in Matlab 2016. Here "1" is defined as authority–user, and "0" as user–authority, so the change between these two formats can be observed clearly in the change of the (0,1) pair.

Algorithm

Neural network is widely used as optimal algorithm in many existing studies (*37–39*). In this paper, an improved wavelet transform is used to accelerate the convergence rate of the WNN calibration procedure and avoid a local optimal solution of total travel time for the entire network.

The dynamic function represented by Equation 6 is approximated using the wavelet transform function and wavelet coefficient expressed in a general form as in Equation 7. More details about the wavelet transform can be found in Antonini et al. (*40*):

$$y(t) = f(\mathrm{y}^{t-1}) + \mathrm{g}(\mathrm{x}^{t}) + E(\mathrm{t}) \tag{6}$$

$$\overline{f(\mathrm{y})} = \sum_{i} w_i \sum_{a,b} \psi_{a,b}(\mathrm{y}) \tag{7}$$

$$i = 1, \ldots, M \ \ \mathrm{a, b} \in \Re \ \psi \in \mathrm{L}^2(\Re)$$

In Equation 6, where x^{t} and $\mathrm{y}^{\mathrm{t}-1}$ represent the current green time ratio status vector and feedback travel time input vector at time t, respectively, f(.) and g(.) are scalar nonlinear mapping functions, and $E(t)$ is the error between the actual and predicted values of future travel time output y(t). In Equation 7, M is the number of wavelets to be discussed in the next section, $\overline{f(\mathrm{y})}$ represents the approximation of f(y), w_i represents the discrete wavelet transform coefficient; and $\psi_{a,b}(\mathrm{y})$ is a two-dimensional wavelet expansion function obtained from the basic wavelet function $\varphi(\mathrm{t})$ by simple scaling and translation which is shown as follows:

$$\psi_{a,b}(\mathrm{y}) = \frac{1}{\sqrt{|a|}} \varphi\left(\frac{y-b}{a}\right) \tag{8}$$

in which the parameter vectors a≠0 and b denote the frequency (or scale) and the time (or space) location vectors corresponding to the multidimensional input vector y, respectively, and $\Re$ is the set of real numbers. The notation $\mathrm{L}^2(\Re)$ represents the square summable constructed state space vectors.

Equations 6, 7, and 8 are used to modify the shift factor and the constriction factor of WNN. The transfer function between input layer and hidden layer is set as follows:

$$h(\mathrm{j}) = h_j \left(\frac{\sum_{i=1}^{k} w_{\mathrm{ij}} x_i - b_j}{a_j} \right) j = 1, 2, \ldots, l \tag{9}$$

in which h(j) is the output of jth node in hidden layer, h_j is a basis transform, w_{ij} is the connection-weight between the input layer and the hidden layer, b_j is the shift factor of h(j), and a_j is the constriction factor of h(j). Meanwhile, the transfer function between the hidden layer and the output layer is presented in Equation 10:

$$\mathrm{p(k)} = \sum_{i=1}^{l} w_{\mathrm{ik}} \ \mathrm{p(i)} \tag{10}$$

in which w_{ik} is the weight between the hidden layer and the output layer, p(k) is the output of the kth node in the output layer, p(i) is the output of the ith node in the hidden layer, l is the number of nodes in the hidden layer, and m is the number of nodes in the output layer. The main steps of the WNN are listed as follows:

- Calculate the prediction error of WNN

$$e = \sum_{k=1}^{m} yn(\mathrm{k}) - \mathrm{y(k)} \tag{11}$$

- where yn(k) is the expected green time ratio and y(k) is the prediction value of the green time ratio.
- Correct the weights based on the prediction error

$$\begin{aligned}
w_{n,k}^{(i+1)} &= w_{n,k}^{i} + \Delta w_{n,k}^{(i+1)} \\
a_k^{(i+1)} &= a_k^{i} + \Delta a_k^{(i+1)} \\
b_k^{(i+1)} &= b_k^{i} + \Delta b_k^{(i+1)} \\
\Delta w_{n,k}^{(i+1)} &= -\eta \frac{\partial e}{\partial \Delta w_{n,k}^{(i)}} \\
\Delta a_k^{(i+1)} &= -\eta \frac{\partial e}{\partial \Delta a_k^{(i)}} \\
\Delta b_k^{(i+1)} &= -\eta \frac{\partial e}{\partial \Delta b_k^{(i)}}
\end{aligned} \tag{12}$$

where η is the learning rate of WNN.

Numerical Experiments

To test the validity of the proposed model, a numerical example is given. The Cournot game, Stackelberg games, and Monopoly games are taken as the benchmarks for LC Stackelberg. Meanwhile, a fixed-time control policy, which is based on classical Webster formulation, is also utilized for averaging traffic flow. The network shown in

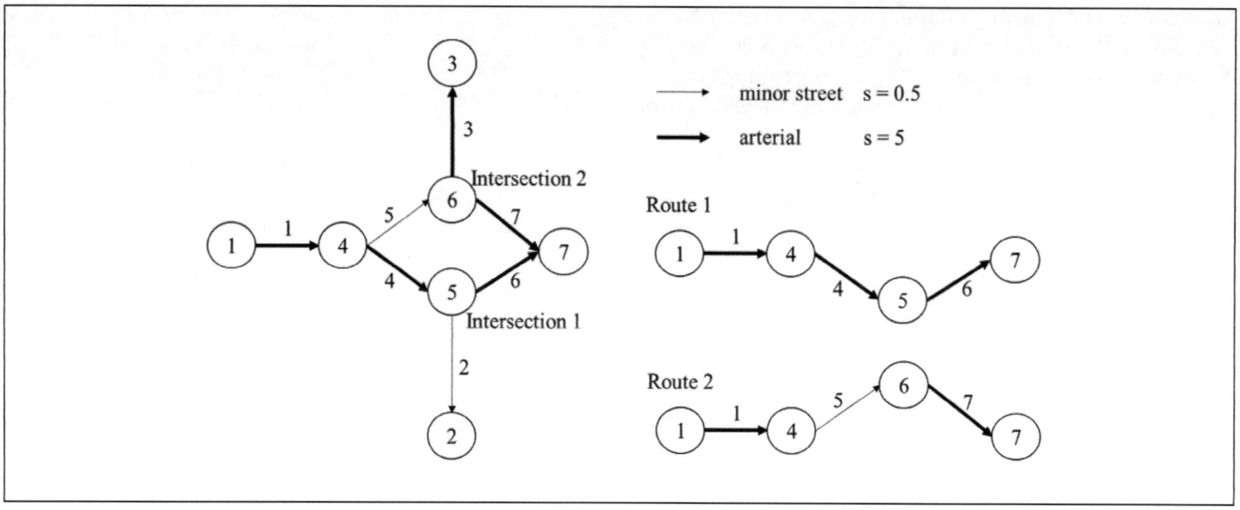

Figure 3. The network used in numerical example.

Figure 3 consists of seven nodes, seven links, and two signalized intersections. The entire network, which was successfully tested in Chen's research (*41*), includes three O-D pairs with nodes 1, 2, and 3 as the origin nodes, respectively, and node 7 as their common destination.

For simplicity of calibration, all the link costs consist of two parts in the assumption: a constant cruise time and a recurrent delay. The recurrent delay is concerned with signalized delay, which is modeled as a deterministic queuing delay in this study. The link cost function used in the numerical experiments is shown below:

$$\tau_a(t) = \tau_{a,0} + \frac{x_a(t)}{s_a \lambda_a(t)} \quad (13)$$

Where $\tau_a(t)$ is travel time on link a for traffic flow entering the link at time t.

$\tau_{a,0}$ is the constant cruise time for link a.

$x_a(t)$ is the number of vehicles on link a at time t.

s_a is the saturation flow rate of link a.

$\lambda_a(t)$ is the green time ratio for link a at time t.

O-D pair 1–7 consists of two selectable paths: route 1 and route 2. Route 1 consists of links 1, 4, and 6, and route 2 consists of links 1, 5, and 7. Links 1, 3, 6, 7 are set as arterial streets with the same capacity (s = 5) and the same cruise time ($\tau_{1,0} = \tau_{3,0} = \tau_{6,0} = \tau_{7,0} = 1$). Link 4 is a long-distance arterial, with saturation flow $s_4 = 5$ and cruise time $\tau_{4,0} = 20$. Link 2 and link 5 are two minor streets with the same capacity (s = 0.5) and the same cruise time ($\tau_{2,0} = \tau_{5,0} = 1$).

According to the route-choice model used in this paper, the probabilities of selecting route 1 (P_1) and route 2 (P_2) are calculated for each time t. When the absolute value of the difference between P_1 and P_2 is more than 0.4, that is:

$$|P_1 - P_2| > 0.4 \quad (14)$$

the level-change function GJR-GARCH is triggered and the change signal is captured by the MPC controller. The present authority–user (user–authority) will be replaced by user–authority (authority–user).

The total rolling horizon of MPC is from 0 to 30 (total time length = 30), and traffic flow of the time-dependent O-D demand is given as follows:

$$D^{1,7}(t) = \begin{cases} 20 + 3t & 0 < t \leq 15 \\ 65 - 3t & 15 < t \leq 30 \end{cases} \quad (14)$$

$$D^{2,7}(t) = 10 \quad 0 \leq t \leq 30 \quad (15)$$

$$D^{3,7}(t) = \begin{cases} 2 + 2t & 0 < t \leq 15 \\ 32 - 2t & 15 < t \leq 30 \end{cases} \quad (16)$$

Table 1. Comparison of All Model Performances

Controls	Total travel time ($*10^3$ seconds)	Optimality gap (%)
Webster	142.5	51.92
Cournot	107.3	14.39
Stackelberg	105.6	12.58
LC Stackelberg	93.8	0.00
Monopoly	103.8	

Results and Discussion

Table 1 presents the total travel time calibrated by five control strategies. The LC Stackelberg model proposed in this paper represents the optimal solution. The Monopoly game is the system-optimal structure, so it achieves the second most optimal performance. The

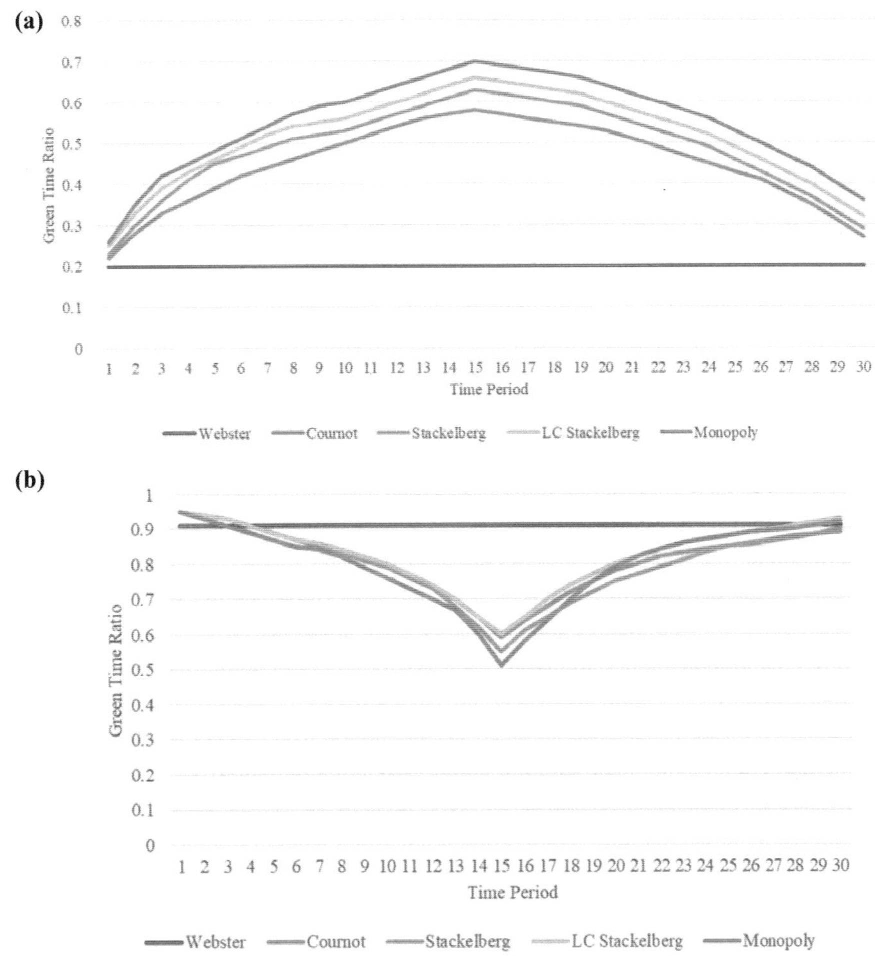

Figure 4. Green time ratio on intersections: (*a*) intersection 1; (*b*) intersection 2.

optimality gaps of other models with respect to LC Stackelberg model are also listed in Table 1. All four model solutions—Cournot, Stackelberg, LC Stackelberg, and Monopoly—decrease total travel time significantly when compared with the Webster method. With the level-change procedure, LC Stackelberg reduces the total travel time by 12.58%. Therefore, the performance of the entire network can potentially be improved by the method presented here. Table 1 also indicates that the LC Stackelberg equilibrium is better than that of Cournot, Stackelberg, or Monopoly, revealing that the level-change function used in the MPC controller is effective. Figure 4 shows the green time ratio settings for two intersections in the network.

Figure 4 shows the green time ratio settings for two intersections in the network. Figure 4*a* shows the green time ratio for link 4, $\lambda_4(t)$, of five different control strategies. A small green time ratio ($g_4 = 0.2$) is allocated in the Webster method. For the other four control methods, all the green time ratios are time-variant and reach their maximum value at t = 15 when traffic flow demand reaches its peak. Figure 4*b* presents the green time ratio for intersection 2. Link 5 is allocated with a large green time ratio ($g_5 = 0.9$) during the time period under the Webster control policy. The value of the green time ratio first decreases and then increases in the other four methods.

Figure 5 shows the level-change procedure and probability of user route choice calibrated in the MPC controller during each time period. Figure 5*a* shows which format (authority–user/user–authority) is selected by MPC at time t. Figure 5*b* shows the values of P_1 and P_2 calculated by route-choice model. These two show clearly that the level-change happens at times 7, 10, 13, 16, and 22. Take time 7 and time 8 as examples. At time 7, the "unbalanced" condition that is defined in section 3.3 has

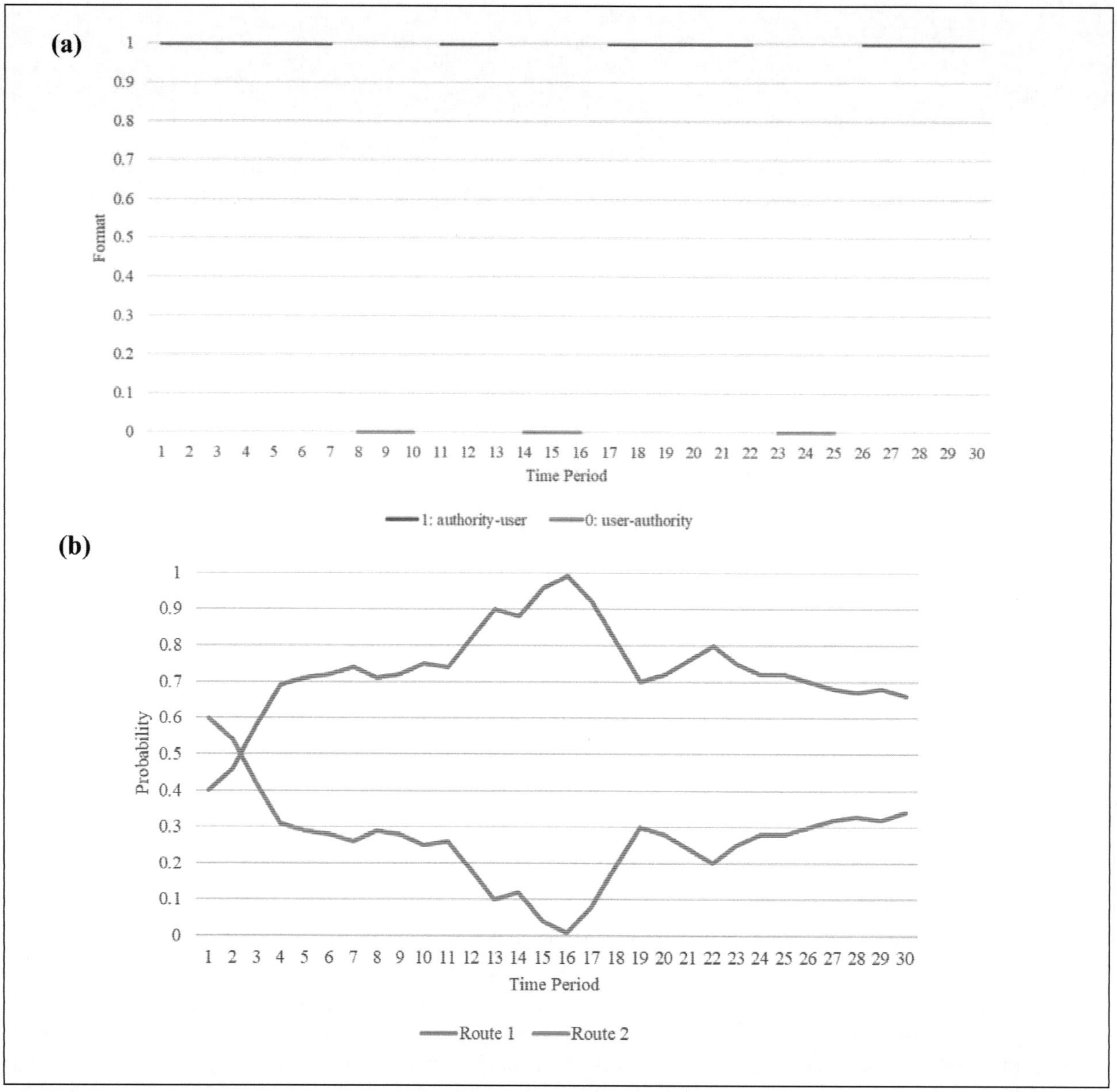

Figure 5. (*a*) level-change procedure during time period; (*b*) route-choice probability on two paths during time period.

lasted for three continuous time points (P_1 = 0.71, 0.72 and 0.74 in time 5, 6, and 7, respectively), so the level-change function is triggered by MPC at the end of time 7. The authority–user format is therefore replaced by user–authority format at the beginning of time 8.

P_1 reaches its highest value, which is very close to 100%, at time 15 and 16 when the demand is in peak value. It reveals that users are more likely to switch from Route 2 to Route 1 when traffic flow increases. When the time approaches time 15, user–authority is chosen by the MPC, which means that authorities are required to follow users' route choices and their corresponding responses to the network at that period of time. In peak time, authority–user is selected. Authority is in the upper level during this period of time, and it should set a new signal control strategy and predict users' responses. After the peak time, P_1 gradually decreases to about 0.65, and level-change in times 16 and 22 may cause this drop. Another finding is that the value of P_1 always experiences a slight decrease when level-change happens.

This observation reveals that the level-change procedure can potentially balance the traffic flow on two parallel links and improve the optimal solution for the entire network. This potential for improvement results mainly from the fact that level-change exists between authority and users in real life; it is not merely a theoretical construct. Its influence is strongest at times of near-peak, peak hour, and congestion dissipation. Therefore, level-change can happen when traffic states change, and the effects of users' route-choice behavior and authority's signal control decisions vary over time. Authority cannot always foresee users' behavior based solely on traffic flow predictions.

Conclusions and Future Work

This paper proposed an extension form of the classical Stackelberg formulations, LC Stackelberg model, which consists mainly of two parts: the authority–user and user–authority formats. These two formats fulfill the level-change procedure in the MPC controller through a level-change function. With a modified WNN algorithm, the green time ratio on each link is simulated and the optimal total travel time solution is obtained for the entire network. The effectiveness of the proposed method is verified and the following conclusions can be drawn:

- The level-change procedure of the LC Stackelberg model is successfully implemented and evaluated using the advanced modeling features provided by the MPC module in Matlab 2016. The MPC controller provided by Matlab allows users to develop and test various control algorithms, especially in modeling DTA and signal control strategies. With the modified WNN algorithm, the optimal solutions are obtained under five strategies.
- Compared with Webster, Cournot, Stackelberg, and Monopoly control strategies, the LC Stackelberg model produced the optimal total travel time solution, as well as the best equilibrium traffic flow pattern. The level-change function proves more effective in the LC Stackelberg model than the classical Stackelberg formulations.
- User is in the upper level when it is close to peak time. While in peak duration, authority plays the leading role in the entire network. The level-change that happens after the peak time potentially reduces P_1 value and then eases traffic burden on the path. Furthermore, after each level-change, the P_1 value drops slightly, which means the level-change may guide users to a relatively uncongested path.

The work done in this paper provides insights into the game between authority and user. The threshold of level-change should be reset when the actual data are available in specific area. Meanwhile, more factors that may influence users' route choice should be considered in future research, and the searching rate of the algorithm needs further acceleration when dealing with a multi-path case or a non-enumerated case. Further study is currently underway to include an actual road network, which consists of expressway systems and surface streets in Shanghai, China, and to test integration strategies with LC Stackelberg on a simulation platform with TransModeler.

Author Contributions

The authors confirm contribution to the paper as follows: study conception and design: Bing Wu, Hang Yang, Zhongyu Wang; data collection: Hang Yang, Yajie Zou; analysis and interpretation of results: Hang Yang, Yajie Zou; draft manuscript preparation: Bing Wu, Hang Yang; draft manuscript revision: Zhongyu Wang, Xuesong Wang. All authors reviewed the results and approved the final version of the manuscript.

References

1. Gartner, N., and C. Stamatiadis. Integration of Dynamic Traffic Assignment with Real-time Traffic Adaptive Control System. *Transportation Research Record: Journal of the Transportation Research Board*, 1998. 1644: 150–156.
2. Allsop, R. E. Some Possibilities for Using Traffic Control to Influence Trip Distribution and Route Choice. *Proc., 6th International Symposium on Transportation and Traffic Theory* (Buckley, D. J., ed.), Elsevier, New York, NY, 1974.
3. Smith, M. J. Traffic Control and Traffic Assignment in a Signal-Controlled Network with Queueing. *Proc., 12th International Symposium on the Theory of Traffic Flow and Transportation*, Berkeley, CA, 1993, pp. 61 –77.
4. Van Vliet, D. SATURN-A Modern Assignment Model. *Traffic Engineering & Control*, Vol. 23, No. 12, 1982, pp. 578–581.
5. Al-Malik, M., and N. H. Gartner. Development of a Combined Traffic Signal Control-Traffic Assignment Model. In *Urban Traffic Networks* (Al-Malik, M., N. H. Gartner, and G. Improta, eds.), Springer, Berlin, 1995, pp. 155–186.
6. Friesz, T. L., J. Luque, R. L. Tobin, and B.-W. Wie. Dynamic Network Traffic Assignment Considered as a Continuous Time Optimal Control Problem. *Operations Research*, Vol. 37, No. 6, 1989, pp. 893–901.
7. Gartner, N. Area Traffic Control and Network Equilibrium. *Lecture Notes in Economics and Mathematical Systems*, Vol. 3, No. 176, 1976, p. 274–297.
8. Tian, Z., K. Balke, R. Engelbrecht, and L. Rilett. Integrated Control Strategies for Surface Street and Freeway Systems. *Transportation Research Record: Journal of the Transportation Research Board*, 2002. 1811: 92–99.
9. Mahmassani, H. S. Dynamic Network Traffic Assignment and Simulation Methodology for Advanced System Management Applications. *Networks and Spatial Economics*, Vol. 1, No. 3, 2001, pp. 267–292.
10. Papageorgiou, M. Dynamic Modeling, Assignment, and Route Guidance in Traffic Networks. *Transportation*

Research Part B: Methodological, Vol. 24, No. 6, 1990, pp. 471–495.

11. Yang, H., and S. Yagar. Traffic Assignment and Traffic Control in General Freeway-arterial Corridor Systems. *Transportation Research Part B: Methodological*, Vol. 28, No. 6, 1994, pp. 463–486.
12. Papageorgiou, M., C. Diakaki, V. Dinopoulou, A. Kotsialos, and Y. Wang. Review of Road Traffic Control Strategies. *Proceedings of the IEEE*, Vol. 91, No. 12, 2003, pp. 2043–2067.
13. Ban, X. J., J.-S. Pang, H. X. Liu, and R. Ma. Modeling and Solving Continuous-time Instantaneous Dynamic User Equilibria: A Differential Complementarity Systems Approach. *Transportation Research Part B: Methodological*, Vol. 46, No. 3, 2012, pp. 389–408.
14. Ban, X. J., H. X. Liu, M. C. Ferris, and B. Ran. A Link-node Complementarity Model and Solution Algorithm for Dynamic User Equilibria with Exact Flow Propagations. *Transportation Research Part B: Methodological*, Vol. 42, No. 9, 2008, pp. 823–842.
15. Yu, H., R. Ma, P. Liu, and H. M. Zhang. Optimal Traffic Signal Control Considering DUE Route Choice and Queue Spillbacks. Presented at 94th Annual Meeting of the Transportation Research Board, Washington, D.C., 2015.
16. Liu, R., and M. Smith. Route Choice and Traffic Signal Control: A Study of the Stability and Instability of a New Dynamical Model of Route Choice and Traffic Signal Control. *Transportation Research Part B: Methodological*, Vol. 77, 2015, pp. 123–145.
17. Meneguzzer, C. Review of Models Combining Traffic Assignment and Signal Control. *Journal of Transportation Engineering*, Vol. 123, No. 2, 1997, pp. 148–155.
18. Chen, O., and M. Ben-Akiva. Game-theoretic Formulations of Interaction between Dynamic Traffic Control and Dynamic Traffic Assignment. *Transportation Research Record: Journal of the Transportation Research Board*, 1998. 1617: 179–188.
19. Jayakrishnan, R., W. K. Tsai, and A. Chen. A Dynamic Traffic Assignment Model with Traffic-Flow Relationships. *Transportation Research Part C: Emerging Technologies*, Vol. 3, No. 1, 1995, pp. 51–72.
20. Maher, M. J., X. Zhang, and D. Van Vliet. A Bi-level Programming Approach for Trip Matrix Estimation and Traffic Control Problems with Stochastic User Equilibrium Link Flows. *Transportation Research Part B: Methodological*, Vol. 35, No. 1, 2001, pp. 23–40.
21. Fisk, C. Game theory and Transportation Systems Modelling. *Transportation Research Part B: Methodological*, Vol. 18, No. 4, 1984, pp. 301–313.
22. Chiu, Y.-C., and H. Mahmassani. Hybrid Real-time Dynamic Traffic Assignment Approach for Robust Network Performance. *Transportation Research Record: Journal of the Transportation Research Board*, 2002. 1783: 89–97.
23. Chiu, Y.-C., and H. S. Mahmassani. Toward Hybrid Dynamic Traffic Assignment-models and Solution Procedures. *Proc., 2001 IEEE, Intelligent Transportation Systems*, Oakland, CA, IEEE, New York, NY, 2001, pp. 245–250.
24. Haddad, J., M. Ramezani, and N. Geroliminis. Cooperative Traffic Control of a Mixed Network with Two Urban Regions and a Freeway. *Transportation Research Part B: Methodological*, Vol. 54, 2013, pp. 17–36.
25. Gartner, N. H. Development of Demand-responsive Strategies for Urban Traffic Control. *Proc., 11th IFIP Conference on System Modelling and Optimization*, Springer, New York, NY, 1984, pp. 166–174.
26. Traffic, P. *UTOPIA/SPOT-Technical Reference Manual*. Peek Traffic, Amersfoort, The Netherlands, 2002.
27. Kotsialos, A., M. Papageorgiou, M. Mangeas, and H. Haj-Salem. Coordinated and Integrated Control of Motorway Networks Via Non-linear Optimal Control. *Transportation Research Part C: Emerging Technologies*, Vol. 10, No. 1, 2002, pp. 65–84.
28. Diakaki, C., V. Dinopoulou, K. Aboudolas, and M. Papageorgiou. *Signal Management in Real Time for Urban Traffic Networks*. Deliverable 9: Final System Development Report, IST-2000-28090, Technical University of Crete, Chania, Greece, 2002.
29. Bellemans, T., B. De Schutter, and B. De Moor. Model Predictive Control for Ramp Metering of Motorway Traffic: A Case Study. *Control Engineering Practice*, Vol. 14, No. 7, 2006, pp. 757–767.
30. Hegyi, A. *Model Predictive Control for Integrating Traffic Control Measures*. PhD thesis. University of Technology, Delft, The Netherlands, 2004.
31. Hegyi, A., B. De Schutter, and H. Hellendoorn. Model Predictive Control for Optimal Coordination of Ramp Metering and Variable Speed Limits. *Transportation Research Part C: Emerging Technologies*, Vol. 13, No. 3, 2005, pp. 185–209.
32. Smith, M. Traffic Control and Route-choice; A Simple Example. *Transportation Research Part B: Methodological*, Vol. 13, No. 4, 1979, pp. 289–294.
33. Bollerslev, T. Generalized Autoregressive Conditional Heteroskedasticity. *Journal of Econometrics*, Vol. 31, No. 3, 1986, pp. 307–327.
34. Glosten, L. R., R. Jagannathan, and D. E. Runkle. On the Relation between the Expected Value and the Volatility of the Nominal Excess Return on Stocks. *The Journal of Finance*, Vol. 48, No. 5, 1993, pp. 1779–1801.
35. Yang, H., Y. Zou, Z. Wang, and B. Wu. A Hybrid Method for Short-term Freeway Travel Time Prediction Based on Wavelet Neural Network and Markov Chain. *Canadian Journal of Civil Engineering*, Vol. 45, No. 2, 2017, pp. 77–86.
36. Yang, H., Z. Wang, Z. Yajie, B. Wu, and X. Wang. A Hybrid Method for Short-Term Freeway Travel Time Prediction Based on Percentage Error Correction and Noise Extraction. Presented at 97th Annual Meeting of the Transportation Research Board, Washington, D.C., 2018.
37. Tang, J., F. Liu, W. Zhang, R. Ke, and Y. Zou. Lane-changes Prediction Based on Adaptive Fuzzy Neural Network. *Expert Systems with Applications*, Vol. 91, 2018, pp. 452–463.
38. Tang, J., F. Liu, Y. Zou, W. Zhang, and Y. Wang. An Improved Fuzzy Neural Network for Traffic Speed Prediction Considering Periodic Characteristic. *IEEE*

Transactions on Intelligent Transportation Systems, Vol. 18, No. 9, 2017, pp. 2340–2350.

39. Tang, J., S. Zhang, X. Chen, F. Liu, and Y. Zou. Taxi Trips Distribution Modeling Based on Entropy-Maximizing Theory: A Case Study in Harbin City—China. *Physica A: Statistical Mechanics and its Applications*, Vol. 493, 2018, pp. 430–443.
40. Antonini, M., M. Barlaud, P. Mathieu, and I. Daubechies. Image Coding Using Wavelet Transform. *IEEE Transactions on Image Processing*, Vol. 1, No. 2, 1992, pp. 205–220.
41. Chen, O. J. *Integration of Dynamic Traffic Control and Assignment*. PhD thesis. Massachusetts Institute of Technology, Cambridge, MA, 1998.

The Standing Committee on Transportation Network Modeling (ADB30) peer-reviewed this paper (18-01632).

Article

TRR
JOURNAL OF THE TRANSPORTATION RESEARCH BOARD

Transportation Research Record
2018, Vol. 2672(48) 36–44

DOI: 10.1177/0361198118774236
journals.sagepub.com/home/trr
SAGE

A Greedy Path-Based Algorithm for Traffic Assignment

Jun Xie[1], Yu (Marco) Nie[2], and Xiaobo Liu[1]

Abstract

This paper presents a new path-based algorithm for the static user equilibrium traffic assignment problem. Path-based algorithms are generally considered less efficient than bush-based counterparts, such as Algorithm B, traffic assignment by paired alternative segments (TAPAS), and iTAPAS, an improved version of TAPAS, because explicitly storing and manipulating paths appears wasteful. However, our numerical experiments indicate that the proposed path-based algorithm can outperform TAPAS or iTAPAS by a wide margin. The proposed algorithm, sharing the same Gauss-Seidel decomposition scheme with existing path-based algorithms, delivered a surprising performance, most likely due to its two main features. First, it adopts a greedy method to solve the restricted subproblem defined on each origin–destination (O-D) pair. Second, instead of sequentially visiting every O-D pair in each iteration, it introduces an intelligent scheme to determine which O-D pairs need more or less work. The proposed algorithm is also more straightforward to implement than bush-based algorithms.

The user equilibrium (UE) static traffic assignment problem (TAP) has long been used as a standard tool to predict network flows (*1*). Under mild assumptions, the UE-TAP can be formulated as a convex optimization program (*2–4*). Designing efficient solution algorithms for this problem in large regional-scale networks has been a recurring research theme in transportation science and has attracted much attention in the past decades.

The Frank and Wolfe (FW) algorithm (*5*) was the first widely applied convergent algorithm for UE-TAP. Easy to implement and fast to converge in early iterations, the FW algorithm has emerged as the primary TAP-solver since the 1970s. However, the FW algorithm is not designed to achieve high-quality solutions that are often deemed necessary in modern planning practice (*6*). A number of strategies have been proposed to accelerate the FW algorithm. Well-known refinements include the parallel tangent algorithm (*7*), the aggregated simplicial decomposition algorithm (*8*) and the conjugate Frank–Wolfe algorithm (*9*). Since FW and its improved variants all operate in the space of link flows, they are usually called link-based algorithms.

The ability to solve UE-TAP with high precision has been enhanced substantially with the development of bush-based algorithms in the past decade. Simply speaking, a bush is an acyclic subnetwork that intends to encompass all the UE shortest paths from a given origin. Typically, a bush-based algorithm constructs and maintains a bush for each origin (or destination) and restricts the assignment operation only to these bushes. The first algorithm of this class is the origin-based algorithm (OBA) (*10*), which sequentially solves a list of node-based subproblems defined on the bush rooted at each origin by a gradient projection (GP) method. As a variant of OBA, the local user cost equilibrium (LUCE) (*11*) solves the node-based subproblems with a greedy method instead of GP. Algorithm B (*12*) simultaneously finds the min- and max-path trees for a given origin; it then equalizes the bush rooted at the origin by shifting flows from the maximum-cost paths to the minimum-cost ones. In traffic assignment by paired alternative segments (TAPAS) (*13*), a pair of alternative segments (PAS) is defined as two completely disjoint path segments with the same starting and ending nodes. Thus, a PAS marks the distinct parts of two or more paths connecting one origin–destination (O-D) pair. The notable innovation of TAPAS is to equalize path costs through identifying, storing, and operating a list of PASs, which

[1]School of Transportation and Logistics, Southwest Jiaotong University, Chengdu, China
[2]Department of Civil and Environmental Engineering, Northwestern University, Evanston, IL

Corresponding Author:
Address correspondence to Jun Xie: jun.xie@swjtu.edu.cn

avoids unnecessary network scanning. An improved version of TAPAS, named iTAPAS (*14*), employs a new PAS identification method and associates each PAS with only one origin to speed up the convergence efficiency. For a more comprehensive comparison of bush-based algorithms the reader is referred to Nie, and Xie and Xie (*15*, *16*).

UE-TAP can also be solved by path-based algorithms. This class of algorithms decomposes the original TAP into a list of subproblems by O-D pairs, which are then solved sequentially in the space of path flows. Whenever one subproblem is solved, the path flows from all other O-D pairs are usually fixed. Typically, a path-based algorithm needs to store and maintain a set of paths for each O-D pair, which can be iteratively generated through a process known as column generation. Several path-based algorithms have been proposed in the literature, which differ from each other mainly in the solution method for the restricted subproblem. Dafermos and Sparrow proposed to solve the subproblems by shifting flows from the longest path to the current shortest path in the path set (*17*); Larsson and Patriksson defined and solved an auxiliary optimization problem for each O-D pair (*18*); Jayakrishnan et al. and Florian et al. adapted the GP and projected gradient (PG) methods respectively to solve the subproblems (*19*, *20*). Two limitations had probably hindered the application of the path-based algorithms in solving large-scale problems in the past: on the one hand, storing and manipulating paths for each O-D pair needs much more random access memory (RAM) than link-based algorithms, which is a crucial issue when RAM is scarce; on the other hand, the decomposition scheme by O-D pairs generates a large number of subproblems, and sequentially solving them with equivalent effort usually leads to a relatively low overall convergence efficiency.

The research motivation of this study was three-fold. First, with the rapid advance in computing power and storage capacity, RAM usage of path-based algorithms is no longer a critical issue for modern-size regional planning networks. Thus, the path-storing requirement should not limit the development of faster traffic assignment algorithms. Second, despite their promising efficiency, most bush-based algorithms are complex in terms of both methodology and implementation, which makes them less accessible to researchers and practitioners. In contrast, path-based algorithms rely on a decomposition scheme that is easy to understand and implement, and they are equally capable of achieving high-precision solutions. Third, much has been learned from developing bush-based algorithms in the past decade, and these lessons can now be adapted to reinventing path-based algorithms. Indeed, recent research (*21*) observed that the efficiency of the path-based algorithm can be improved by only skipping column generation in certain iterations. In a nutshell, the time is ripe to give the path-based algorithm a fresh look.

The new path-based algorithm developed in this paper shares the same Gauss-Seidel decomposition scheme with existing path-based algorithms, in which O-D pairs are visited sequentially. Yet, two main innovative features differentiate it from the path-based algorithms in the literature. First, the greedy method, which serves as the solver of node-based subproblems in LUCE, is adopted to solving the restricted subproblem defined on each O-D pair. The greedy method is generally faster than GP in solving the O-D subproblems (*22*). Second, instead of sequentially visiting each O-D pair in each iteration, it introduces an intelligent scheme to determine which O-D pairs need more or less work. Specifically, an inner loop is added to the main loop to make sure that the frequency of path flow adjustment is higher than that of column generation. Moreover, the algorithm tracks the maximum path cost difference for each O-D pair and repeats the path flow adjustment more frequently on those O-D pairs with a larger cost difference. We also improve the computational efficiency by properly setting a maximum convergence precision in each iteration,that is, skipping the path flow adjustment for the O-D pairs with a cost difference smaller than the current relative convergence gap. Our numerical experiments show that the proposed path-based algorithm is significantly faster than the state-of-the-art bush-based algorithms.

The remainder of this paper is organized as follows. Section 2 presents the notation and formulation required. Section 3 describes the algorithm details and implementation procedures. Section 4 demonstrates the high efficiency of the proposed algorithm by comparing it with several of the latest bush-based algorithms. The last section concludes the paper and provides directions for future research.

Notation and Formulation

Consider a directed transportation network $G(N, A)$ where N and A denote the sets of nodes and links, respectively. Suppose that G is strongly connected, that is, there is at least one path between any two nodes in G. N consists of a set of origins R and a set of destinations S. The travel demand between an origin $r \in R$ and a destination $s \in S$ is denoted by d_{rs}. Let H_{rs} denote the set of simple paths connecting O-D pair (r, s) and f_h the flow on path $h \in H_{rs}$. A link is defined by (i,j) with i and j being the tail and head node respectively. The general cost of traversing link (i,j), denoted as $t_{ij}(x_{ij})$, is assumed to be a separable, strictly positive, and monotonically increasing function of x_{ij}, the flow on link (i,j).

Under the above assumptions the TAP satisfying the UE conditions (or Wardrop's first principle) can be formulated mathematically as the following convex program:

$$\min z(f) = \sum_{(i,j)\in A} \int_0^{x_{ij}} t_{ij}(w)dw \tag{1}$$

subject to

$$\sum_{h\in H_{rs}} f_h = d_{rs}, \quad \forall r \in R, \forall s \in S, \tag{2}$$

$$f_h \geq 0, \quad \forall h \in H_{rs}, \forall r \in R, \forall s \in S, \tag{3}$$

$$x_{ij} = \sum_{r\in R}\sum_{s\in S}\sum_{h\in H_{rs}} \delta_{ij}^h f_h, \quad \forall (i,j) \in A, \tag{4}$$

where $\delta_{ij}^h = 1$ if path h passes link (i,j); and 0 otherwise. It is well known that the UE-TAP is strictly convex given the above mild assumption on $t_{ij}(x_{ij})$ and hence admits a unique solution in the space of link flows. However, its path flow solution is not unique.

Central to the development of path-based algorithms is a sequential (known as Gauss-Seidel) decomposition of the above problem with respect to O-D pairs. As the decomposed subproblems are solved sequentially, it suffices for our purpose to focus on a single O-D subproblem, which may be formulated as follows:

$$\min z_{rs}(f) = \sum_{(i,j)\in A} \int_0^{x_{ij}^- + x_{ij}^{rs}} t_{ij}(w)dw \tag{5}$$

subject to

$$\sum_{h\in H_{rs}} f_h = d_{rs}, \tag{6}$$

$$f_h \geq 0, \quad \forall h \in H_{rs}, \tag{7}$$

$$x_{ij}^{rs} = \sum_{h\in H_{rs}} \delta_{ij}^h f_h, \quad \forall (i,j) \in A, \tag{8}$$

where $x_{ij}^- = \sum_{o\in R, o\neq r}\sum_{q\in S, q\neq s} x_{ij}^{oq}$ are flows contributed by all O-D pairs other than (r,s). To solve the above nonlinear program, one can construct a quadratic approximation for it to provide a descent direction. At the current solution $\{g_h, \forall h \in H_{rs}, \forall r \in R, \forall s \in S\}$, the objective function (5) can be approximated by the second-order Taylor expansion as follows:

$$\hat{z}_{rs}(f) \simeq z_{rs}(g) + \sum_{h\in H_{rs}} \Big(\frac{\partial z_{rs}(g)}{\partial f_h}(f_h - g_h) + 0.5\frac{\partial z_{rs}^2(g)}{\partial^2 f_h}(f_h - g_h)^2\Big). \tag{9}$$

Let $v_h^g = \frac{\partial z_{rs}(g)}{\partial f_h}$, $s_h^g = \frac{\partial z_{rs}^2(g)}{\partial^2 f_h}$ and remove the constants from formulation (9), the quadratic approximation of the single O-D assignment problem can then be formulated as follows:

$$\min \hat{z}_{rs}(f) = \sum_{h\in H_{rs}} ((v_h^g - s_h^g g_h)f_h + 0.5 s_h^g f_h^2) \tag{10}$$

subject to

$$\sum_{h\in H_{rs}} f_h = d_{rs}, \tag{11}$$

$$f_h \geq 0, \quad \forall h \in H_{rs}. \tag{12}$$

Let w_{rs} be the multiplier associated with O-D pair (r,s), then the Karush-Kuhn-Tucker conditions of problem (10–12) indicate,

$$v_h^g + s_h^g(f_h - g_h) - w_{rs} \geq 0, \tag{13}$$

$$f_h(v_h^g + s_h^g(f_h - g_h) - w_{rs}) = 0. \tag{14}$$

Here w_{rs} should be interpreted as the minimum travel cost between r and s, and the quadratic approximation problem in fact approximates v_h with the following linear function at the current solution g,

$$\hat{v}_h = v_h^g + s_h^g(f_h - g_h). \tag{15}$$

where v_h^g and s_h^g can be easily calculated precisely as follows:

$$v_h^g = \frac{\partial z_{rs}(g)}{\partial f_h} = \sum_{(i,j)\in A} t_{ij}(x_{ij}^g)\delta_{ij}^h, \tag{16}$$

$$s_h^g = \frac{\partial z_{rs}^2(g)}{\partial^2 f_h} = \sum_{(i,j)\in A} \frac{\partial t_{ij}(x_{ij}^g)}{\partial x_{ij}^g}\delta_{ij}^h. \tag{17}$$

Many methods can be used to solve the quadratic approximation program (10–12). In the literature, the Goldstein–Levitin–Polyak GP method and Rosen's PG – known respectively as the GP algorithm and the PG algorithm in traffic assignment – were adapted respectively by Jayakrishnan et al. and Florian et al. (*19*, *20*). In the next section, we will introduce the greedy method to solve the above quadratic program, followed by an implementation.

Algorithm Details

Adaption of the Greedy Method

The greedy method was first used in the hyper-path algorithm for transit assignment (*23*, *24*), and then adapted by Gentile in the LUCE algorithm (*11*) to solve the node-based assignment problems. In what follows we adopt this method to solve the single O-D assignment problem.

Let us first simplify our notation by introducing the following constant:

$$c_h^g \equiv \; v_h^g - s_h^g g_h. \tag{18}$$

Suppose we know the set of all used paths $\hat{H}_{rs}$ between r and s; then the conditions (13)–(14) can be translated to the following equation system.

$$c_h^g + s_h^g f_h = \bar{w}_{rs}, \quad \forall h \in \hat{H}_{rs}, \tag{19}$$

$$\sum_{h \in \hat{H}_{rs}} f_h = d_{rs}. \tag{20}$$

From the above system, one can solve the minimum average path travel time $\bar{w}_{rs}$ and f_h as follows:

$$\bar{w}_{rs} = \frac{d_{rs} + \sum_{h \in \hat{H}_{rs}} c_h^g / s_h^g}{\sum_{h \in \hat{H}_{rs}} 1/s_h^g}, \tag{21}$$

$$f_h = \bar{w}_{rs}/s_h^g - c_h^g/s_h^g, \quad \forall h \in \hat{H}_{rs}. \tag{22}$$

All paths in $H_{rs} \backslash \hat{H}_{rs}$ should receive zero flow. To determine $\hat{H}_{rs}$, the procedure described in Algorithm 1 is used.

Algorithm 1 consists of three parts: initialization (cf. lines 3–9), main loop (cf. lines 11–17), and flow update (cf. lines 19–31). In the initialization part, all paths in the set H_{rs} should be sorted according to the increasing order of c_h^g, and then $\bar{w}_{rs}$ is initialized using the path with the least c_h^g (cf. Line 8). In the worst case, the main loop part will find the exact solution to the quadratic program (10–12) after all paths in H_{rs} are visited, which is faster than the GP method when there are more than two used paths in H_{rs}. Note that the typical GP [e.g. the one presented in Jayakrishnan et al. (*19*)] can only give an approximate solution to the quadratic program in this situation. The reader is referred to Xie et al. (*22*) for a comparison between the two methods in solving a quadratic program. Finally, the flow update part updates the flow of the paths in H_{rs}, as well as the flow and cost of the links on those paths. Note that, in our current implementation, only the links on the paths subject to flow change are updated (cf. Line 26 of Algorithm 1). Whether these paths are overlapping or not is not considered. Furthermore, at the end of Algorithm 1 the paths receiving zero flow are all removed from the path set H_{rs} (cf. Line 31 of Algorithm 1).

Algorithm 1 Adaption of the greedy method for a single O-D assignment problem

1: **Input:** The current solution $\{g_h, \forall h \in H_{rs}\}$ is given.
2: **Initialization:** lines 3–9.
3: **for** each $h \in H_{rs}$ **do**
4: Compute v_h^g and s_h^g according to formulation (16) and formulation (17), respectively.
5: Compute c_h^g according to formulation (18).
6: **end for**
7: Sort all paths $h \in H_{rs}$ according to the increasing order of c_h^g, i.e., $H_{rs} = \{1, 2, 3, \ldots\}$ with $c_1^g \leqslant c_2^g \leqslant c_3^g \leqslant \ldots$.
8: Let $B = 1/(s_1^g d_{rs})$, $C = c_1^g/(s_1^g d_{rs})$ and $\bar{w}_{rs} = (1.0 + C)/B$.
9: Set $h = 2$, $\hat{H}_{rs} = \{1\}$.
10: **Main Loop:** lines 11–17.
11: **while** $h \leqslant |H_{rs}|$ and $c_h^g < \bar{w}_{rs}$ **do**
12: Set $C = C + c_h^g/(s_h^g d_{rs})$.
13: Set $B = B + 1/(s_h^g d_{rs})$.
14: Set $\bar{w}_{rs} = (1.0 + C)/B$.
15: Let $\hat{H}_{rs} = \hat{H}_{rs} \cup \{h\}$.
16: Set $h = h + 1$.
17: **end while**
18: **Flow Update:** lines 19–31.
19: **end for**
20: **for** each $h \in \hat{H}_{rs}$ **do**
21: Set $f_h = (\bar{w}_{rs} - c_h^g)/s_h^g$.
22: **for** each $h \in H_{rs} \backslash \hat{H}_{rs}$ **do**
23: Set $f_h = 0$.
24: **end for**
25: **for** each $h \in H_{rs}$ **do**
26: **if** $f_h \neq g_h$ **then**
27: Update the link flow by $x_{ij} = x_{ij} + (f_h - g_h), \forall (i,j) \in h$.
28: Update $t_{ij}(x_{ij})$ and $\frac{\partial t_{ij}(x_{ij})}{\partial x_{ij}}$ for each link $(i,j) \in h$.
29: **end if**
30: **end for**
31: Let $H_{rs} = \hat{H}_{rs}$.
32: **Output:** A new solution $\{f_h, \forall h \in H_{rs}\}$.

Algorithm Implementation

The implementation of the greedy path-based algorithm is described in Algorithm 2. In addition to the greedy-method solver for O-D subproblems, another noteworthy improvement was an efficient implementation scheme, which was designed to expedite the overall convergence of path-based algorithms mainly from three aspects. First, an inner loop (cf. lines 18–34 of Algorithm 2) was added to make sure that the frequency of path flow adjustment (by Algorithm 1) was higher than that of column generation. Second, it tracked the maximum path cost difference for each O-D pair, that is, $\Delta_{rs} = \max\{v_h\} - \min\{v_h\}, \forall h \in H_{rs}$, and repeated the path flow adjustment more frequently on the O-D pairs with larger Δ_{rs} (cf. lines 22–29 of Algorithm 2). Third, it improved computational efficiency by properly setting a maximum convergence precision in each iteration, that is, skipping the path flow adjustment for those O-D pairs with $\Delta_{rs} < RG^{k-1}/2.0$, where RG^{k-1} is the relative convergence gap in the last iteration (cf. lines 25–29 of Algorithm 2). The above three improvements in implementation ensured that more computational resources were allocated to handle those O-D pairs that can lead to more reduction of the objective value.

Algorithm 2 The greedy path-based algorithm

1: **Initialize**: lines 2–7
2: **for** each O-D pair (r, s) **do**
3: Compute the shortest path $\hat{h}$ from r to s.
4: Assign all d_{rs} to $\hat{h}$ and push it into H_{rs}.
5: **end for**
6: Update link flows by $x_{ij} = \sum_{r \in R} \sum_{s \in S} \sum_{h \in H_{rs}} f_h \delta_{ij}^h, \quad \forall (i,j) \in A$.
7: Update the link cost $t_{ij}(x_{ij})$ and its derivative $\frac{\partial t_{ij}(x_{ij})}{\partial x_{ij}}$ for each link $(i,j) \in A$.
8: **Main Loop**: lines 9–34
9: **for** each $r \in R$ **do**
10: Compute the shortest path tree from r to all its destinations S_r.
11: **for** each $s \in S_r$ **do**
12: Build the shortest path $\hat{h}$ from r to s.
13: Push $\hat{h}$ into H_{rs} if $\hat{h} \notin H_{rs}$; otherwise delete $\hat{h}$.
14: Perform path flow adjustment for H_{rs} by Algorithm 1.
15: **end for**
16: **end for**
17: **Inner Loop**: lines 18–34
18: Set $I = 0, MaxI = 1000, FC = 0$.
19: **while** $I < MaxI$ **do**
20: Let $I = I + 1$ and $FC = 0$.
21: **for** each O-D pair (r, s) **do**
22: **if** $I\%100 = 0$ **then**
23: Compute $\Delta_{rs} = \max\{v_h\} - \min\{v_h\}, \forall h \in H_{rs}$.
24: **end if**
25: **if** $\Delta > RG^{k-1}/2.0$ **then**
26: Let $FC = FC + 1$.
27: Perform path flow adjustment for H_{rs} by Algorithm 1.
28: Update x_{ij}, $t_{ij}(x_{ij})$ and $\frac{\partial t_{ij}(x_{ij})}{\partial x_{ij}}$ for $\forall (i,j) \in h, \forall h \in H_{rs}$.
29: **end if**
30: **end for**
31: **if** $FC = 0$ **then**
32: Break the inner loop.
33: **end if**
34: **end while**
35: **Convergence Check:** If the convergence criterion is achieved, stop; otherwise, repeat the Main Loop.

In the current implementation, several parameters were set according to the limited experience gained from the computational experiments on the networks shown in the next section. First, in Line 26 of Algorithm 2, we set a maximum convergence precision for each O-D subproblem as $RG^{k-1}/2.0$, which was sufficient to ensure a consecutive decrease of the objective value in each main loop. Second, in Line 19 of Algorithm 2, the maximum allowed inner loop iteration was set at 1,000, trying to enforce all O-D pairs to reach the desired convergence precision. Third, in lines 22–24, the Δ_{rs} for each O-D pair was updated only when the inner loop index I was out by a factor of 100; otherwise, Δ_{rs} was kept as the value obtained in the last update. Fourth, we created a parameter FC to track how many O-D pairs were updated by Algorithm 1 in each inner loop (cf. lines 18–20 and 26 in Algorithm 2); if $FC = 0$ (cf. lines 31–33 in Algorithm 2), which generally means that all O-D pairs have achieved the desired convergence precision in the current main loop, then we should break the inner loop.

Finally, a very simple data structure was employed in the current implementation to keep and operate paths. Basically, we maintained a vector H_{rs} for each O-D pair $(r, s), \forall r \in R, s \in S$, which was initialized with the shortest path from r to s in the beginning (cf. Line 4 of Algorithm 2); and then it was expanded when there was a new path generated in the column generation (cf. Line 13 of Algorithm 1), or reduced when there were paths receiving zero flow in the flow assignment (cf. lines 22–24 and Line 31 of Algorithm 1).

Numerical Results

In this section, we compare the convergence of the proposed greedy path-based algorithm with that of TAPAS, iTAPAS, GP, and an improved GP (iGP for short) on computing four testing networks. GP was implemented exactly following the procedure provided in Mitradjieva and Lindberg (*19*); the iGP was an implementation of GP plus the proposed intelligent scheme, that is, a replication of Algorithm 2 by replacing the greedy solver (cf. Line 14) with a GP solver; TAPAS and iTAPAS were implemented following the procedures in Xie and Xie (*14*). All these algorithms were coded using the network modeling toolkit, a C++ class library specialized in modeling transportation networks. Different algorithms shared as many subroutines as they could to ensure a fair comparison. All numerical results reported in this section were produced on a Windows 10 64-bit workstation with Intel Xeon central processing unit (CPU) E3-1225 V3 3.3 GHz CPU and 16G RAM.

The algorithms were applied to the four testing networks obtained from http://www.bgu.ac.il/?bargera/tntp/. The properties of the four networks are given in Table 1. The convergence indicator used in these experiments was the relative gap (RG), which was calculated as follows:

$$RG = 1 - \frac{\sum_{rs} \mu_{rs} d_{rs}}{\sum_{(i,j) \in A} x_{ij} t_{ij}(x_{ij})} \tag{23}$$

where μ_{rs} is the minimum travel cost between O-D pair (r, s). Finally, the Bureau of Public Roads function was used to calculate link costs.

The computational results for all the testing networks with the five algorithms are shown in Figure 1. All the testing algorithms were stopped when $RG = 10^{-14}$ was achieved or 1 h had passed.[1] In these plots, the horizontal

Table 1. Test Networks Used in the Numerical Experiments

Networks	Nodes	Links	Zones	O-D Pairs	V/C
Chicago Sketch	933	2950	386	1,260,910	0.54
PRISM	14,639	33,937	898	609,670	0.26
Chicago Regional	12,982	39,018	1771	3,136,441	0.60
Philadelphia	13,389	40,223	1525	1,150,722	0.75

Note: V/C = volume/capacity.

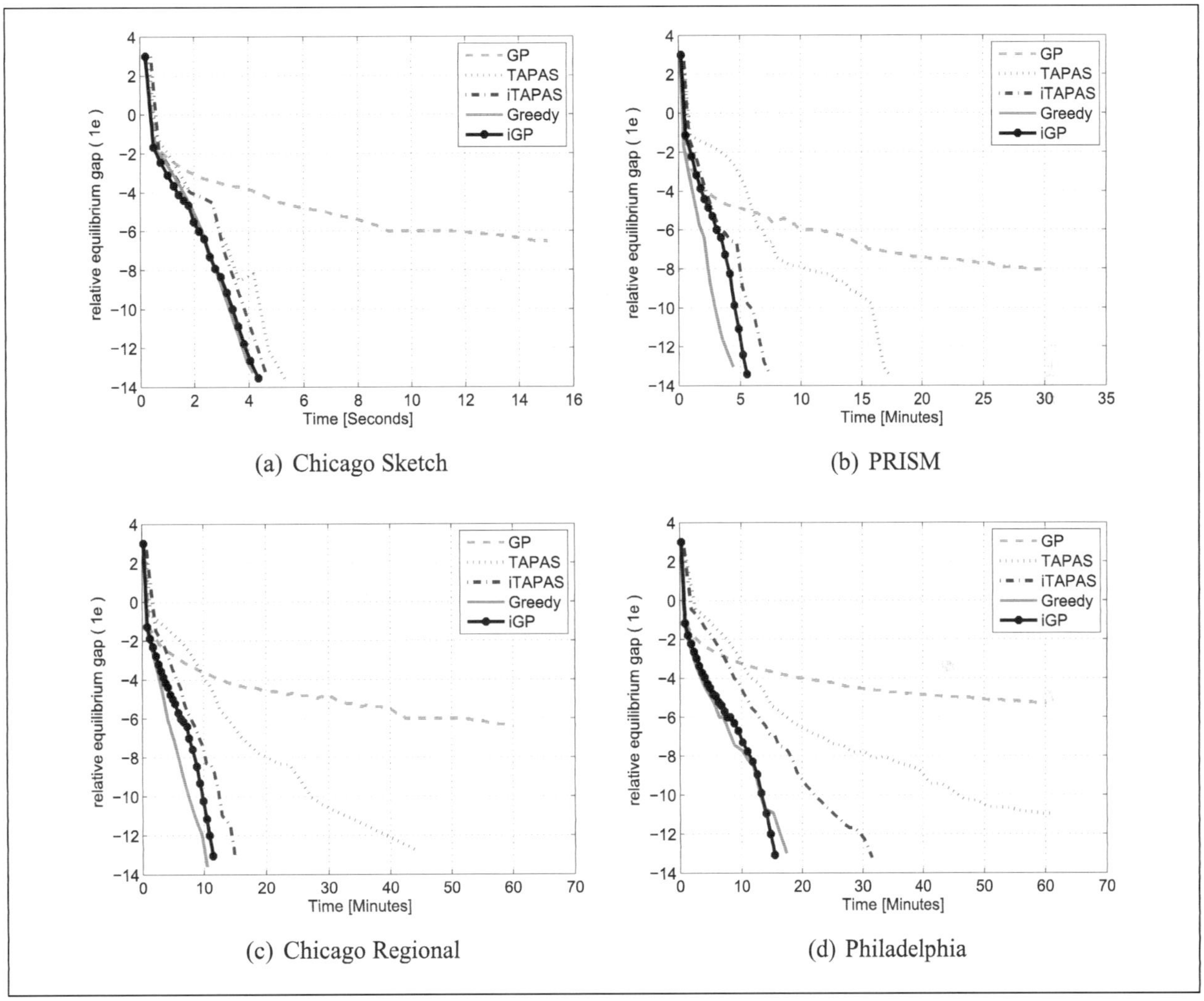

Figure 1. Convergence performance comparison on computing selected test networks.

axis represents CPU times, and the vertical axis represents the RG. Generally, the proposed greedy algorithm could outperform iTAPAS, TAPAS, and GP with a wide margin in all testing networks. Specifically, for the Chicago Regional network (cf. Figure 1*c*), TAPAS and iTAPAS took about 38 min and 14 min respectively to achieve $RG = 10^{-12}$, and the greedy algorithm took only 9 min to converge to the same precision, which is about 55% faster than iTAPAS and four times faster than TAPAS. In contrast, the GP algorithm only converged to about $RG = 10^{-7}$ within 1 h.[2] Overall, the same comparison result that the greedy algorithm is faster than the GP, TAPAS, and iTAPAS algorithms was obtained from the other three testing networks. These results generally

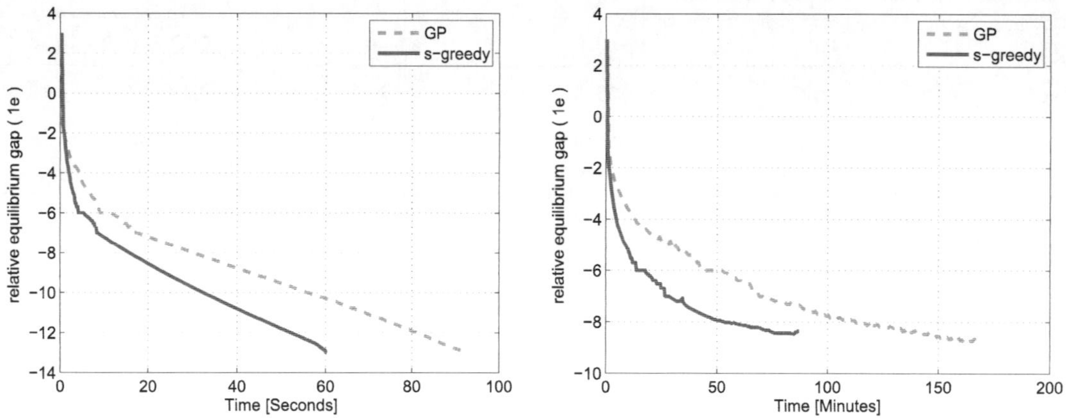

(a) Chicago Sketch (b) Chicago Regional

Figure 2. Comparison on the convergence performance of s-greedy and GP.

Table 2. Path Information for Different Algorithms

Networks	Information	GP	greedy	iGP
Chicago Regional	path number	1,985,744	1,991,055	1,920,298
	memory size	489 M	491 M	473 M
Philadelphia	path number	1,712,502	1,709,818	1,376,289
	memory size	583 M	562 M	437 M

Note: M = megabyte.

indicate that the proposed path-based algorithm is faster than the state-of-the-art bush-based algorithms.

The proposed intelligent scheme for path-based algorithms (cf. Algorithm 2) plus the GP solver, iGP, was also tested in these numerical examples. Interestingly, iGP and the greedy algorithm performed very similarly in all experiments. Taking a closer look, the greedy algorithm performed a little better than iGP in computing the PRISM (this model of the Birmingham City Region in England is a modified version of the Prism model, described at http://www.prism-wm.com/) and the Chicago Regional, but iGP was still ahead of the other three algorithms in the whole convergence process for all experiments. Recall that the major difference between the GP method and the greedy method for solving a quadratic program lies in the fact that the former can obtain an exact solution, whereas the later can only get an approximate solution when there are more than two used paths in the feasible set. Yet, we noticed that GP was also able to reach the same convergence precision as the greedy method does through repeating itself multiple iterations, that is, performing the flow shifting operations as many times as required to converge to the desired level of precision.[3] Given the above background, a reasonable explanation for the similar performance between iGP and the greedy method may be that, the proposed intelligent scheme, which is designed to allow the GP method or the greedy method to repeat multiple times to reach the desired convergence precision, thoroughly makes up for the inefficient convergence of the GP method.

To further support this explanation, we implemented the greedy method in the same scheme with GP, that is, a procedure of Algorithm 2, removing the inner loop. Such a simplified greedy algorithm is named s-greedy. The convergence performance of s-greedy and GP in solving Chicago Sketch and Chicago Regional is shown in Figure 2. Clearly, s-greedy is more efficient than GP in these two examples. For the Chicago Sketch network, s-greedy was over 60% faster than GP to achieve $RG = 10^{-12}$; and for the Chicago Regional network, s-greedy was about twice as fast as GP to achieve $RG = 10^{-8}$. As a result, it is safe to conclude that both the implementation strategies proposed in this paper contribute positively to enhancing the computational performance of the path-based algorithms.

Table 2 reports the number of paths generated by each path-based algorithm for the two regional testing networks and the corresponding memory (M represents megabyte) required to save them in the computer. Overall, all three path-based algorithms generated less than 2 million used paths in the two networks, which required less than 600 M memory to save in the computer. Such a requirement on memory is clearly no longer a hurdle for most PCs.

Conclusion

An efficient implementation of a path-based algorithm is proposed for the static UE-TAP. It retains the simplicity of path-based algorithms, while taking advantage of the lessons learned from developing bush-based algorithms in the past decade. The proposed algorithm adapts a greedy method to solve the restricted subproblem for each O-D pair, and introduces an intelligent scheme to determine which O-D pairs need more or less work in each iteration. These features make the new algorithm about 55% faster than iTAPAS, or more than four times faster than TAPAS in solving large regional-scale problems, according to our numerical experiments.

Our experience has revealed several promising generic rules for designing efficient path-based UE-TAP algorithms. First, the frequency of path flow adjustment should be significantly higher than that of column generation. Second, one should try to repeat more flow adjustment on the less converged O-D pairs. Third, keep in mind that each subproblem is only subject to a local optimal solution, hence do not waste time seeking high-precision solutions to subproblems in early iterations.

As the potential of path-based algorithms for the standard static UE-TAP has been well demonstrated here, a future research direction could be to extend this to solving more sophisticated models, such as stochastic, multi-class, and asymmetric TAPs.

Acknowledgments

The work was conducted when the first author visited Northwestern University as a visiting postdoctoral researcher. He was funded by Chinese National Nature Science Foundation (Grant No. 71501129) and Chinese International Postdoctoral Exchange Fellowship Program (No. 20150045). The work was also partially funded by the United States National Science Foundation under the award number CMMI-1402911.

Author Contributions

The authors confirm contribution to the paper as follows: study conception and design: J.X.; data collection: J.X.; analysis and interpretation of results: J.X., Y.M.N., X.L.; draft manuscript preparation: J.X., Y.M.N., X.L. All authors reviewed the results and approved the final version of the manuscript.

Notes

1. For the Chicago Sketch and PRISM networks, different convergence time intervals, that is, 15 s and 30 min respectively, are shown in the figures to highlight the difference.
2. Note that the GP algorithm can converge better with time elapses.
3. Refer to Section 6 of Xie et al. (*22*) for more details on the comparison of GP and greedy.

References

1. Wardrop, J. G. Some Theoretical Aspects of Road Traffic Research. *Proceedings of the Institution of Civil Engineers*, Vol. 1, No. 3, 1952, pp. 325–362.
2. Beckmann, M., C. McGuire, and C. B. Winsten. *Studies in the Economics of Transportation*. Yale University Press, New Haven, 1956.
3. Sheffi, Y. *Urban Transportation Network*. Pretince Hall, Englewood Cliffs, N.J., 1985.
4. Patriksson, M. *The Traffic Assignment Problem: Models and Methods*. Courier Dover Publications, New York, 2015.
5. Frank, M., and P. Wolfe. An Algorithm for Quadratic Programming. *Naval Research Logistics Quarterly*, Vol. 3, No. 1–2, 1956, pp. 95–110.
6. Boyce, D., B. Ralevic-Dekic, and H. Bar-Gera. Convergence of Traffic Assignments: How Much is Enough? *Journal of Transportation Engineering*, Vol. 130, No. 1, 2004, pp. 49–55.
7. Florian, M., J. Guálat, and H. Spiess. An Efficient Implementation of the Partan Variant of the Linear Approximation Method for the Network Equilibrium Problem. *Networks*, Vol. 17, No. 3, 1987, pp. 319–339.
8. Lawphongpanich, S., and D. W. Hearn. Simplical Decomposition of the Asymmetric Traffic Assignment Problem. *Transportation Research Part B: Methodological*, Vol. 18, No. 2, 1984, pp. 123–133.
9. Mitradjieva, M., and P. O. Lindberg. The Stiff is Moving-conjugate Direction Frank-Wolfe Methods with Applications to Traffic Assignment. *Transportation Science*, Vol. 47, No. 2, 2013, pp. 280–293.
10. Bar-Gera, H. Origin-Based Algorithm for the Traffic Assignment Problem. *Transportation Science*, Vol. 36, No. 4, 2002, pp. 398–417.
11. Gentile, G. Local User Cost Equilibrium: A Bush-Based Algorithm for Traffic Assignment. *Transportmetrica A: Transport Science*, Vol. 10, No. 1, 2014, pp. 15–54.
12. Dial, R. B. A Path-Based User-Equilibrium Traffic Assignment Algorithm that Obviates Path Storage and Enumeration. *Transportation Research Part B: Methodological*, Vol. 40, No. 10, 2006, pp. 917–936.
13. Bar-Gera, H. Traffic Assignment by Paired Alternative Segments. *Transportation Research Part B: Methodological*, Vol. 44, No. 8, 2010, pp. 1022–1046.
14. Xie, J., and C. Xie. New Insights and Improvements of Using Paired Alternative Segments for Traffic Assignment. *Transportation Research Part B: Methodological*, Vol. 93, 2016, pp. 406–424.
15. Nie, Y. M. A Class of Bush-Based Algorithms for the Traffic Assignment Problem. *Transportation Research Part B: Methodological*, Vol. 44, No. 1, 2010, pp. 73–89.
16. Xie, J., and C. Xie. Origin-Based Algorithms for Traffic Assignment: Algorithmic Structure, Complexity Analysis,

and Convergence Performance. *Transportation Research Record: Journal of the Transportation Research Board*, 2015. 2498: 46–55.
17. Dafermos, S. C., and F. T. Sparrow. The Traffic Assignment Problem for a General Network. *Journal of Research of the National Bureau of Standards B*, Vol. 73, No. 2, 1969, pp. 91–118.
18. Larsson, T., and M. Patriksson. Simplicial Decomposition with Disaggregated Representation for the Traffic Assignment Problem. *Transportation Science*, Vol. 26, No. 1, 1992, pp. 4–17.
19. Jayakrishnan, R., W. T. Tsai, J. N. Prashker, and S. Rajadhyaksha. *A Faster Path-Based Algorithm for Traffic Assignment*. University of California Transportation Center, 1994.
20. Florian, M., I. Constantin, and D. Florian. A New Look at Projected Gradient Method for Equilibrium Assignment. *Transportation Research Record: Journal of the Transportation Research Board*, 2009. 2090: 10–16.
21. Galligari, A., and M. Sciandrone. A Convergent and Fast Path Equilibration Algorithm for the Traffic Assignment Problem. *Optimization Methods and Software*, Vol. 33, No. 2, 2017, pp. 354–371.
22. Xie, J., Y. M. Nie, and X. Yang. Quadratic Approximation and Convergence of Some Bush-Based Algorithms for the Traffic Assignment Problem. *Transportation Research Part B: Methodological*, Vol. 56, 2013, pp. 15–30.
23. Nguyen, S., and S. Pallottino. Equilibrium Traffic Assignment for Large Scale Transit Networks. *European Journal of Operational Research*, Vol. 37, No. 2, 1988, pp. 176–186.
24. Spiess, H., and M. Florian. Optimal Strategies: A New Assignment Model for Transit Networks. *Transportation Research Part B: Methodological*, Vol. 23, No. 2, 1989, pp. 83–102.

The Standing Committee on Transportation Network Modeling (ADB30) peer-reviewed this paper (18-03128).

Article

TRR
JOURNAL OF THE TRANSPORTATION RESEARCH BOARD

Transportation Research Record
2018, Vol. 2672(48) 45–57

DOI: 10.1177/0361198118795006
journals.sagepub.com/home/trr
SAGE

Impact of Extreme Events on Transportation Infrastructure in Iowa: A Bayesian Network Approach

Abhijit R. Kulkarni[1] and Behrouz Shafei[1]

Abstract

Iowa's roadway network is an important part of the state's transportation infrastructure and plays a critical role in the functionality and economic development of the entire state. This network primarily consists of three interstate highways that pass through Iowa, connecting it to the neighboring states and eventually Canada. Various businesses are located near this roadway network and rely on it for everyday operation. In recent years, however, the growth of agricultural and biofuel industries has intensified the demand on the roads and bridges in Iowa. The state's roads and bridges have also witnessed a number of flooding events, which have caused extensive traffic disruptions and economic losses. Thus, it is imperative to develop a fundamental approach to evaluate the impact of extreme events on the transportation infrastructure of Iowa and other similar states. Towards this goal, the current study investigates the existing condition of Iowa's transportation infrastructure, possibility of occurrence of extreme weather events, and scenarios that may lead to the failure of transportation infrastructure components. For this purpose, the capabilities of Bayesian belief networks are utilized to quantify the effects of extreme precipitation and extreme temperature on the performance of transportation infrastructure and then predict the probability of damage to roads and bridges. This will be achieved through the identification of the most influential factors using a set of sensitivity analyses, assessment of overall vulnerability with evidence-based propagation analyses, and quantification of response to extreme weather events, taking into consideration climate projections.

The network of Iowa's roads and bridges plays an important role in the state's economic development and growth, which rely on various sectors, including agriculture, manufacturing, and renewable energy. The state of Iowa, similar to other neighboring states, however, faces a number of challenges in providing a transportation network that meets constantly evolving safety and efficiency requirements (*1*). Vulnerability of the existing roadway network to extreme weather events is particularly a growing concern, given the widespread consequences and losses associated with them (*2*). This is in part due to the fact that the occurrence of such events has become more intense and frequent in the Midwest region during the past two decades. In addition, aging effects have reduced the structural capacity of physical infrastructure components currently in service (*3–8*). This complex situation highlights the importance of systematic studies to evaluate the performance of transportation infrastructures under both normal and distressed conditions. Among various extreme events, the climatic ones commonly affect large geographic areas, disrupt necessary services, damage existing infrastructure components, and require costly maintenance and repair activities (*9*). This is clearly reflected in the extent of physical damage and service disruptions to the transportation networks during past weather events. Considering the current vulnerabilities and future risks (*10*), the objective of this study is to investigate the extreme weather scenarios and establish a fundamental framework to understand the causes and consequences of them. For this purpose, a Bayesian belief network (BBN) is employed to understand the uncertainties and interdependencies that exist in the response of a transportation network to an extreme event. BBNs are proven to be capable of predicting the probability of occurrence of a given scenario and the outcome associated with it if the contributing factors are properly introduced and quantified (*11, 12*). The current study provides an overview of the condition of transportation

[1]Institute for Transportation, Department of Civil, Construction, and Environmental Engineering, Iowa State University, Ames, IA

Corresponding Author:
Address correspondence to Behrouz Shafei: shafei@iastate.edu

infrastructure in Iowa, as well as possible weather events in the region, including extreme precipitation and extreme temperature. Taking into consideration the variety of factors that influence the performance of roadway networks, a BBN is developed to systematically capture the causes and consequences of various climate scenarios. The developed BBN is then employed to investigate the likelihood of destruction of roads and collapse of bridges, as ultimate consequences of extreme events. Through a set of sensitivity analyses, the factors that demonstrate the highest contribution to the probability of damage are identified. This covers not only individual factors, but also sequences of them. In the next step, evidence-based propagation analyses are performed in both forward and backward fashion. This is to understand how the collection of necessary evidence can help refine predictions on the vulnerability of the network. This study is concluded by the investigation of nine climate scenarios, which reflect the most extreme events. The outcome of this study is expected to pave the way for transportation authorities to benefit from advanced statistical inference methods with the ultimate goal of improving the preparedness to respond to extreme events by focusing on the most critical causes and consequences based on existing and new sources of information.

Transportation Infrastructure in Iowa

Iowa's public transportation infrastructure consists of more than 114,200 miles of roads (*13*) with an average Infrastructure Condition Evaluation (ICE) rating between 65 and 75. Of the 114,200 miles of roadways, there are 9,400 miles of primary roads, 89,900 miles of secondary roads, and 14,900 miles of municipal roads. The Iowa Department of Transportation (DOT) manages close to 10% of the total roadways (mainly primary roads), which carry more than 60% of all the state's vehicular traffic. The primary system also carries more than 90% of the total large truck traffic. The large truck traffic on the primary roadway system has witnessed a 123% increase over the last 30 years, partly due to the growth of agricultural and biofuel industries in Iowa. This is expected to grow further by 66% in the next 20 to 25 years (*14*). Despite the importance of primary roadways, approximately 40% of them have been rated fair to poor in 2012. Similar to the roadways, the bridges are in need of attention (*15*). Of 24,184 bridges in Iowa, 4,968 bridges have been reported structurally deficient. The combination of increasing traffic and aging effects has expedited the deterioration of the current roadway network. Such a situation is expected to increase the vulnerability of the transportation infrastructure to extreme events, as well as the likelihood of traffic congestions, delays, and economic losses in the state.

On the other hand, a growing demand for fuels produced from renewable resources has encouraged the biofuel industry to invest in Iowa. Since 1980, ethanol production in Iowa has witnessed a significant growth (almost 160-fold). This has accelerated the construction of new biorefineries and expansion of existing ones. However, the transportation infrastructure that is essential to support the operation of biofuel industry has not received the same level of development. Iowa produced more than 4.4 billion gallons of renewable fuel from its ethanol and biodiesel refineries in 2016 (*16*). The estimated total of 1.7 million truckloads for supplying feedstock, 0.52 million truckloads for supplying by-product to livestock facilities, and 0.56 million truckloads for supplying the finished biofuel to fueling and distribution centers add up to approximately 2.78 million truckloads in Iowa every year. With increasing the number of ethanol plants and biorefineries, the truckloads are anticipated to further increase, demanding a more reliable and efficient transportation infrastructure. However, recent extreme events, particularly the floods that occurred in 2008, 2010, 2011, and 2016, have raised some concerns about the vulnerability of not only existing transportation infrastructure components, but also the industries that greatly depend on them.

Impact of Extreme Events on Transportation Infrastructure in Iowa

The main focus of this study is on extreme events caused by weather conditions in the state of Iowa (Table 1). The 2008 flood in Iowa, for example, was one of the historical flooding events in the Midwest region. The flood was originated from a mixture of extreme weather conditions, including record snowfall and ice accumulation, below normal temperature, and high precipitation. As a consequence, Des Moines, Cedar, and Iowa Rivers, as well as their tributaries, witnessed an extreme flood with stream flows exceeding the 500-year (0.2% annual probability of occurrence) predictions (*17*). Because of this extreme event, more than 80 of Iowa's 99 counties had disaster declarations with an estimated damage of more than $10 billion across the state. On primary highways, 125 miles were completely washed out and 464 miles had to be closed. The closure of roads forced a 115-mile detour from Davenport to Des Moines. In addition, over 1,500 miles of roads were estimated to be in need of replacement. Over 300 bridges were also damaged or completely destroyed. An investment of $80 million was assessed to be required for the purpose of repairing or replacing the damaged roads and bridges (*18*).

Flooding in Iowa in 2011 was triggered by three main factors: melting of above normal snowpack in plains from March to April 2011, melting of above normal

Table 1. Extreme Flooding Events in Iowa Since 1952

Year	Flooded area(s)	Causes
1952	Sioux City	Cold and snowy winter during 1951–52, followed by a dramatic warm up in March 1952
1965	Minnesota Border to Quad Cities	Heavy snowfall, extreme cold which prevented gradual melting of snow, spring warm up, and heavy rainfalls
1993	Entire Iowa	Record rains in fall 1992, cold and snowy winter during 1992–93, heavy early-spring rainfalls followed by multiple days of rainfall during summer 1993
2008	Eastern Iowa, Cedar, and Iowa River	Heavy snowfalls in 2007, deep snowpack during 2007–08, and heavy rainfalls during the first half of June 2008
2010	Ames, Colfax, and Des Moines	Heavy rainfalls, leading to major flooding
2011	Missouri and Souris River Basin	Record snowfall in the Rocky Mountains of Montana, leading to highest runoff on record since 1898
2016	Northeast Iowa	Heavy rainfalls, leading to major flooding

snowpack in mountains from May to July 2011, and heavy rainfall in the upper Missouri river basin from March to October 2011 (*19*). Due to the cumulative effect of these three factors, the average runoff was about 246% of the normal runoff for the state. During the time span of March to July 2011, the storage capacity of all the dams and reservoirs in the region was exceeded. This led the authorities to have to release record volumes of water, which resulted in a major flooding of cities and towns along the Missouri river. Due to this flooding event, interchanges along I-29 and hundreds of miles of the secondary roadway system in six counties were closed. A $63.5 million investment was found to be required to restore the damaged roads and bridges to a working condition.

A systematic study of damage to transportation infrastructure caused by flooding events was carried out after the 2011 flood in Iowa (*19*). The study consisted of visual observations and field tests. Shoulder erosion, formation of voids, complete washout of pavement, and erosion of subbase layers were reported as the most common forms of damage to the pavement. As for bridges, there were reports that important bridge components experienced various levels of damage, leading to the complete washout of the entire bridge superstructure on certain occasions. Erosion of approach backfill and embankment foreslopes were also noted in various locations. In addition, several culverts were found damaged due to flooding. Weep holes, which indicate the erosion of culvert backfill, were observed in a number of cases. Water blockage and separation of culverts were the main types of damage recorded during visual inspections.

Several studies have investigated future climate conditions in Iowa (*2, 20–22*). Among them, Iowa Climate Change Adaptation and Resilience Report (*23*) provided an overview of changes likely to occur in the state during the next 50 to 60 years. The anticipated changes particularly include an increase in springtime precipitation, as well as more extreme low and high temperatures. Such scenarios are expected to result in heavier rains, earlier snow meltdowns, higher stream runoffs, and eventually increased frequency and intensity of floods. Thus, a higher risk of landslide, slope failure, bridge washout, and road closure are among the consequences that must be taken into consideration for the long-term design of new transportation infrastructure components. This aspect is equally important for the assessment of existing transportation infrastructure components since many of them have not been designed for the projected level of environmental stressors.

According to the climate models developed for the Midwest region (*9, 21*), heat waves are also anticipated to become more frequent, more intense, and longer lasting in the next 50 to 60 years. Expedited deterioration of physical infrastructure components is one of the main consequences of heat waves and extreme temperatures (*21, 24*). As a result of changes in the extreme low and high temperatures, freeze and thaw cycles are anticipated to increase. This can cause structural damage to the road and bridge surfaces. Changes in the temperature range may also affect construction materials, as a premature failure of them has been observed in several instances. In addition, temperature-based expansions and contractions can adversely affect subgrades, embankments, and levees. In bridge structures, extreme thermal effects can cause excessive movements of bridge girders, decks, and bearings, for which the bridge has not been originally designed (*25*). Such additional loading demands, paired with increasing traffic loads, can lead to an increased vulnerability and even early failure of transportation infrastructure components.

Bayesian Belief Networks

A holistic assessment of the impact of extreme events on the performance of roadway networks requires an investigation of multiple factors related to climate conditions (e.g., precipitation and temperature), interdependencies between causes and consequences of extreme events, and current condition states of infrastructure components in service. Inherent variations and uncertainties associated with contributing factors, however, pose certain challenges to a proper evaluation of the risk of damage and probability of failure. Conventional probabilistic risk assessment methods are often sufficient for the evaluation of an individual infrastructure component, but they are proven to fall short in capturing uncertain scenarios for complex infrastructure systems (*26*). In such situations, a BBN can be considered as a promising alternative, given that it is capable of (a) representing complex interactions between variables, (b) capturing their dependencies using conditional probabilities, and (c) updating the probabilities as soon as new information becomes available (*27, 28*). In addition, BBN is known to provide an efficient platform in situations where different types of variable and sources of knowledge must be integrated for decision-making purposes (*29, 30*).

BBNs are directed acyclic graphs that provide a graphical representation for probabilistic relationships between a set of random variables (*29, 31*). The graphical representation, which depicts causal relationships of the involved variables, is found to be useful, especially if various levels of data describing the real-world problem are present (*32*). BBNs are based on the Bayes theorem, which adds the capability of updating the probability of occurrence based on new evidence. This can be immediately employed to calculate the posterior probability of variables that are dependent on other variables whose probabilistic information is extracted over time. A BBN, in general, consists of a set of nodes and directed links between them. Each node, which represents a random variable, can be a parent or child, depending on the structure of the model. Child nodes (or variables) are caused by parent nodes (or variables). The nodes that have both parent and child roles are referred to as intermediate nodes. The links are used to model dependencies between variables, capturing causal relationships. This happens in a "directed" and "acyclic" manner, indicating that the links are not allowed to form two-way or loop-like relationships in the BBN.

For a BBN of variables (A_1, A_2, A_3, ..., A_n), the joint probability of occurrence of all the involved variables can be expressed as:

$$p(A_1, A_2, A_3, \ldots, A_n) = p(A_1|A_2, A_3, \ldots, A_n)\, p(A_2|A_3, \ldots, A_n)p(A_3|A_4, A_5, \ldots, A_n) \quad (1)$$

By denoting a set of parent nodes, $PP(A_i)$, the joint probability equation can be written as:

$$p(A_1, A_2, A_3, \ldots, A_n) = \prod_{i=1}^{n} p(A_i|PP(A_i)) \quad (2)$$

The random variable A_i can be either in a Boolean (Yes/No, True/False) or in a qualitative format. Further to their predictive and explanatory capabilities, the possibility of including random variables of different nature makes BBNs particularly useful for the performance assessment of roadway networks. The Iowa DOT's Bridge Inspection Manual (*33*), for example, defines 11 condition ratings for decks, superstructures, and substructures. The ratings cover a spectrum of Excellent, Very Good, Good, Satisfactory, Fair, Poor, Critical, Imminent Failure, and Failed, further to Not Applicable. Similarly, the road condition index offers three categories: Good, Fair, and Poor. The Iowa DOT's Bridge Inspection Manual provides a detailed description of each category, which helps utilize the qualitative information collected through inspections for preventive and/or corrective actions.

BBNs for Roadway Networks

For the performance assessment of roadway networks, the required information can be obtained using (i) frequentist or (ii) knowledge-based approaches. Through the frequentist approach, statistical measures, such as mean, coefficient of variation, and marginal probability distribution, are extracted from available data (*27*). In some cases, however, no reliable sources of data are available. This requires using a knowledge-based approach to formulate the conditional probabilities needed for statistical inference. This is commonly achieved through expert elicitation. Based on a detailed review of relevant papers and reports (*2, 13, 20, 21, 34*), a BBN is developed in the current study for the investigation of the impact of extreme events on the roadway network of Iowa (with the possibility of extension to other similar states in the Midwest region and beyond). The following steps are taken to create this BBN: (a) gathering the necessary information on the intensity and frequency of occurrence of extreme weather events, (b) generating various exposure scenarios, reflecting extreme precipitation and extreme temperature, and (c) evaluating the consequences of such scenarios on transportation infrastructure components. Figure 1 shows the developed BBN, which contains 24 random variables.

For the state of Iowa, extreme precipitation and extreme temperature are employed as the two main environmental stressors. Extreme precipitation initiates landslide, mudslide, and flooding, whereas extreme temperature causes early melting of snow, premature deterioration of pavement, expansion and contraction of

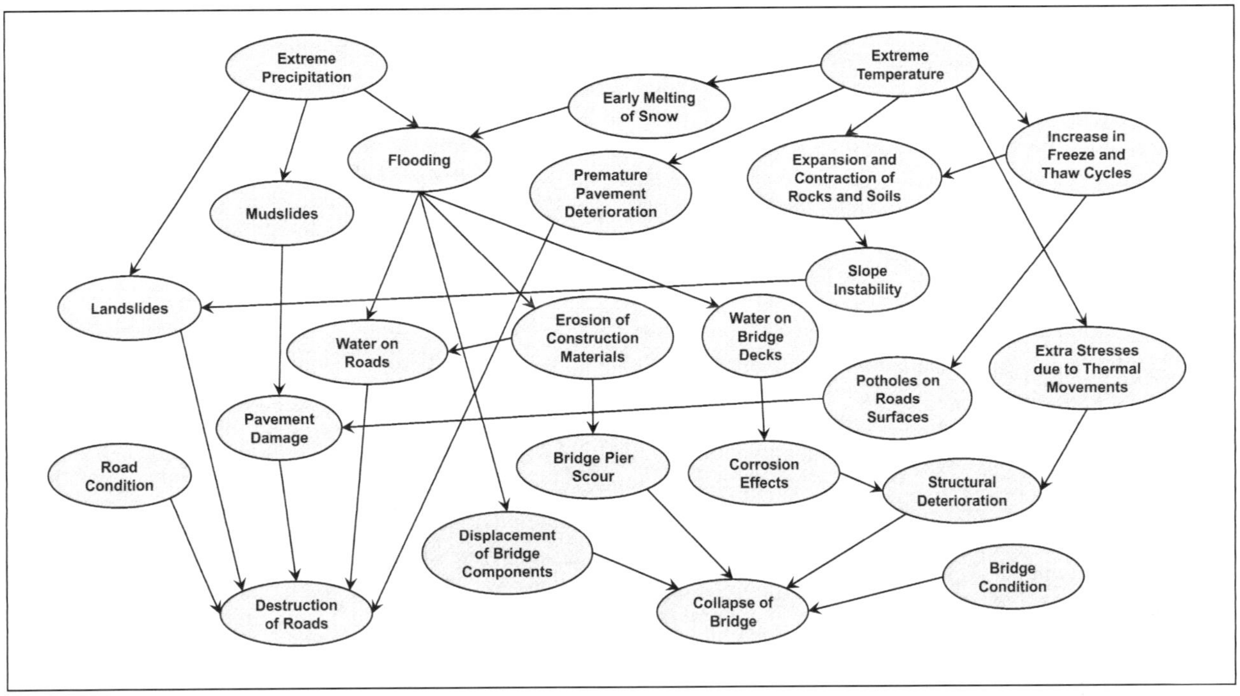

Figure 1. BBN model developed in the current study to investigate the impact of extreme weather events on transportation infrastructure components.

rocks and soils, extra stresses due to thermal movements, and increase in freeze and thaw cycles. Among the most important variables, flooding is influenced by two parent nodes, which are extreme precipitation and early melting of snow. Because of flooding, a range of immediate consequences are anticipated, including water accumulation on roads, erosion of construction materials, submersion of bridge decks, and displacement of main bridge components. As a result of extreme precipitation, mudslides can directly cause damage to the pavement, leading to the destruction of the entire road. On the other hand, extreme temperature can increase the freeze and thaw cycles, which cause more frequent expansion and contraction of rocks and soils. This can be potentially translated to slope instability, and ultimately, landslides. Increased freeze and thaw cycles can also create potholes on road surfaces, leading to damage to the pavement.

A similar causal relationship can be established for bridges, in which scouring and corrosion effects are the key contributors that make a bridge vulnerable to collapse. Further to all the identified variables, it must be noted that the current condition of roads and bridges plays a pivotal role to determine how they survive possible extreme events. Separate random variables are included in the developed BBN to capture this important factor.

Table 2. Conditional Probabilities Assumed for "Destruction of Roads"

Landslides	Water on roads	Premature pavement deterioration	Pavement damage	Road condition	Yes	No
Yes	Yes	Yes	Yes	Good	0.45	0.55
				Fair	0.60	0.40
				Poor	0.70	0.30
Landslides	Water on roads	Premature pavement deterioration	Pavement damage	Road condition	Yes	No
No	No	No	No	Good	0.01	0.99
				Fair	0.05	0.95
				Poor	0.075	0.925

Table 3. Conditional Probabilities Assumed for "Collapse of Bridges"

Bridge pier scour	Structural deterioration	Displacement of bridge components	Bridge condition	Yes	No
Yes	Yes	Yes	Excellent	0.8	0.2
			Very good	0.825	0.175
			Good	0.85	0.15
			Satisfactory	0.875	0.125
			Fair	0.9	0.1
			Poor	0.925	0.075
			Serious	0.95	0.05
			Critical	0.975	0.025
			Imminent failure	0.99	0.01
No	No	No	Excellent	0.01	0.99
			Very good	0.05	0.95
			Good	0.075	0.925
			Satisfactory	0.1	0.9
			Fair	0.125	0.875
			Poor	0.15	0.85
			Serious	0.175	0.825
			Critical	0.2	0.8
			Imminent failure	0.225	0.775

Table 4. Conditional Probabilities of Variables with One Parent Node

		Corrosion effects				Potholes on road surfaces	
		Yes	No			Yes	No
Water on Bridge decks	Yes	0.70	0.30	Increase in freeze and thaw cycles	Yes	0.75	0.25
	No	0.10	0.90		No	0.20	0.80
		Early melting of snow				Extra stresses due to thermal movements	
		Yes	No			Yes	No
Extreme temperature	Yes	0.40	0.6	Extreme temperature	Yes	0.85	0.15
	No	0.01	0.99		No	0.15	0.85
		Premature pavement deterioration				Increase in freeze and thaw cycles	
		Yes	No			Yes	No
Extreme temperature	Yes	0.70	0.30	Extreme temperature	Yes	0.80	0.20
	No	0.15	0.85		No	0.10	0.90
		Mudslides				Pavement damage	
		Yes	No			Yes	No
Extreme Precipitation	Yes	0.80	0.20	Mudslides	Yes	0.80	0.20
	No	0.20	0.80		No	0.20	0.80
		Water on bridge deck				Slope instability	
		Yes	No			Yes	No
Flooding	Yes	0.99	0.01	Expansion and contraction of rocks & soils	Yes	0.80	0.20
	No	0.20	0.80		No	0.20	0.80
		Erosion of construction materials				Bridge pier scour	
		Yes	No			Yes	No
Flooding	Yes	0.99	0.01	Erosion of construction materials	Yes	0.90	0.10
	No	0.20	0.80		No	0.25	0.75

Table 5. Conditional Probabilities of Variables with Two Parent Nodes

Extreme precipitation	Early melting of snow	Flooding Yes	Flooding No	Extreme temperature	Increase in freeze and thaw cycles	Expansion and contraction of rocks & soils Yes	Expansion and contraction of rocks & soils No
Yes	Yes	0.634	0.366	Yes	Yes	0.35	0.65
Yes	No	0.550	0.450	Yes	No	0.20	0.80
No	Yes	0.002	0.998	No	Yes	0.05	0.95
No	No	0.001	0.999	No	No	0.01	0.99
Extreme precipitation	**Slope instability**	**Landslides Yes**	**Landslides No**	**Extreme precipitation**	**Slope instability**	**Pavement damage Yes**	**Pavement damage No**
Yes	Yes	0.80	0.20	Yes	Yes	0.75	0.25
Yes	No	0.70	0.30	Yes	No	0.20	0.80
No	Yes	0.11	0.89	No	Yes	0.6	0.40
No	No	0.01	0.99	No	No	0.10	0.90
Flooding	**Erosion of construction materials**	**Water on roads Yes**	**Water on roads No**	**Corrosion effects**	**Extra stresses due to thermal movements**	**Structural deterioration Yes**	**Structural deterioration No**
Yes	Yes	0.95	0.05	Yes	Yes	0.90	0.10
Yes	No	0.75	0.25	Yes	No	0.90	0.10
No	Yes	0.65	0.35	No	Yes	0.10	0.90
No	No	0.10	0.90	No	No	0.10	0.90

The probability of occurrence, U, of extreme precipitation is derived from the following equation (*35*):

$$U = 1 - \left(1 - \frac{1}{T}\right)^r. \tag{3}$$

where T is the return period and r is the number of years. Assuming a 100-year return period, the unconditional probability of extreme precipitation in 100 years is 0.63 (Yes) and 0.37 (No). In the absence of a similar equation for extreme temperature, a knowledge-based approach is employed and the unconditional probability of extreme temperature in 100 years is estimated as 0.20 (Yes) and 0.80 (No) based on the projections made by the climate models developed for the Midwest region.

A review of the nature of the variables included in the developed BBN indicates that many of them can be sufficiently represented by Boolean variables. The exceptions are road and bridge condition states, which commonly have a qualitative format. Following a combination of frequentist and knowledge-based approaches, the conditional probabilities for all the Boolean variables are obtained. While efforts have been made to identify the most representative conditional probabilities, taking into account how the child nodes are affected by parent nodes based on the available data and observations, the conditional probabilities can be further adjusted, depending on the region of interest and type of infrastructure under consideration. Tables 2 to 5 provide an overview of the conditional probabilities assumed for the random variables identified in the current study. For statistical inference, the GeNIe Modeler package (*36*) has been employed. This package offers a range of capabilities, including parameter and structure learning. Furthermore, it supports various inference algorithms that can facilitate understanding causal relationships between the involved variables.

Results and Discussion

Using the BBN developed in the current study, a set of investigations are performed to understand the sensitivity of ultimate consequences, such as destruction of roads and collapse of bridges, to various contributing factors. This is expected to reflect real-world situations, in which uncertainty exists in the occurrence of events. For comparison purposes, a matrix of occurrence is introduced for extreme precipitation (EP) and extreme temperature (ET) with the following four cases: (A1) EP = Yes, ET = Yes, (A2) EP = Yes, ET = No, (A3) EP = No, ET = Yes, and (A4) EP = No, ET = No. From each sensitivity analysis, the top 10 scenarios that contribute most to the probability of destruction of roads and collapse of

Figure 2. Sensitivity analysis for the probability of (*a*) destruction of roads and (*b*) collapse of bridges, assuming extreme precipitation = yes and extreme temperature = yes.

bridges are identified and then the probability range associated with them is determined. Figure 2 shows the outcome of sensitivity analysis for Case A1 in a tornado graph format. As can be seen in this figure, the expected probability of destruction of roads and collapse of bridges is 0.48 and 0.59, respectively. Considering that such tornado graphs highlight the most influential scenarios, they can be considered as an efficient tool to

Table 6. Probability of Destruction of Roads and Collapse of Bridges under Six Evidence-Based Scenarios

Extreme precipitation	Extreme temperature	Early melting of snow	Erosion of construction materials	Bridge pier scour	Displacement of bridge components	Destruction of roads	Collapse of bridges
Yes	—	—	—	—	—	39%	58%
Yes	Yes	—	—	—	—	48%	59%
Yes	Yes	Yes	—	—	—	48%	60%
Yes	Yes	Yes	Yes	—	—	53%	79%
Yes	Yes	Yes	Yes	Yes	—	53%	83%
Yes	Yes	Yes	Yes	Yes	Yes	54%	91%

prioritize the pre-event efforts needed to strengthen the transportation roadway network against extreme events. The sensitivity analysis also provides invaluable information about the vulnerability of various transportation infrastructure components to various sources of extreme events. In the current study, the probability of destruction of roads is found to be 0.48, 0.37, 0.20, and 0.15 for Case A1 through A4, respectively. The probability of collapse of bridges, however, is predicted as 0.59, 0.58, 0.40, and 0.39 for Case A1 through A4, respectively. This leads to two main findings: (a) extreme precipitation poses a higher risk to the roadway network than extreme temperature, and (b) bridges are more vulnerable to extreme events than roads. It must be noted that the current findings are only intended to demonstrate the capabilities of the developed BBN based on the assumed conditional probabilities and do not reflect the actual condition of roads and bridges in the state of Iowa.

The BBN has the capability to update the marginal probabilities of the involved variables when new evidence is presented. This can be achieved with forward and backward propagation analyses, which update the marginal probabilities of unobserved variables. Figure 3 shows the updated BBN after EP = Yes and ET = Yes are provided as evidence (separately). This figure clearly highlights the outcome of forward propagation analysis, which directly influences the conditional probabilities of not only contributing variables, but also ultimate consequences. For further comparisons, new evidence is added to the BBN following the sequence listed in Table 6. In the first case, only extreme precipitation is set as evidence (EP = Yes). In the second case, both extreme precipitation and extreme temperature are set as evidence (EP = Yes and ET = Yes). In the remaining four cases, early melting of snow, erosion of construction materials, bridge pier scour, and displacement of bridge components are introduced one by one as new evidence. Assuming a poor condition for roads and bridges, the probabilities of destruction of roads and collapse of bridges are extracted from the BBN. According to Table 6, as more evidence of extreme conditions is added, the probability of destruction of roads and collapse of bridges increases from 0.39 to 0.54 and from 0.58 to 0.91, respectively. This indicates the unique feature of BBNs to update posterior probabilities as soon as evidence is collected from an affected region.

The developed BBN offers the flexibility needed to predict the condition of roads and bridges based on future climate projections. In the Midwest region, for example, an increase in extreme precipitation and extreme temperature has been anticipated (*2, 9, 14, 21*). Thus, in order to assess the vulnerability of roadway networks to growing environmental stressors, nine climate scenarios are investigated in the current study. These scenarios are obtained from a matrix of three extreme precipitation probabilities, that is, 0.63, 0.75, and 0.90, and three extreme temperature probabilities, that is, 0.20, 0.35, and 0.50. Among the listed scenarios, the first scenario, which reflects a probability of occurrence of 0.63 for extreme precipitation and 0.20 for extreme temperature, is identical to the climate scenario considered for the previous analyses. Figure 4 shows the marginal probabilities obtained from the BBN for roads and bridges under the nine climate scenarios and six evidence cases (following Table 6). As can be seen in this figure, the climate parameters are most consequential in the first three evidence cases, in which only limited evidence exists. Depending on climate scenarios, the marginal probability of destruction of roads and collapse of bridges is found to increase up to 3% and 6%, respectively. However, as the number of pieces of evidence of failure grows, the marginal probability approaches its highest level and becomes less sensitive to climate variables. This clearly reflects how the response of a roadway network to an extreme event is influenced by environmental and physical factors.

Conclusions

This study provided a holistic framework to evaluate the vulnerability of roadway networks subjected to extreme weather events. For this purpose, the predictive and

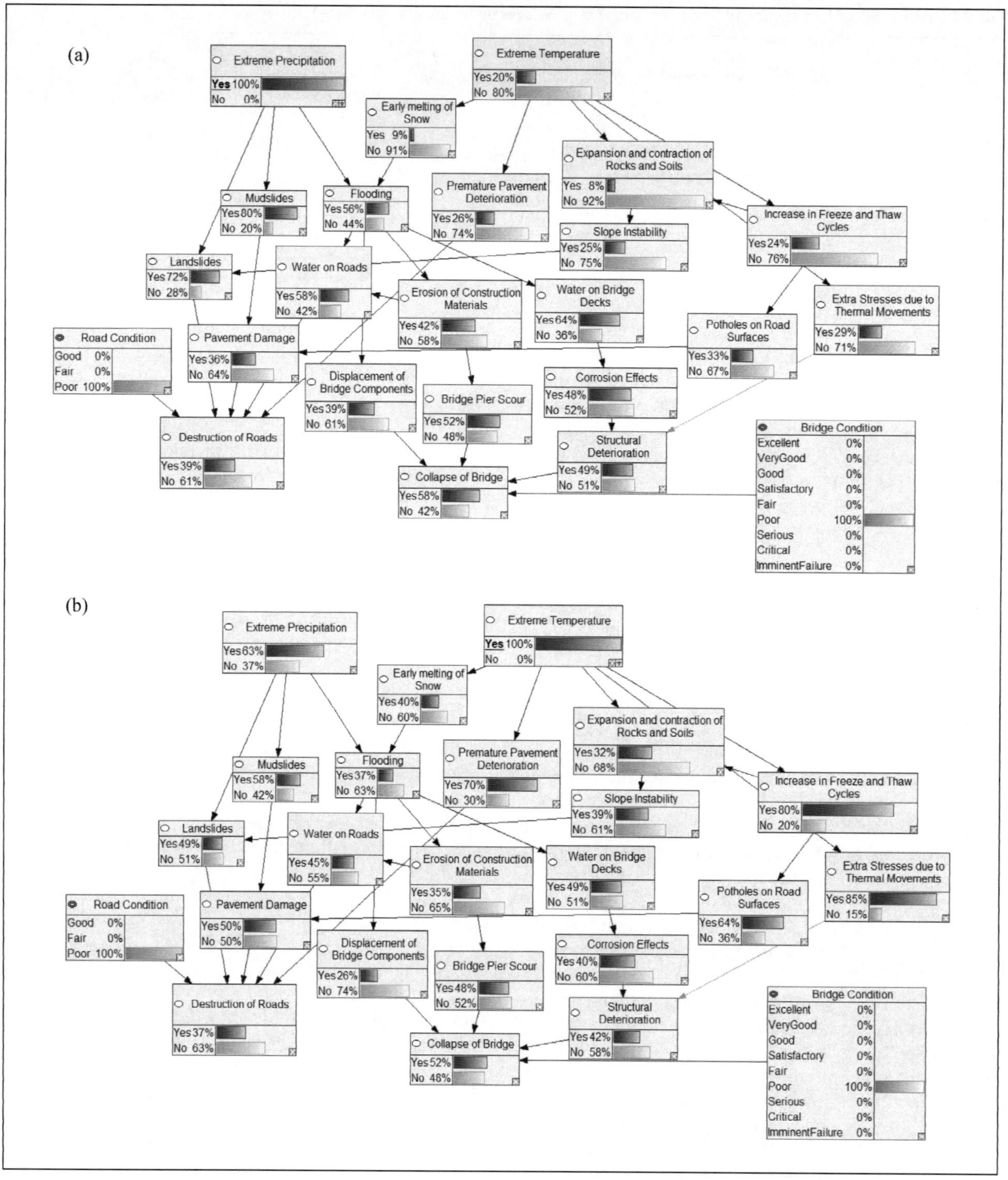

Figure 3. BBN after propagation analysis for (*a*) extreme precipitation = yes and (*b*) extreme temperature = yes.

explanatory capabilities of BBNs were utilized to understand the main causes and consequences of damage to roads and bridges. The developed BBN considered extreme precipitation and extreme temperature as the two primary stressors on the roadway network of Iowa, which was investigated as a case study. Based on historical extreme events and future climate predictions, these two stressors were proven to be able to affect the

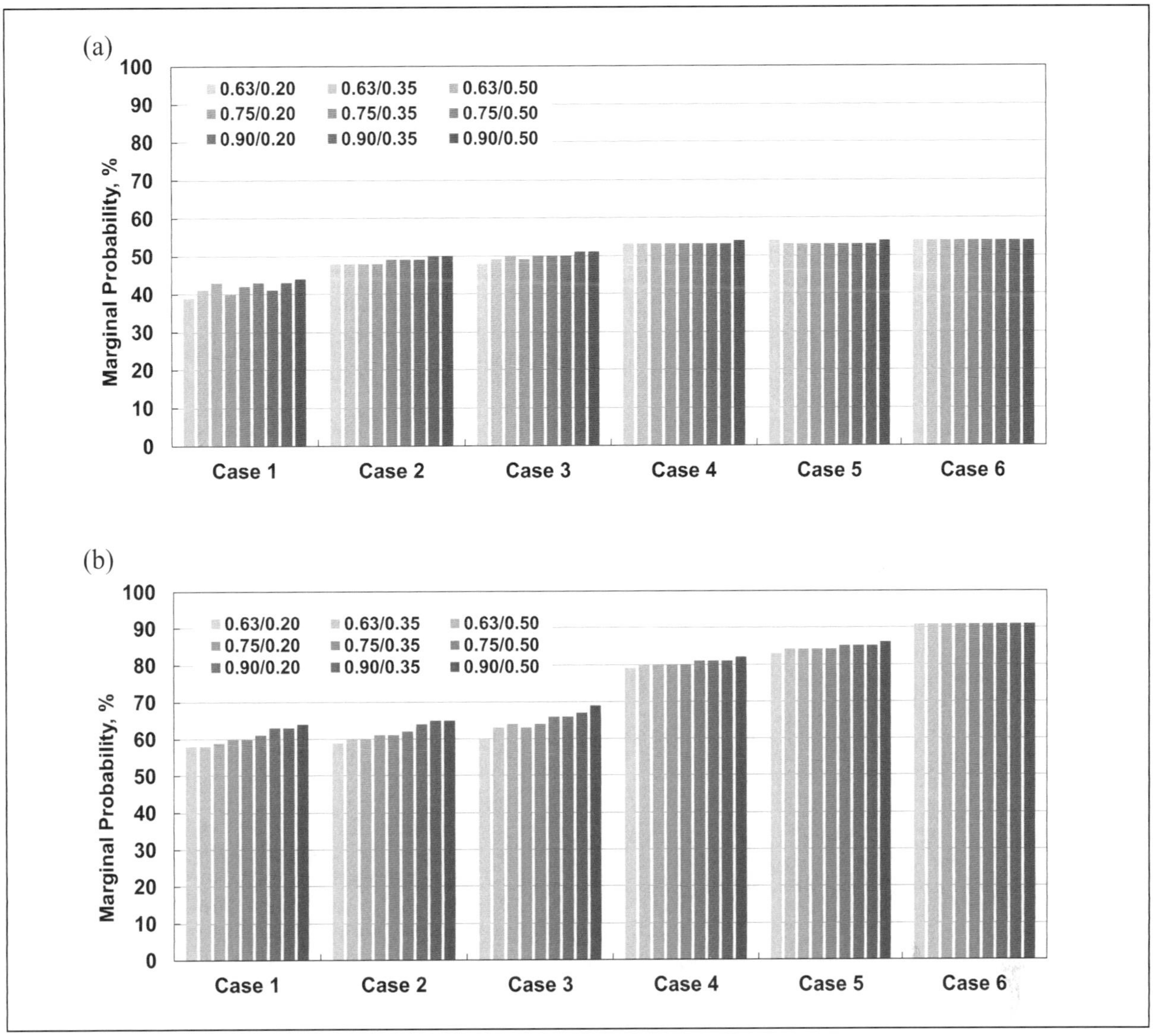

Figure 4. Effect of various climate scenarios on six evidence cases considered for (*a*) destruction of roads and (*b*) collapse of bridges.

performance of transportation infrastructure in the state of Iowa and other neighboring states in the Midwest region. The impact of such events can be in various forms, ranging from flooding and landslides to increase in freeze and thaw cycles. Based on a detailed review of relevant papers and reports, the key contributing factors were identified to formulate the variables included in the BBN. Following a combination of frequentist and knowledge-based approaches, the conditional probabilities were then assigned to parent and child nodes. This enabled the developed BBN to assess the vulnerability of the roadway network under consideration, in terms of destruction of roads and collapse of bridges, as two ultimate consequences. Despite the focus of the current study on physical infrastructure components, it was found that the capabilities of this framework can be extended to evaluate both direct and indirect losses. Through a set of sensitivity analyses and evidence-based propagation analyses, the current study showed how the most influential factors and the effects associated with them can be systematically studied using BBNs. While this critical aspect has been quantified with marginal probabilities, the predicted risks are absolutely not representative of the actual vulnerability of Iowa's transportation infrastructure and are only intended to demonstrate how such statistical inference methods can be employed for generating the information necessary for resource allocation and decision-making purposes.

Author Contributions

The authors confirm contribution to the paper as follows: study conception and design: Behrouz Shafei; data collection: Abhijit Kulkarni; analysis and interpretation of results: Abhijit Kulkarni and Behrouz Shafei; draft manuscript preparation: Abhijit Kulkarni. Both authors reviewed the results and approved the final version of the manuscript.

References

1. The Road Information Program. *Iowa's Top Transportation Challenges: Meeting the State's Need for Safe and Efficient Mobility*. Washington, D.C., 2015.
2. Meyer, M. D., A. F. Choate, and E. Rowan. Adapting Infrastructure to Extreme Weather Events: Best Practices and Key Challenges. In *AASHTO Workshop*, Traverse City, MI, Vol. 20, 2012.
3. Khatami, D., B. Shafei, and O. Smadi. Management of Bridges under Aging Mechanisms And Extreme Events: A Risk-Based Approach. *Transportation Research Record: Journal of the Transportation Research Board*, 2016. 2550: 89–95.
4. Alipour, A., B. Shafei, and M. Shinozuka. Capacity Loss Evaluation of Reinforced Concrete Bridges Located in Extreme Chloride-Laden Envrionments. *Journal of Structure and Infrastructure Engineering*, Vol. 9, No. 1, 2013, pp. 8–27.
5. Shafei, B., and A. Alipour. Application of Large-Scale Non-Gaussian Stochastic Fields for the Study of Corrosion-Induced Structural Deterioration. *Journal of Engineering Structures*, Vol. 88, 2015, pp. 262–276.
6. Shafei, B., and A. Alipour. Estimation of Corrosion Initiation Time in Reinforced Concrete Bridge Columns: How To Incorporate Spatial and Temporal Uncertainties. *ASCE Journal of Engineering Mechanics*, Vol. 141, No. 10, 2015, p. 04015037.
7. Shafei, B., A. Alipour, and M. Shinozuka. A Stochastic Computational Framework To Investigate the Initial Stage of Corrosion in Reinforced Concrete Superstructures. *Journal of Computer-Aided Civil and Infrastructure Engineering*, Vol. 28, No. 7, 2013, pp. 482–494.
8. Shafei, B., A. Alipour, and M. Shinozuka. Prediction of Corrosion Initiation in Reinforced Concrete Members Subjected to Environmental Stressors: A Finite-Element Framework. *Journal of Cement and Concrete Research*, Vol. 42, No. 2, 2012, pp. 365–376.
9. Meyer, M. D., E. Rowan, C. Snow, and A. Choate. Impacts of Extreme Weather on Transportation. In *2013 Extreme Weather Events Symposium*, Washington, D.C., 2013.
10. Baglin, C. *Response to Extreme Weather Impacts on Transportation Systems*. National Cooperative Highway Research Program, Washington, D.C., 2014.
11. Jha, M. K. Applying Bayesian Networks to Assess Vulnerability of Critical Transportation Infrastructure. *Proc., 19th International Conference on Applications of Advanced Technology in Transportation (AATT)*, Chicago, IL, 2006, pp. 1–6.
12. Jha, M. K., and R. A. Keele. Using Dynamic Bayesian Networks for Investigating the Impacts of Extreme Events. In *Bayesian Networks*, InTech, London, UK, 2012, pp. 36–48.
13. Iowa Department of Transportation. *Iowa Infrastructure Condition Evaluation*. Iowa Department of Transportation, Ames, IA, 2015–2016.
14. Iowa Department of Transportation. *Iowa In Motion—Planning Ahead*. Iowa Department of Transportation, Ames, IA, 2012.
15. U.S. Department of Transportation. *Deficient Bridges by Highway System 2016*. https://www.fhwa.dot.gov/bridge/nbi/no10/defbr16.cfm#a. Accessed December 12, 2016.
16. Iowa Renewable Fuels Association. *Iowa Renewable Fuels Association: Statistics*. 2016. http://iowarfa.org/resource-center/statistics/. Accessed July 30, 2017.
17. Zogg, J. *The Top Five Iowa Floods*. National Weather Services, WFO, Des Moines, IA, 2014.
18. U.S. Department of Commerce. *Central Iowa Floods of 2008*. National Weather Service, Des Moines, IA, 2009.
19. Vennapussa, P., D. White, and K. Miller. *Western Iowa Missouri River Flooding: Geo-Infrastructure Damage Assessment, Repair and Mitigation Strategies*. Institute for Transportation, Iowa State University, Ames, IA, 2013.
20. Peterson, T. C., M. McGuirk, T. G. Houston, A. H. Horvitz, and M. F. Wehner. *Climate Variability and Change with Implications for Transportation*. Transportation Research Board, Washington D.C., 2008.
21. Meyer, M., M. Flood, J. Keller, J. Lennon, G. McVoy, C. Dorney, K. Leonard, R. Hyman, and J. Smith. *Strategic Issues Facing Transportation, Volume 2: Climate Change, Extreme Weather Events and the Highway System: Practitioner's Guide and Research Report*. Transportation Research Board, Washington, D.C., 2014.
22. Anderson, C., D. Claman, and R. Mantilla. *Iowa's Bridge and Highway Climate Change and Extreme Weather Vulnerability Assessment Pilot*. 2015. http://lib.dr.iastate.edu/intrans_techtransfer/74. Accessed July 30, 2017.
23. United States Environmental Protection Agency. *Iowa Climate Change Adaptation and Resilience Report*. United States Environmental Protection Agency (EPA), Washington, D.C., 2011.
24. Committee on Climate Change and U.S. Transportation. *Potential Impacts of Climate Change on U.S. Transportation*. Transportation Research Board, Washington, D.C., 2008.
25. Moorty, S., and C. W. Roeder. Temperature-Dependent Bridge Movements. *Journal of Structural Engineering*, Vol. 118, No. 4, 1992, pp. 1090–1105.
26. Yodo, N., and P. Wang. Resilience Modeling and Quantification for Engineered Systems Using Bayesian Networks. *Journal of Mechanical Design*, Vol. 138, 2016, p. 031404.
27. Tesfamariam, S., and B. Martin-Perez. Bayesian Belief Network Too Assess Carbonation-Induced Corrosion in Reinforced Concrete. *Journal of Materials in Civil Engineering, ASCE*, Vol. 20, No. 11, 2008, pp. 707–717.
28. Wang, F., and Q.-L. Zhang. Systemic Estimation of Dam Overtopping Probability: Bayesian Networks Approach.

Journal of Infrastructure Systems, Vol. 23, No. 2, 2017, p. 04016037.
29. Pearl, J. *Probabilistic Reasoning in Intelligent Systems: Networks of Plausible Inference*. Morgan Kaufmann Publishers, San Mateo, CA, 1988.
30. Jensen, F. V. *An Introduction to Bayesian Networks*. Springer, New York, NY, 1996.
31. Krieg, M. *A Tutorial on Bayesian Network*. DSTO Electronics and Surveillance Research Laboratory, Edinburgh, South Australia, 2001.
32. Hosseini, S., and K. Barker. Modeling Infrastructure Resilience Using Bayesian Networks: A Case Study of Inland Waterway Ports. *Journal of Computers and Industrial Engineering*, Vol. 93, 2016, pp. 252–266.
33. Iowa Department of Transportation. *Bridge Inspection Manual*. Office of Bridges and Structures, Ames, IA, 2015.
34. Leviäkangas, P., A. Tuominen, R. Molarius, H. Kojo, J. Schabel, S. Toivonen, J. Keränen, J. Ludvigsen, A. Vajda, H. Tuomenvirta, I. Juga, P. Nurmi, J. Rauhala, F. Rehm, T. Gerz, T. Muehlhausen, J. Schweighofer, S. Michaelides, M. Papadakis, N. Dotzek, and P. Groenemeijer. *Extreme Weather Impacts on Transport Systems*. VTT Technical Research Centre of Finland, Oulu, Finland, 2011.
35. Hershfield, D. M. On the Probability of Extreme Rainfall Event. *American Meteorological Society*, Vol. 54, No. 10, 1973, pp. 1013–1018.
36. BayesFusion, LLC. *GeNIe Modeler*. BayesFusion, LLC, 1998.

The Standing Committee on Transportation Network Modeling (ADB30) peer-reviewed this paper (18-03968).

Article

TRR
JOURNAL OF THE TRANSPORTATION RESEARCH BOARD

Transportation Research Record
2018, Vol. 2672(48) 58–68

DOI: 10.1177/0361198118782041
journals.sagepub.com/home/trr
SAGE

A Generalized Single-Level Formulation for Origin–Destination Estimation under Stochastic User Equilibrium

Wei Ma[1] and Zhen (Sean) Qian[2]

Abstract

Origin–destination (OD) demand is an indispensable component for modeling transportation networks, and the prevailing approach to estimating OD demand using traffic data is through bi-level optimization. A bi-level optimization approach considering equilibrium constraints is computationally challenging for large-scale networks, which prevents the OD estimation (ODE) being scalable. To solve for ODE in large-scale networks, this paper develops a generalized single-level formulation for ODE incorporating stochastic user equilibrium (SUE) constraints. Two single-level ODE models are specifically discussed and tested. One employs a SUE based on the satisfaction function, and the other is based on the Logit model. Analytical properties of the new formulation are analyzed. The estimation methods are proven to be unbiased. Gradient-based algorithms are proposed to solve for this formulation. Numerical experiments are conducted on a small network and a large network, along with sensitivity analysis on sensor locations, historical OD information and measurement error. Results indicate that the new single-level formulation, in conjunction with the proposed solution algorithms, can achieve accuracy comparable with the bi-level formulation, while being much more computationally efficient for large networks.

For decades, origin–destination (OD) demand has played a vital role in traffic planning and operation (*1–5*). Estimating OD demand has become one of the most fundamental problems in transportation engineering and has been extensively studied over the years. The prevailing approach to estimating OD demand using traffic data is through a bi-level optimization framework. The bi-level approach considering user equilibrium (UE) constraints is computationally challenging for large-scale networks, which prevents the OD estimation (ODE) being scalable. In view of this, this research develops a new single-level ODE method that incorporates general stochastic user equilibrium (SUE) constraints. The single-level approach would be more efficient and scalable than a bi-level approach.

Estimating OD demands is notoriously difficult. Census provides accurate results, but they are costly and can take a long time. Various estimation methods have been proposed to infer OD from traffic data. In the early stage of ODE methods, the road networks were assumed to be uncongested, therefore a pre-determined route choice matrix can be used in the estimation process (*6–10*). While the uncongested assumption is usually violated in urban networks, researchers combined UE conditions together with ODE models to search for ODs that satisfy the UE conditions (*10–13*). The UE conditions can help estimate the route choice matrix within the ODE process. This combined model is known as the bi-level ODE formulation, also known as mathematical programming with equilibrium constraints (MPEC) in other fields (*14*).

The bi-level models integrating equilibrium constraints and ODE formulation (with historical OD information) are widely used for practical applications (*15*). To our best knowledge, there are two types of solution algorithms for the bi-level formulations. The first type of method derives gradient of traffic equilibrium and performs a gradient descent on the ODE problem (*1*). The gradient is constrained in the equilibrium feasible domain, so the lower level constraint is satisfied. The computed gradient is usually based on the sensitivity analysis of equilibrium conditions, which usually requires

[1]Department of Civil and Environmental Engineering, Carnegie Mellon University, Pittsburgh, PA
[2]Department of Civil and Environmental Engineering and Heinz College, Carnegie Mellon University, Pittsburgh, PA

Corresponding Author:
Address correspondence to Zhen (Sean) Qian: seanqian@cmu.edu

solving a complex optimization problem (*16*). The second method is a heuristic method that computes the OD demand by iteratively solving the traffic equilibrium model and the ODE model (*13*). The heuristic method is also computationally intensive on large networks as solving the equilibrium model is costly. The bi-level formulations for the ODE problem are naturally non-convex and non-continuous (*14*), therefore both methods have limited potentials for scalability.

In recent years, a single-level ODE method considering UE constraints was proposed (*17*, *18*). The intuition behind the new method is to relax the lower level equilibrium constraint as a penalization term in the upper-level objective function, such that the estimation method is able to optimize the ODE model and traffic equilibrium model simultaneously (*17*). Shen and Wynter (*19*) show that (*17*) does not always obtain a desired estimation, and proposed a fix to enhance the accuracy of the single-level ODE under large-scale networks.

Shen and Wynter (*19*) offer a satisfying approach to efficiently solving ODE problems under UE constraints. However, UE models assume users having full knowledge of network conditions, which is evidently not realistic in the real world (*20*, *21*). On the contrary, the SUE models can handle perception errors of travelers and are generally preferred in practice. For a detailed review of SUE models, we refer to (*22*). Many studies (*23–25*) have proposed SUE-based ODE formulations, most of which are formulated as a bi-level optimization problem. The bi-level SUE-based ODE formulation is relatively inefficient and cannot be applied to large networks. Therefore, there is a lack of ODE methods considering SUE constraints that can be efficiently applied on large scale networks. This research fills the gap by generalizing the single-level ODE to accommodate SUE, and thoroughly testing and validating the proposed methods using real-world networks and data.

In this paper, we generalize the previously studied single-level ODE method (*19*) to a generalized formulation applicable to all the equilibrium constraints. We especially examine the case of SUE and propose different variants for single-level ODE methods considering SUE constraints. In this paper, we select the satisfaction-function-based SUE (*20*) and the Logit SUE to formulate the model. The accuracy, efficiency and sensitivity of the proposed methods are tested on small and large networks. The major contributions of this paper are summarized as follows:

1) It proposes a generalized form for single-level ODE formulation with general equilibrium constraints and revisits each term in the formulation.
2) It derives two specific SUE-based ODE formulations and proves the unbiasedness of the OD estimator.
3) The proposed methods are tested on a large-scale real network to gain insights from solutions. The computational efficiency of the solution algorithms is also examined.

The rest of paper is organized as follows. We first review the ODE problems and propose the generalized formulation for single-level ODE problem with equilibrium constraints. We then present the new ODE formulations under SUE constraints. The solution algorithms and framework are proposed, and then examined using numerical experiments on both small and large networks. Finally, we conclude the findings.

Overview of ODE Model

We review variants of ODE formulations such as the traditional formulation, the bi-level formulation and the single-level formulation under UE constraints (*19*). We use the following notations throughout this paper:

A: the set of all links
A^o: the set of links with flow observations
K_q: the set of all OD pairs
K_{rs}: the set of all paths between OD pair r-s
q: the vector of OD demands
q_{rs}: OD demand for OD pair r-s
q^H_{rs}: historical OD demand for OD pair r-s
p^k_{rs}: the probability of choosing path k in all paths between OD pair r-s
p: OD to path flow matrix, consisting of all p^k_{rs}
c: the vector of path costs for all OD pairs
x: the vector of link flow
x_a: link flow on link a
x^o_a: measured link flow on link a
f: the vector of path flow
f^k_{rs}: path flow for path k in OD pair r-s

Overview of the ODE Problem

Traditional bi-level ODE formulates its upper-level to minimize the square error between estimated and observed traffic counts, and the lower level formulation searches for the UE traffic pattern.

Upper Level:

$$\begin{aligned} &\min_q \frac{1}{2}\sum_{a\in A^o}\left(x_a - x^o_a\right)^2 \\ &s.t.\ q_{rs} \geq 0 \quad \forall rs \in K_q \\ &x \in \arg\min_x z(q) \end{aligned} \tag{1}$$

Lower Level z(q):

$$\min_{x} \sum_{a \in A} \int_{0}^{x_a} t_a(w)dw$$
$$\text{s.t. } q_{rs} \geq 0 \forall rs \in K_q$$
$$\sum_{k \in K_{rs}} f_{rs}^k = q_{rs} \quad (2)$$
$$\sum_{rs \in W} \sum_{k \in K_{rs}} \delta_{rs}^{ka} f_{rs}^k = x_a, \forall a \in A$$
$$f_{rs}^k \geq 0, \forall rs \in K_q \; k \in K_{rs}$$

where $t_a(\cdot)$ is the link performance function for any link a.

There are three important issues for a typical ODE method. First, the uniqueness of the estimated OD demand is usually not guaranteed (*26*). The historical OD information $\sum_{rs \in K_q} (q_{rs} - q_{rs}^H)^2$ is usually introduced to the upper-level objective function as a prior information, which narrows the searching domain. When the actual ODs are close to historical ODs, the bi-level programming can converge to the actual ODs accurately.

The second issue is the computational complexity of the bi-level formulation. To address this, (*17*) proposed a relaxed ODE formulation which combines both ODE and UE in a single-level formulation. Later, (*19*) pointed out that the formulation in (*17*) does not satisfy the Karush-Kuhn-Tucker (KKT) condition of the lower level, causing an underestimation of OD demand. (*19*) proposed a modified method which is consistent with the UE constraint:

$$\min_{q} \frac{1}{2} \sum_{a \in A^o} (x_a - x_a^o)^2 + \frac{w_q}{2} \sum_{rs \in K_q} (q_{rs} - q_{rs}^H)^2 + \frac{1}{\theta} \left(\sum_{a \in A} \int_{0}^{x_a} t_a(w)dw - \sum_{rs \in K_q} \hat{c}_{rs} q_{rs} \right) \quad (3)$$
$$\text{s.t. } (q, x, f) \in H$$

where w_q is a weight parameter representing the confidence level of historical OD information and θ is the dispersion factor representing the importance of the UE constraint. The polyhedron H in (q, x, f) is defined by the following relationships:

$$\sum_{k \in K_{rs}} f_{rs}^k = q_{rs}, \forall rs \in K_q \quad (4)$$

$$\sum_{rs \in K_q} \sum_{k \in K_{rs}} \delta_{rs}^{ka} f_{rs}^k = x_a, \forall a \in A \quad (5)$$

$$q_{rs} \geq 0, \forall rs \in K_q \quad (6)$$

$$f_{rs}^k \geq 0, \forall rs \in K_q, k \in K_{rs} \quad (7)$$

Formulation (3) addresses the first two issues in a traditional ODE model, while still using UE constraints. The UE constraints assume all travelers are aware of the complete information about the network conditions. Therefore, the third issue is that UE constraints are not realistic for the real-world environment (*21*). A good alternative to UE can be SUE that assumes travelers have perception errors on their knowledge about network conditions and use route choice model to compute the route choice probabilities. Till now, how to incorporate SUE constraints into a single-level framework of (*19*) has not been explored. We will generalize the single-level ODE framework by (*19*) and propose new single-level ODE formulations considering SUE constraints.

A Generalized Formulation with Equilibrium Constraints

The main idea of the single-level ODE framework in (*19*) relies on the objective to optimize the measurement mismatch error and equilibrium condition mismatch error simultaneously. The framework uses an adjustment factor to prevent one objective dominating the other. The adjustment factor leverages between measurement mismatch error and equilibrium condition mismatch error such that both errors can reach optimal points at the same time when their summation reaches optimal. The adjustment factor for different equilibrium constraints can be different. We first generalize the single-level ODE formulation, and then we derive the corresponding forms for variants of SUE.

Following the concept of (*19*), the generalized ODE formulation with equilibrium constraint can be formulated as follows:

$$\min_{q,x,f} \epsilon_m + w_e(\epsilon_e + \kappa)$$
$$\text{s.t. } (q, x, f) \in H \quad (8)$$

where

$$\epsilon_m = \frac{1}{2} \sum_{a \in A^o} (x_a - x_a^o)^2 + \frac{w_q}{2} \sum_{rs \in K_q} (q_{rs} - q_{rs}^H)^2 \quad (9)$$

ϵ_m represents the measurement mismatch error which comes from the traditional ODE formulation and ϵ_e represents the equilibrium constraint mismatch error. For the case of UE constraint, $\epsilon_e = \frac{1}{\theta} \sum_{a \in A} \int_{0}^{x_a} t_a(w)dw$. We optimize both objectives simultaneously by adding them together, however the optimal solution (q, x, f) for $\epsilon_m + \epsilon_e$ is usually not the optimal solution for ϵ_e and ϵ_m separately, so the adjustment factor κ and weight factor w_e are included to ensure both objectives achieve the optimal at the same time during the optimization, to avoid one objective dominating the other.

For different equilibrium constraints, the challenge lies in κ as it is not always available. Formulation (8) can be viewed as a Lagrangian relaxation of the bi-level formulation, and κ represents the dual variables from constraints. The dual variables might not be derived analytically before solving the bi-level formulation.

The single-level ODE formulation in (*19*) for UE constraints well-fits the generalized form (8) by defining $\kappa = -\sum_{rs \in K_q} \hat{c}_{rs} q_{rs}$. The formulation of κ is derived from the KKT conditions of the original bi-level formulation.

A New Single-Level ODE Formulation With SUE Constraints

In this section, two SUE models are employed as the constraints for the new single-level ODE method. One SUE model is the satisfaction-function-based unconstrained model and the other is the Logit-based constrained model. To accommodate both formulations to the generalized formulation (8), we explore ϵ_e and κ for different SUE models separately. The proposed formulation can be potentially applied to other variants of the SUE model, such as c-Logit, but it is not the focus of this paper.

Satisfaction-Function-Based SUE Model

The SUE model in (*20*) is formulated to minimize $z_1(x)$, which is presented follows:

$$z_1(x) = -\sum_{rs} q_{rs} S_{rs}[c^{rs}(x)] + \sum_a x_a t_a(x_a) - \sum_a \int_0^{x_a} t_a(\omega) d \tag{10}$$

where

$$(q, x, f) \in H \tag{11}$$

$$S_{rs}[c^{rs}(x)] = E\left[\min_{k \in K_{rs}} \{C_r^{rs}\} | c^{rs}(x)\right] \tag{12}$$

$z_1(x)$ is the objective of the satisfaction-function-based SUE model. Having a good choice of satisfaction function $S_{rs}[c^{rs}(x)]$, the optimization problem (10) can be cast to an unconstrained problem without constraint (11). Then by chain rule, the derivative of $z_1(x)$ by path flow f_{rs}^k is:

$$\begin{aligned} \frac{\partial z_1(x)}{\partial f_{rs}^k} &= \sum_b \left(\frac{\partial z_1(x)}{\partial x_b} \frac{\partial x_b}{\partial f_{rs}^k} \right) \\ &= \sum_b \delta_{rs}^{kb} \left[-\sum_{rs'} \sum_{k'} q_{rs'} P_{rs'}^{k'} \delta_{k'b}^{rs'} + x_b \right] \frac{dt_b}{dx_b} \end{aligned} \tag{13}$$

Since $\frac{\partial z_1(x)}{\partial f_{rs}^k}$ is zero when the traffic conditions achieve equilibrium, both ϵ_m and ϵ_e achieve minimum simultaneously without any adjustment, therefore $\kappa = 0$. And ϵ_e is the objective of the satisfaction-function-based SUE model itself, which is presented in Formulation (14).

$$\epsilon_e = -\sum_{rs} q_{rs} S_{rs}[c^{rs}(x)] + \sum_a x_a t_a(x_a) - \sum_a \int_0^{x_a} t_a(\omega) d\omega \tag{14}$$

Proposition 1. (Unbiasedness). The optimization problem defined by Equations 8, 9, 14 and $\kappa = 0$ is unbiased if x^o and q^H are unbiased.

Proof. Consider the KKT conditions for the optimization problem, the complementary slackness, primal feasibility and dual feasibility are easy to satisfy. If we write the stationarity property of f:

$$0 \in \sum_{a \in A^o} \Delta^T(\Delta f - x^o) + w_q M^T(Mf - q_{rs}^H) + dz \tag{15}$$

where

$$\begin{aligned} dz_{rs}^k = \frac{\partial z_1(x)}{\partial f_{rs}^k} &= \sum_b \left(\frac{\partial z_1(x)}{\partial x_b} \frac{\partial x_b}{\partial f_{rs}^k} \right) \\ &= \sum_b \delta_{rs}^{kb} \left[-\sum_{rs'} \sum_{k'} q_{rs'} P_{rs'}^{k'} \delta_{k'b}^{rs'} + x_b \right] \frac{dt_b}{dx_b} \end{aligned} \tag{16}$$

Since x^o and q^H is unbiased, and dz is zero when the network condition (f, x) achieves stochastic equilibrium (*20*). Therefore Equation 15 is satisfied if and only if f is unbiased.

Logit-Based SUE Model

Another well-known SUE model is the Logit-based model, which assumes travelers' perception errors follow Gumbel distribution, and therefore their choices follow the Logit function. Logit-based SUE is formulated as follows:

$$z_2(x) = \sum_{a \in A} \int_0^{x_a} t_a(w) dw + \frac{1}{\Theta} \sum_{rs \in W} \sum_{k \in K_{rs}} f_{rs}^k \log f_{rs}^k \tag{17}$$

where

$$(q, x, f) \in H \tag{18}$$

Constraint (18) is always active since $z_2(x)$ goes to infinity without constraint (18). Therefore, the adjustment κ

is not zero in this case. The KKT condition of the Logit-constrained problem can be written as:

$$c_{rs}^{k} + \frac{1}{\Theta}\left(\log f_{rs}^{k} + 1\right) - \pi_{rs} = 0 \tag{19}$$

ϵ_m and ϵ_e cannot achieve the minimal points simultaneously, so the adjustment factor κ needs to reproduce the KKT condition of Equation 19 during the optimization. Comparing with the KKT conditions for the original bi-level formulation, we have the following formulations of ϵ_m and κ in Equations 20, 21.

$$\epsilon_e = \sum_{a \in A} \int_0^{x_a} t_a(w)dw + \frac{1}{\Theta} \sum_{rs \in W} \sum_{k \in K_{rs}} f_{rs}^{k} \log f_{rs}^{k} \tag{20}$$

$$\kappa = -\sum_{rs \in W} q_{rs} \hat{\pi}_{rs} \tag{21}$$

$\hat{\pi}_{rs}$ is the expected cost from r to s in a Logit-based SUE traffic pattern, and it cannot be derived analytically. We will discuss how to compute it using an iterative ODE framework in the following section.

Proposition 2. (Unbiasedness). The optimization problem defined by Equations 8, 9, 20 and 21 is unbiased if x^o and q^H are unbiased.

Proof. Following similar steps, we write down the stationarity condition of the KKT conditions of the optimization problem.

$$0 \in \sum_{a \in A^o} \Delta^T(\Delta f - x^o) + w_q M^T\left(Mf - q_{rs}^{H}\right) + dz \tag{22}$$

where

$$dz_{rs}^{k} = c_{rs}^{k} + \frac{1}{\Theta}\left[\log(f_{rs}^{k}) + 1\right] - \hat{\pi}_{rs} \tag{23}$$

If the network conditions achieve stochastic equilibrium, then Equation 19 holds, thus dz equals zero. Again if x^o and q^H are unbiased and other KKT conditions hold naturally, then f is unbiased.

Solution Algorithm

We present a gradient-based algorithm to solve both satisfaction-function- and Logit-based formulations. The gradients of ϵ_m can be derived from sensitivity analysis of equilibrium conditions, which has been extensively studied over years (*1*). According to those sensitivity studies, the gradients of z_1 and z_2 are derived in this section. By applying gradient projection or Frank-Wolfe algorithms, generalized single-level optimization formulation under SUE constraint (*8*) can be solved.

Since our ODE formulations are path-based, the path size increases exponentially when the network grows. For small networks, path enumeration is possible. If the networks are large, we simply enumerate K shortest paths (*27*) for each OD pair and therefore shrink the path size. To be more advanced, we can fit the ODE formulation in a column generation framework (*15*, *28*).

Gradient-Based Algorithms

The gradients of both optimization formulations with different SUE constraints are derived in this section. Gradients for such equilibrium-constrained formulation are composed of two parts: ϵ_m and $\epsilon_e + \kappa$. Gradient for ϵ_m is identical for all formulations, while the gradients for $\epsilon_e + \kappa$ vary across different formulations.

For satisfaction-function-based ODE formulation, the derivative of the second part is written as follows:

$$\left(\frac{\partial z_1(x)}{\partial f_{rs}^{k}}\right)^{(i)} = \sum_b \delta_{rs}^{kb} \left[-\sum_{rs'}\sum_{k'} q_{rs'}{}^{(i-1)} P_{rs'}^{k'}{}^{(i-1)} \delta_{k'b}^{rs'} + x_b^{(i-1)}\right] \frac{dt_b\left(x_b^{(i-1)}\right)}{dx_b} \tag{24}$$

where $q_{rs'}{}^{(i-1)}$ and $P_{rs'}^{k'}(i-1)$ are derived from $f_{rs}^{k}(i-1)$, and t_b is the link performance function of link b.

For Logit-based ODE formulation, the derivative of second part can be derived as follows:

$$\left(\frac{\partial z_2(x)}{\partial f_{rs}^{k}}\right)^{(i)} = \left(c_{rs}^{k}\right)^{(i-1)} + \frac{1}{\Theta}\left(\log f_{rs}^{k\,(i-1)} + 1\right) - \pi_{rs}{}^{(i-1)} \tag{25}$$

To determine $\hat{\pi}_{rs}$ in the optimization problem, we borrow a similar idea from (19) and devise an iterative updating framework to update $\hat{\pi}_{rs}$ at each iteration. For each iteration i, $\hat{\pi}_{rs}^{(i)}$ is determined as:

$$\hat{\pi}_{rs}^{(i)} = \frac{1}{|K_{rs}|} \sum_{k \in K_{rs}} c_{rs}^{k}(i-1) \tag{26}$$

Solution Procedure

The solution procedure is summarized as follows:

Step 0: *(Initialization)*. Iteration i = 1, generate a path set for each OD pair. Set path flow $f_{rs}^{k(i)}$ and link flow $x_a{}^{(i)}$ as zero.

Step 1: *(Path cost update)*. Compute $c_{rs}^{k(i)}$ by the new network conditions $x_a{}^{(i)}$ and $f_{rs}^{k(i)}$.

Step 2: *(Update adjustment factor)*. If using satisfaction-function-based formulation, skip; if using Logit-based formulation, use Equation 26 to update adjustment factor κ.

Step 3: *(Gradient descent)*. Perform a one-step gradient projection descent or Frank–Wolfe descent.

Step 4: *(Network Update)*. Update path flow $f_{rs}^{k(i)}$ and link flow $x_a^{(i)}$.

Step 5: *(Convergence check)*. Check the difference of link flow f, if the convergence criterion is met, go to Step 6; if not, i = i + 1, go to Step 1.

Step 6: *(Output)*. Output (f, q).

We use the K-shortest path set on the empty network as the fixed path set. For congested networks, a column generation scheme on a larger K will be preferred. We use $\frac{1}{\sqrt{i}}$ as the step size for the gradient descent at iteration i. Other advanced line search methods can also be applied. The convergence criterion is based on the estimated (f, q). When (f, q) converges over the iteration, we terminate the algorithm.

Numerical Experiments

We test the proposed single-level ODE methods on both small network and real-world large scale network. We adopt a validation framework commonly used in ODE studies. Given a road network, we construction the OD demand based on the previous studies or historical OD and then treat it as the "true" OD demand. Then we generate the observation data by running the SUE model and add some perturbations on the results of the equilibrium models. The performance of the ODE formulations is based on the comparison between true ODs and estimated ODs.

To demonstrate the efficiency and accuracy of our formulations, we compare our formulations with two other commonly used ODE methods: the sensitivity-based approach for bi-level optimization and the traditional approach for non-equilibrium-based optimization. First we solve these four formulations on a small network and illustrate how our formulations work. The sensitivity of sensor location, observation noise and historical OD for four formulations are also tested and analyzed. Then we compare four formulations on a large-scale network and demonstrate the efficiency of our formulations.

A Small Network

A toy network with four links, three paths, and one OD pair is used to get a sense of how the estimation works. The toy network is presented in Figure 1. The Bureau of Public Roads (BPR) link travel time function is adopted,

$$t_a(X_a) = t_a^0\left[1 + \alpha\left(\frac{X_a}{cap_a}\right)^{\beta}\right] \tag{27}$$

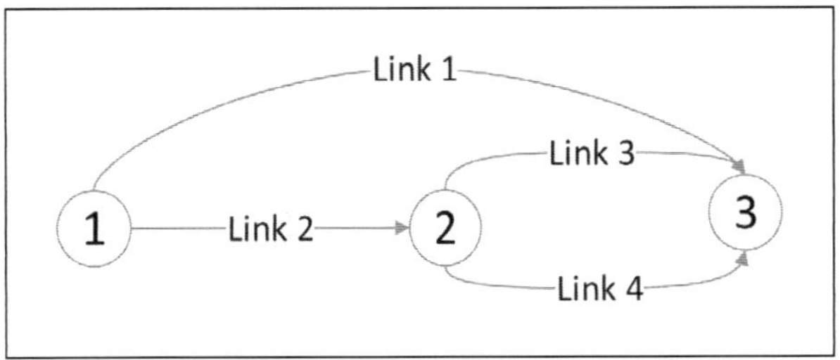

Figure 1. A three-path small network.

where t_0^a is the free-flow travel time on link $a \in A$, $\beta = 4$, $\alpha = 0.15$ are constant parameters. cap_a denotes the capacity of link a. The parameters in the BPR function are set as $t_1^0 = 20, t_2^0 = t_3^0 = 10$ and $t_4^0 = 8$, $cap_a = 360, \forall a \in A$, $q_{true} = 1000$.

A Moving Successive Average (MSA) based algorithm is used to solve the stochastic traffic equilibrium (MSA is used for demonstration purpose; more advanced algorithms can be found in (*20*)). The stochastic traffic equilibrium conditions are $x_1 = 360$, $x_2 = 640$, $x_3 = 317$ and $x_4 = 323$. To evaluate different ODE methods, three metrics are adopted to measure the difference between estimated ODs and actual ODs:

Percentage root mean square error (PRMSE):

$$\text{PRMSE} = 100\% \times \sqrt{\frac{\sum_{rs\in K_q}\left(q_{rs} - q_{rs}^{true}\right)^2}{|K_q|}} \times \frac{|K_q|}{\sum_{rs\in K_q} q_{rs}^{true}} \tag{28}$$

Mean absolute relative error (MARE):

$$\text{MARE} = 100\% \times \frac{\sum_{rs\in K_q}\frac{|q_{rs} - q_{rs}^{true}|}{q_{rs}^{true}}}{|K_q|} \tag{29}$$

Percentage of total demand not captures (PTD):

$$\text{PTD} = 100\% \times \left(1 - \frac{\sum_{rs\in K_q} q_{rs}}{\sum_{rs\in K_q} q_{rs}^{true}}\right) \tag{30}$$

Basic Results. Assuming link 1 is observed without any noise, that is $x_1^o = 360$. We compare different ODE formulations with and without the accurate historical OD information. Table 1 presents the estimation results with and without accurate historical OD information. (Note: we drop the decimal part of the link flow, so the table may look slightly inconsistent.)

The results with historical OD seem straightforward because the historical OD contains perfect information. Yet it validates the Proposition 1 and 2, which proves that estimated OD of all four formulations are unbiased.

Table 1. Comparison between Different ODE Formulations With and Without Accurate Historical OD Information

	Methods	Estimated link flows				Estimated OD metrics		
	Link 1 observed	Link 1	Link 2	Link 3	Link 4	PRMSE (%)	MARE (%)	PTD (%)
With historical OD	Traditional ODE	360	640	320	320	0.00	0.00	0.00
	Bi-level ODE	360	640	320	320	0.00	0.00	0.00
	Satisfaction-function-based ODE	360	640	317	323	0.01	0.01	–0.01
	Logit-based ODE	360	639	320	320	0.06	0.06	0.06
Without historical OD	Traditional ODE	360	0	0	0	63.99	63.99	63.99
	Bi-level ODE	360	640	316	324	0.02	0.02	0.02
	Satisfaction-function-based ODE	360	643	318	325	0.48	0.48	−0.48
	Logit-based ODE	360	399	199	200	24.17	24.17	24.17

The results without historical OD are what we are really concerned with. As can be seen, the traditional ODE is not able to capture the link information on unobserved links since it does not contain an equilibrium constraint; bi-level ODE performs best among all algorithms, but it is not applicable on large networks. Both of our formulations are single-level, so they can be applied on large networks. Satisfaction-function-based ODE outperforms the Logit-based ODE, since Logit-based ODE needs to iteratively update a dual variable π_{rs}. A wrong estimation of π_{rs} may lead the gradients to a wrong direction. However, quantifying the satisfaction function requires more computations than the Logit-based ODE, so the Logit-based ODE has trade-off between its accuracy and efficiency.

Impact of Sensor Locations. To test the sensitivity of our formulations on sensor location, we compare both new single-level formulations with the bi-level ODE formulation. The traditional ODE is ruled out due to its poor performance without historical OD. We assume different link flows are observed without noise in each case and compare the estimated link flows and ODs, the results are presented in Table 2.

As a benchmark of the ODE with different observed link flows, the bi-level ODE formulation performs reliably. The estimation error is always below 1%. Logit-based ODE performs slightly worse than satisfaction-function-based ODE, while it always outperforms the traditional ODE formulation.

We take a close look at the Logit-based ODE. For example, the observed link is Link 4, and the estimated flow on Link 3 is 44. The Logit-based ODE searches for the gradients of equilibrium, but they can be inaccurate and non-convergent. As a result, the Logit-based ODE always outperforms the traditional formulation, but cannot compare to the bi-level formulation.

As for the satisfaction-based ODE, it closely approximates the performance of the bi-level formulation. Overall the satisfaction-based ODE would be the best choice for ODE on small and medium scale networks.

Impact of Observation Noise. So far, we have assumed that the observations are free of noise, but the actual measurements are always corrupted with noise. In this section, we assume Link 1 is observed with noise, and the quantity of noise is proportional to the actual link flow. The relative deviation is used to measure the noise. If the relative deviation is 0.1, it means the observed link flow can be as large as 110% of the actual link flow. We set the relative deviation from −0.3 to 0.3 and conduct the experiments, then plot the results in Figure 2.

With the help of historical OD, all four formulations achieve high accuracy under the observational noise. The PRMSE of estimated OD is at most 5%. However, without the historical OD, all formulations become more sensitive to the observational noise. The Logit-based ODE outperforms traditional one but is still not good enough. Both the bi-level and satisfaction-function-based ODE methods remain stable under the observational noise.

The Logit-based ODE relies on the historical OD more than the satisfaction-function-based ODE. With better information on historical OD, Logit-based ODE can outperform the satisfaction-function-based ODE. The satisfaction-function-based ODE is more robust without the historical information, as its accuracy is roughly linear to the observational noise.

Impact of Historical OD Demand. Previous experiments are based on perfect historical OD or no OD information. In fact, a historical OD with error is usually the case in practice. Figure 3 explores the impact of historical OD.

All four ODE formulations are comparably reliable with respect to the historical OD deviation; namely the sensitivity is approximately linear to it. The bi-level and satisfaction-function-based ODE are less sensitive than the other two formulations. Regardless of the relative sensitivity, the accuracy of historical OD is of great

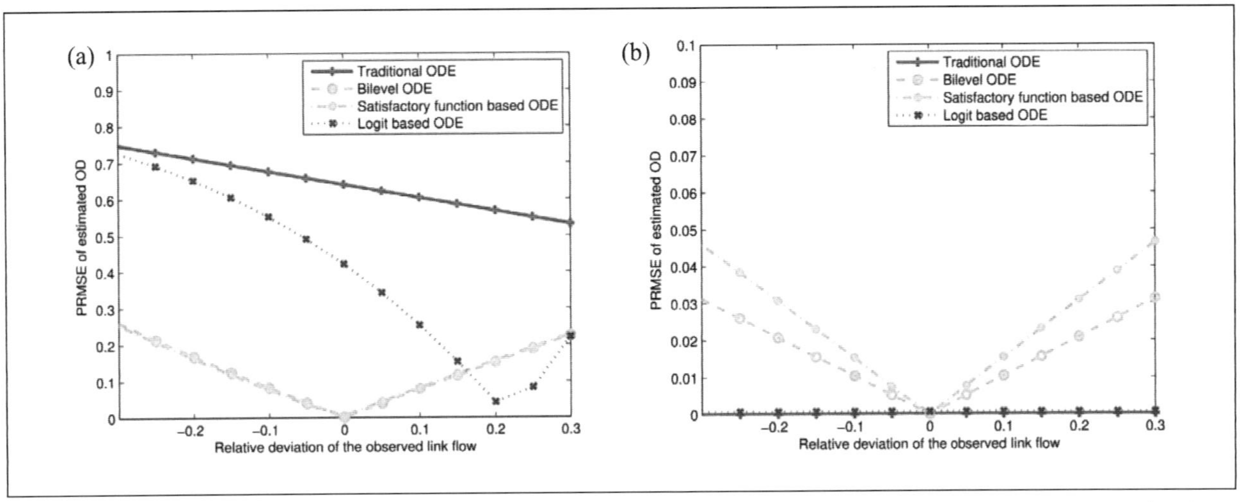

Figure 2. Change in PRMSE of estimated OD respect to measurement deviation: (*a*) without historical OD information, and (*b*) with perfect historical OD information.

Table 2. Different ODE Methods on Different Observed Links

		Estimated link flows				Estimated OD metrics		
	Observed link number	Link 1	Link 2	Link 3	Link 4	PRMSE (%)	MARE (%)	PTD (%)
Bi-level ODE	1	360	640	316	324	0.02	0.02	0.02
	2	360	640	317	323	0.01	0.01	0.01
	3	360	640	316	324	0.05	0.05	0.05
	4	360	640	316	324	0.05	0.05	0.05
	1,2	360	640	320	320	0.01	0.01	0.01
	1,3	360	640	317	323	0.01	0.01	0.01
	1,4	360	640	317	323	0.01	0.01	0.01
Satisfaction-function-based ODE	1	362	643	318	325	0.48	0.48	−0.48
	2	393	663	328	335	5.58	5.58	−5.58
	3	317	591	297	294	9.22	9.22	9.22
	4	319	594	289	305	8.79	8.79	8.79
	1,2	360	640	317	223	0.15	0.15	0.15
	1,3	361	642	317	325	0.31	0.31	−0.31
	1,4	361	642	318	324	0.30	0.30	−0.30
Logit-based ODE	1	360	399	199	200	24.17	24.17	24.17
	2	1	639	320	319	35.95	35.95	35.95
	3	109	358	316	42	53.25	53.25	53.25
	4	119	367	44	323	51.38	51.38	51.38
	1,2	360	640	320	320	0.12	0.12	0.12
	1,3	360	446	316	130	19.43	19.43	19.43
	1,4	360	465	142	323	17.58	17.58	17.58

importance to all four formulations. Therefore, one should carefully decide whether to use the historical OD in the ODE process.

A Real-World Network

We now compare four formulations on a real-world network to further demonstrate the accuracy and computational efficiency of the proposed method. The SR41 corridor network, in the City of Fresno, California, consists of one major freeway and two parallel arterial roads connected with local streets. It contains 2,413 links and 7,110 OD pairs. Figure 4 is an overview of the SR41 network. Researchers have made efforts on calibrating its OD demand (*29*), which we adopt in this study as the true OD demand.

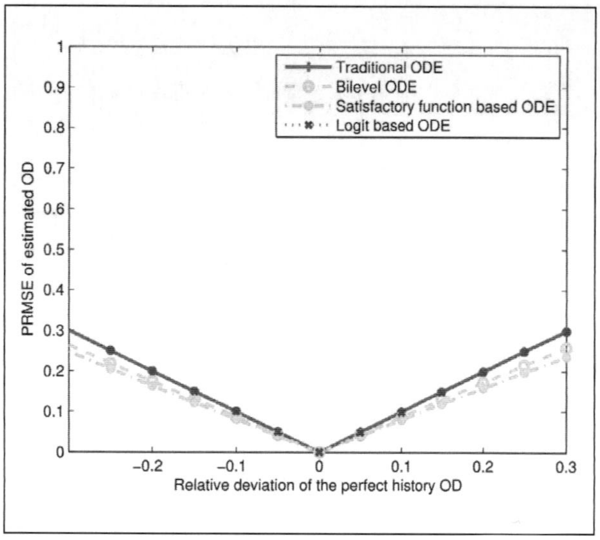

Figure 3. Change in PRMSE of estimated OD respect to historical OD deviation.

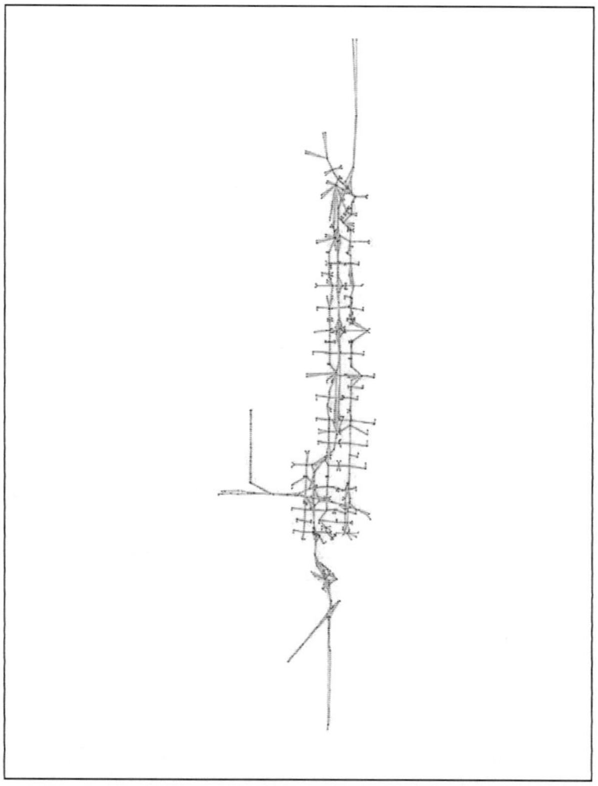

Figure 4. Overview of SR41 network.

We randomly pick 6% of links to be observable and the observations are free of noise. Historical OD information is not included in all four formulations. Four formulations are implemented and tested under the same settings. After 50 iterations on a personal computer, the results are presented in Table 3.

The traditional ODE totally fails without the help of historical OD information; the measurement metrics are out of bound (>1). This indicates traditional ODE method is not able to handle such a large number of ODs without historical ODs. Bi-level ODE outperforms all the other methods in terms of MARE and moderate PRMSE, but it takes almost 20 minutes to finish 100 iterations, thus the computational time will grow very fast as the network scale increases. Both satisfaction-function-based ODE and Logit-based ODE perform well on SR-41 corridor and the time consumption is acceptable.

We also plot the estimated path flows against the real path flows after 50 iterations, as shown in Figure 5.

As a conclusion, the proposed single-level formulations achieve similar results to the bi-level formulation and save approximately 80% of computational budget. Thus, the proposed formulations have potential to scale on large-scale networks and compare to the bi-level methods.

Conclusion

We propose a generalized framework for ODE problems with general equilibrium constraints. Based on the proposed framework, we formulate two single-level ODE methods under SUE constraints. The new formulations address the computational complexity issue from the traditional bi-level formulations. Analytical properties of the new formulation are analyzed, and the unbiasedness of the estimation results is proved.

The new ODE formulations are based on two SUE models, the satisfaction-function-based SUE and the Logit-based SUE. For the satisfaction-function-based SUE constraint, a direct optimization formulation is used to determine the OD estimators; while for the Logit-based SUE, an iterative parameter updating framework is used to approach the OD estimators iteratively.

Gradient-based algorithms are proposed to solve the new formulations. Numerical experiments are also conducted on both small and large networks. Results indicate that under the same settings, our formulations outperform the traditional ODE formulation. In fact, the satisfaction-function-based ODE can oftentimes provide comparable results to the bi-level formulations. The Logit-based ODE also performs well on large networks, and can be more computationally efficient than the satisfaction-function-based ODE. Sensitivity analysis on sensor locations, observation noise and historical ODs are also conducted to show our new formulations are as stable as the bi-level formulation.

Table 3. Comparison of Four ODE Formulations on SR41 Networks after 50 Iterations

Methods	Estimated OD metrics			Time consumption
	PRMSE(%)	MARE(%)	PTD(%)	(s)
Traditional ODE	N/A	N/A	N/A	24
Bi-level ODE	21.47	20.84	0.02	1170
Satisfaction-function-based ODE	21.39	21.17	0.05	214
Logit-based ODE	21.33	23.49	0.04	177

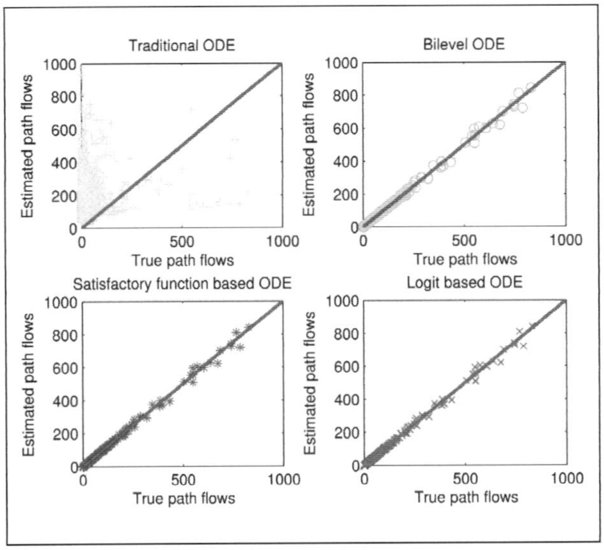

Figure 5. Estimated against true path flows for different ODE formulations.

The effectiveness of the new formulations compares to the bi-level ODE formulation, and the efficiency of the new formulation outperforms the bi-level ODE formulation. Future work includes dealing with the non-equilibrium conditions and extending this formulation to the case of dynamic ODE.

Acknowledgments

This research is funded in part by Traffic 21 Institute and Carnegie Mellon University's Mobility 21, a National University Transportation Center for mobility which is sponsored by the US Department of Transportation.

References

1. Tobin, R. L., and T. L. Friesz. Sensitivity Analysis for Equilibrium Network Flow. *Transportation Science*, Vol. 22, No. 4, 1988, pp. 242–250.
2. Patriksson, M. Sensitivity Analysis of Traffic Equilibria. *Transportation Science*, Vol. 38, No. 3, 2004, pp. 258–281.
3. Mahmassani, H. S. Dynamic Network Traffic Assignment and Simulation Methodology for Advanced System Management Applications. *Networks and Spatial Economics*, Vol. 1, No. 3, 2001, pp. 267–292.
4. Ben-Akiva, M., M. Bierlaire, J. Bottom, H. Koutsopoulos, and R. Mishalani. Development of a Route Guidance Generation System for Real-Time Application. *IFAC Proceedings Volumes*, Vol. 30, No. 8, 1997, pp. 405–410.
5. Chen, A., Z. Zhou, P. Chootinan, S. Ryu, C. Yang, and S. C. Wong. Transport Network Design Problem under Uncertainty: A Review and New Developments. *Transport Reviews*, Vol. 31, No. 6, 2011, pp. 743–768.
6. Robillard, P. Estimating the OD Matrix from Observed Link Volumes. *Transportation Research*, Vol. 9, No. 2–3, 1975, pp. 123–128.
7. Van Zuylen, H. J., and L. G. Willumsen. The Most Likely Trip Matrix Estimated from Traffic Counts. *Transportation Research Part B: Methodological*, Vol. 14, No. 3, 1980, pp. 281–293.
8. Maher, M. J. Inferences on Trip Matrices from Observations on Link Volumes: A Bayesian Statistical Approach. *Transportation Research Part B: Methodological*, Vol. 17, No. 6, 1983, pp. 435–447.
9. Cascetta, E. Estimation of Trip Matrices from Traffic Counts and Survey Data: A Generalized Least Squares Estimator. *Transportation Research Part B: Methodological*, Vol. 18, No. 4, 1984, pp. 289–299.
10. Nguyen, S. *Estimating an OD Matrix from Network Data: A Network Equilibrium Approach*. Universite de Montreal, Centre de Recherche sur les Transports, 1977.
11. Fisk, C. S. Game Theory and Transportation Systems Modelling. *Transportation Research Part B: Methodological*, Vol. 18, No. 4, 1984, pp. 301–313.
12. Yang, H., T. Sasaki, Y. Iida, and Y. Asakura. Estimation of Origin-Destination Matrices from Link Traffic Counts on Congested Networks. *Transportation Research Part B: Methodological*, Vol. 26, No. 6, 1992, pp. 417–434.
13. Yang, H. Heuristic Algorithms for the Bilevel Origin-Destination Matrix Estimation Problem. *Transportation Research Part B: Methodological*, Vol. 29, No. 4, 1995, pp. 231–242.
14. Luo, Z. Q., J. S. Pang, and D. Ralph. *Mathematical Programs with Equilibrium Constraints*. Cambridge University Press, 1996.
15. Bera, S., and K. V. Rao. Estimation of Origin-Destination Matrix from Traffic Counts: The State of the Art. *European Transport, Institute for the Study of Transport within the European Economic Integration*, Vol. 49, 2011, pp. 2–23.

16. Josefsson, M., and M. Patriksson. Sensitivity Analysis of Separable Traffic Equilibrium Equilibria with Application to Bilevel Optimization in Network Design. *Transportation Research Part B: Methodological*, Vol. 41, No. 1, 2007, pp. 4–31.
17. Nie, Y. M., and H. M. Zhang. A Relaxation Approach for Estimating Origin–Destination Trip Tables. *Networks and Spatial Economics*, Vol. 10, No. 1, 2010, pp. 147–172.
18. Meng, Q., H. Yang, and M. G. Bell. An Equivalent Continuously Differentiable Model and a Locally Convergent Algorithm for the Continuous Network Design Problem. *Transportation Research Part B: Methodological*, Vol. 35, No. 1, 2001, pp. 83–105.
19. Shen, W., and L. Wynter. A New One-Level Convex Optimization Approach for Estimating Origin–Destination Demand. *Transportation Research Part B: Methodological*, Vol. 46, No. 10, 2012, pp. 1535–1555.
20. Sheffi, Y. *Urban Transportation Network. Equilibrium Analysis with Mathematical Programming Methods.* Prentice Hall, Englewoods Cliffs, N.J., 1985.
21. Nakayama, S., R. Kitamura, and S. Fujii. Drivers' Route Choice Rules and Network Behavior: Do Drivers Become Rational and Homogeneous Through Learning? *Transportation Research Record: Journal of the Transportation Research Board*, 2001. 1752: 62–68.
22. Meng, Q., W. H. Lam, and L. Yang. General Stochastic User Equilibrium Traffic Assignment Problem with Link Capacity Constraints. *Journal of Advanced Transportation*, Vol. 42, No. 4, 2008, pp. 429–465.
23. Maher, M. J., X. Zhang, and D. Van Vliet. A Bi-Level Programming Approach for Trip Matrix Estimation and Traffic Control Problems with Stochastic User Equilibrium Link Flows. *Transportation Research Part B: Methodological*, Vol. 35, No. 1, 2001, pp. 23–40.
24. Wang, Y., X. Ma, Y. Liu, K. Gong, K. C. Henricakson, M. Xu, and Y. Wang. A Two-Stage Algorithm for Origin-Destination Matrices Estimation Considering Dynamic Dispersion Parameter for Route Choice. *PLoS One*, Vol. 11, No. 1, 2016, p. e0146850.
25. Yang, H., Q. Meng, and M. G. Bell. Simultaneous Estimation of the Origin-Destination Matrices and Travel-Cost Coefficient for Congested Networks in a Stochastic User Equilibrium. *Transportation Science*, Vol. 35, No. 2, 2001, pp. 107–123.
26. Yang, H., Y. Iida, and T. Sasaki. The Equilibrium-Based Origin-Destination Matrix Estimation Problem. *Transportation Research Part B: Methodological*, Vol. 28, No. 1, 1994, pp. 23–33.
27. Eppstein, D. Finding the K Shortest Paths. *SIAM Journal on Computing*, Vol. 28, No. 2, 1998, pp. 652–673.
28. Sherali, H. D., and T. Park. Estimation of Dynamic Origin–Destination Trip Tables for a General Network. *Transportation Research Part B: Methodological*, Vol. 35, No. 3, 2001, pp. 217–235.
29. Liu, H. X., L. Ding, J. X. Ban, A. Chen, and P. Chootinan. A Streamlined Network Calibration Procedure for California SR41 Corridor Traffic Simulation Study. Presented at 85th Annual Meeting of the Transportation Research Board, Washington, D.C., 2006.

The Standing Committee on Transportation Network Modeling (ADB30) peer-reviewed this paper (18-04243).

Article

TRR
JOURNAL OF THE TRANSPORTATION RESEARCH BOARD

Transportation Research Record
2018, Vol. 2672(48) 69–78

DOI: 10.1177/0361198118792336
journals.sagepub.com/home/trr
SAGE

Optimal Information Provision at Bottleneck Equilibrium with Risk-Averse Travelers

Peng Liu[1] and Yang Liu[1]

Abstract

This paper aims to study the effects of travel information provision on risk-averse travelers when travel time is uncertain. A stochastic bottleneck model is examined with risk-averse commuters, in which the free-flow travel time is assumed to be uncertain and follows a uniform distribution. A mean-variance approach is adopted to measure the travel cost under risk. It is proven that the individual travel cost at bottleneck equilibrium monotonically increases with the risk-aversion level. With a higher risk-aversion level, the morning peak hour starts earlier, but the duration of the peak hour remains constant regardless of the risk-aversion level. If improvement in information quality can reduce the travel time uncertainty, risk-averse commuters will benefit from the higher quality of information. Nevertheless, when the cost of information provision is also considered, the optimal information provision strategy is derived to minimize the total system cost. The numerical examples demonstrate the information efficiency and provision strategy. The findings reveal the congestion patterns of a stochastic bottleneck with risk-averse travelers and will guide us to provide appropriate travel information.

As transportation systems can be affected by various internal and external incidents, such as traffic accidents and unpredictable weather, travel time is mostly uncertain (*1*, *2*). Empirical studies have revealed that travelers are not only interested in saving the expected travel time, but also in reducing the travel time variability (or the improvement of travel time reliability) (*3*, *4*). The common measures of travel time variability are the variance and the standard deviation of the probability distribution of travel time. Furthermore, travelers are found to be risk averse when they make travel decisions under risk. Therefore, risk-aversion is considered to be an important dimension in travel preference. Many researchers have studied the impacts of risk-averse attitude in transportation systems (*5*–*8*).

There are limited studies which examine the departure time choices of bottleneck equilibrium with risk-averse commuters. Therefore, it is not intuitive to explain some behavioral realities owing to the lack of analytical models and results. When travel time is uncertain, the traffic information provision plays a vital role in improving travelers' cognition of stochastic travel time and guiding their decision-making. However, the cost of information collection and dissemination should also be considered. This study investigates the effects of travel time information on the risk-averse travelers' departure time choices. The travel cost is explicitly measured under risk by adopting the mean-variance approach. Specifically, the free-flow travel time is assumed as a random variable which follows a uniform probability distribution. Based on these assumptions, the dynamic user equilibrium with risk-averse commuters is analytically solved at different levels of information quality. The cost of information collection and dissemination is formulated as an increasing function of information quality, and the trade-off between the information cost and the commuting cost is analyzed. The optimal information provision strategy which maximizes the social welfare is solved.

The remainder of this paper is organized as follows. The following section reviews the literature. Then, the model settings and assumptions are presented, and the equilibrium departure times and the travel cost under risk are derived. Afterwards, the effects of travel time information on departure time choice, travel cost, and total system cost are further analyzed. Numerical examples demonstrate the information efficiency and

[1]Department of Industrial Systems Engineering and Management, National University of Singapore, Singapore

Corresponding Author:
Address correspondence to Yang Liu: iseliuy@nus.edu.sg

provision strategy. The final section concludes the findings and provides future directions.

Literature Review

Vickrey first proposed the classical bottleneck model to explain the departure time choices in the morning peak hour (*9*). Traffic congestion is formulated as a queuing delay at the bottleneck. Based on this model, Small further highlighted that the schedule delay, that is, the difference between the actual and desired arrival time, obviously affects the work trip schedule (*10*). Therefore, in the classical bottleneck model, all travelers attempt to minimize their individual travel costs. They choose their departure times from their origin by trading off the cost of the travel time against the cost of schedule delay (i.e., early/late arrival penalty incurred when travelers cannot arrive at the destination at their desired time). Since Vickrey proposed the original bottleneck model, the deterministic assumptions have dominated the related studies. In recent decades, traffic bottleneck problems in the stochastic context have received considerable attention (*11–13*). Arnott et al. investigated the effects of information on time-of-day decisions with stochastic bottleneck capacities or elastic traffic demand (*14, 15*).

In the literature, Bates et al. (*16*) and Fosgerau and Karlstrm (*17*) used the bottleneck models to study the value of travel time reliability. Afterward, Batley (*18*), Börjesson et al. (*19*) and Xiao et al. (*20*) found that the valuation of travel time variability is much higher than the cost of schedule delay. That is, the value of schedule delay implied by the bottleneck models may be overestimated when the travel time reliability is treated as the schedule delay. In conclusion, travel time reliability and schedule delay are positively correlated. Schedule delay cannot completely capture the cost arising from travel time variability. Therefore, some scholars have modeled risk-aversion in the stochastic bottleneck model. According to the different types of risk measurement, the existing literature can be divided into two broad categories, which are the concave utility function (*13, 21–24*) and the mean-variance form (*25–30*).

In research published by de Palma and Fosgerau, travelers are considered to be risk averse with a general concave utility function depending on the duration in the queue and exit time from the queue (*21*). Siu and Lo proposed a concept of punctuality reliability in a stochastic bottleneck model (*22*). In this paper, the values of the scheduled delay early and late are respectively decreasing and increasing with respect to the probability of being late. Siu and Lo further extended the punctuality-based bottleneck model to include the stochastic capacity and departure time/route choices in traffic networks (*23, 24*). Based on the study of Friesz et al. (*25*), Yao et al. (*26*) introduced a stochastic congestion toll for a single bottleneck. The proportional standard deviation of a stochastic toll is added in the travel cost to reflect the risk-averse preferences. Subsequently, Yao et al. extended the model to include heterogeneous commuters with different degrees of risk-aversion (*27*). Li et al. considered a uniformly distributed capacity for risk-averse commuters in a single bottleneck setting and derived analytical solutions for the equilibrium departure time choice (*28*). The results showed that travelers depart earlier and the number of early arrivals increases at equilibrium state. Li et al. formulated the schedule delay cost as a linear function of the mean and the standard deviation of travel time when the independence of travel time and departure time is assumed (*29, 30*).

Risk Aversion Bottleneck Model

Overview of Original Bottleneck Model

In the original bottleneck model, a highway with a bottleneck located at the exit of the highway is considered, which connects a residential district with a central business district (CBD) (*9*). During the morning peak-hour, a fixed number of N travelers commute from home to CBD for work with desired arrival times t^*. When commuters arrive at their workplace, their schedule delay will be the difference between the desired and actual arrival times. When the time-dependent travel demand exceeds the capacity of the bottleneck, denoted as s, a queue forms and consequently commuters experience queuing delay. Therefore, the travel time along the highway consists of two parts: the free-flow travel time τ (the travel time with no congestion), and the queuing delay. The individual travel cost $C(t)$ consists of the travel time cost and schedule delay cost (i.e., a penalty of early or late arrival):

$$C(t) = \alpha T(t) + \beta SDE(t) + \gamma SDL(t). \qquad (1)$$

The first term on the right-hand side of Equation 1 is the travel time cost, in which $T(t) = \tau + Q(t)/s$ is the travel time if one departs from home at time t, and α is the value of time. $Q(t)$ is the length of the queue at the departure time t, which can be formulated as:

$$Q(t) = \max\{F(t) - s(t - t_o), 0\}, \qquad (2)$$

in which $F(t) = \int_{t_o}^{t} n(t)dt$ is the cumulative departure rate and t_o is the earliest time with a positive departure rate. And $F(t_e) = N$ holds, in which t_e is the latest time with a positive departure rate.

The second and the third terms in Equation 1 denote the schedule delay cost, in which $SDE(t)$ and $SDL(t)$ are the schedule delay time of an early arrival and late arrival respectively if a commuter departs at time t, expressed by:

$$SDE(t) = \{t^* - t - T(t), 0\}, \quad (3)$$

$$SDL(t) = \{t + T(t) - t^*, 0\}. \quad (4)$$

In addition, β and γ are the unit penalty cost of being early or late to work.

In the classic deterministic traffic bottleneck model, the earliest and latest departure times are:

$$t_o = t^* - \tau - \frac{\gamma}{\beta + \gamma}\frac{N}{s}, \quad (5)$$

$$t_e = t^* - \tau + \frac{\beta}{\beta + \gamma}\frac{N}{s}. \quad (6)$$

Furthermore, at the equilibrium, the individual travel cost is $C = \alpha\tau + \frac{\beta\gamma}{\beta + \gamma}\frac{N}{s}$.

Stochastic Bottleneck Model with Risk-Averse Travelers

Now, Vickrey's deterministic model is extended by assuming the free-flow travel time is a stochastic variable (*15, 31*). τ is still used to represent the stochastic free-flow travel time, and it is assumed that:

(A1) The free-flow travel time τ follows a uniform distribution within interval $[\bar{\tau} - \theta, \bar{\tau} + \theta]$;
(A2) The risk-aversion travel cost is measured by the mean-variance approach;
(A3) The traffic demand N is no less than a threshold $(\frac{\beta + \gamma}{\beta}\theta + \frac{\lambda}{\beta}\frac{\theta^2}{3})s$;
(A4) The relationship $\gamma > \alpha > \beta$ holds.

In assumption A1, $\bar{\tau}$ is the mean of the free-flow travel time and θ denotes the maximum variation from the mean. The stochastic free-flow travel time τ is identical across commuters within a day but fluctuates from day to day. It is exogenous and independent of departures. In assumption A2, the risk is measured by the mean-variance approach (*26–29*). Owing to the uncertain travel time, there could be three departure time intervals: (1) travelers always arrive early; (2) travelers can either arrive early or late depending on actual free-flow travel time; and (3) travelers always arrive late (*32*). The second interval does not emerge when the travel time is deterministic. Assumption A3 ensures that the demand reaches a certain threshold so that all of the three departure intervals occur. The proof of this can be found in the Appendix. Assumption A4 ensures the existence of the equilibrium and the consistency with empirical findings.

Under assumptions A1 and A2, the risk-aversion travel cost $C^R(t)$ can be formulated as:

$$C^R(t) = \mathbb{E}[C(t)] + \lambda\mathbb{V}[C(t)], \quad (7)$$

in which $\mathbb{E}[C(t)]$ is the mean of the travel cost. λ denotes the risk-aversion level, that is, the larger λ indicates the higher level of risk-aversion, and $\mathbb{V}[C(t)]$ is the variance of the travel cost at departure time t. It should be noted that Equation 7 is reduced to the risk-neutral travel cost if λ is zero.

Equilibrium Travel Cost

Under assumption A3, departure time intervals exist in which travelers will always arrive early or always arrive late. Therefore, at the earliest departure time t_o^R, the schedule delay for late arrival is zero, that is, $SDL(t_o^R) = 0$. Similarly, at the latest departure time t_e^R, the schedule delay for early arrival is zero, that is, $SDE(t_e^R) = 0$. Combining Equations 1–4 and 7, gives:

$$\begin{aligned} C^R(t_o^R) &= \alpha\mathbb{E}[T(t_o^R)] + \beta\mathbb{E}[SDE(t_o^R)] + \alpha^2\lambda\mathbb{V}[\tau] \\ &+ \beta^2\lambda\mathbb{V}[SDE(t_o^R)] + 2\alpha\beta\lambda Cov[\tau, SDE(t_o^R)] = \alpha\mathbb{E}[T(t_o^R)] \\ &+ \beta\mathbb{E}[SDE(t_o^R)] + (\alpha - \beta)^2\lambda\mathbb{V}[\tau], \end{aligned} \quad (8)$$

$$\begin{aligned} C^R(t_e^R) &= \alpha\mathbb{E}[T(t_e^R)] + \gamma\mathbb{E}[SDL(t_e^R)] + \alpha^2\lambda\mathbb{V}[\tau] \\ &+ \gamma^2\lambda\mathbb{V}[SDL(t_e^R)] + 2\alpha\gamma\lambda Cov[\tau, SDL(t_e^R)] \\ &= \alpha\mathbb{E}[T(t_e^R)] + \gamma\mathbb{E}[SDL(t_e^R)] + (\alpha + \gamma)^2\lambda\mathbb{V}[\tau]. \end{aligned} \quad (9)$$

As the departure rate $n(t) = 0$ if $t < t_o^R$ and $n(t) = 0$ if $t > t_e^R$, both $F(t_o^R) = 0$ and $F(t_e^R) = N$ hold. It is noted that the queue length is zero when travelers depart at time t_o^R and t_e^R. In addition, the arrival time at work are $t_o^R + \tau$ and $t_o^R + \tau + N/s$ if one departs at t_o^R and t_e^R, respectively. Consequently, it can be derived that:

$$C^R(t_o^R) = \alpha(\bar{\tau} + t_o^R - t) + \beta(t^* - t_o^R - \bar{\tau}) + (\alpha - \beta)^2\lambda\frac{\theta^2}{3}, \quad (10)$$

$$\begin{aligned} C^R(t_e^R) &= \alpha(\bar{\tau} + t_o^R + \frac{N}{s} - t) + \gamma(t_o^R + \bar{\tau} + \frac{N}{s} - t^*) \\ &+ (\alpha + \gamma)^2\lambda\frac{\theta^2}{3}. \end{aligned} \quad (11)$$

Each traveler minimizes their individual risk-aversion travel costs by choosing their departure time. At equilibrium, nobody has an incentive to alter his/her departure time. The equilibrium condition implies that the risk-aversion travel cost is constant across the departure time t when the departure rate is positive, and thus $C^R(t_o^R) = C^R(t_e^R)$ holds. Substituting Equations 10 and 11 into it gives:

$$\beta(t^* - t_o^R - \bar{\tau}) - \gamma(t_o^R + \bar{\tau} + \frac{N}{s} - t^*) = ((\alpha + \gamma)^2 - (\alpha - \beta)^2)\lambda \frac{\theta^2}{3}. \quad (12)$$

Therefore, combined with Equation 12 and the relationship $t_e^R = t_o^R + N/s$, the earliest and latest departure times can be derived as:

$$t_o^R = t^* - \bar{\tau} - \frac{\gamma}{\beta + \gamma} \frac{N}{s} - (2\alpha + \gamma - \beta)\lambda \frac{\theta^2}{3}, \quad (13)$$

$$t_e^R = t^* - \bar{\tau} + \frac{\beta}{\beta + \gamma} \frac{N}{s} - (2\alpha + \gamma - \beta)\lambda \frac{\theta^2}{3}. \quad (14)$$

Equations 13 and 14 reveal the impacts of risk (uncertainty) and risk-aversion levels on the congestion time window. As $2\alpha + \gamma - \beta > 0$ from the assumption A4, at the higher risk (θ-value) in travel time, the congestion period will shift earlier, that is, $\partial t_o^R/\partial\theta < 0$, $\partial t_e^R/\partial\theta < 0$. When travelers are more risk averse (i.e., the higher λ), the congestion period will shift earlier, that is, $\partial t_o^R/\partial\lambda < 0$ and $\partial t_e^R/\partial\lambda < 0$. However, the duration of the congestion hour remains constant, that is, $t_e^R - t_o^R = \frac{N}{s}$. Substituting Equation 13 into Equation 10, the risk-aversion travel cost C^R at equilibrium is obtained as:

$$C^R = \alpha\bar{\tau} + \frac{\beta\gamma}{\beta + \gamma} \frac{N}{s} + (\alpha^2 + \beta\gamma)\lambda \frac{\theta^2}{3}. \quad (15)$$

It is worthwhile noting that the first term and the second term in Equation 15 are the expected travel cost in the risk-neutral case. The third term denotes the additional cost associated with a travelers' risk-aversion preference. It can be easily proved that the travel cost at a bottleneck equilibrium monotonically increases with the uncertainty level, that is, $\partial C^R/\partial\theta > 0$, holding the others constant. The travel cost monotonically increases with the risk-aversion level, that is, $\partial C^R/\partial\lambda > 0$, holding the others constant.

Information Provision

With technologies such as the global navigation satellite system (GNSS), the travel time and traffic information can be collected and disseminated efficiently. Therefore, it is assumed that the morning commuters can obtain the pre-trip information about the free-flow travel time τ within the current day in advance, although τ fluctuates from day to day. In this section, how information quality affects departure time choice, individual travel cost, and system performance is investigated.

Perfect Information

First, a situation was investigated in which all commuters know the true free-flow travel time τ each day before their departures, which is the case with perfect or full information. In this case, commuters will make the same departure choices as in the deterministic case. Therefore, the gaps of the earliest and the latest departure times between the no-information case and the perfect-information case are respectively:

$$t_o^R - t_o = \tau - \bar{\tau} - (2\alpha + \gamma - \beta)\lambda \frac{\theta^2}{3}, \quad (16)$$

$$t_e^R - t_e = \tau - \bar{\tau} - (2\alpha + \gamma - \beta)\lambda \frac{\theta^2}{3}. \quad (17)$$

According to assumption **A1**, τ follows a uniform distribution between interval $[\bar{\tau} - \theta, \bar{\tau} + \theta]$. Furthermore, the expected earliest departure time t_o^P and the expected latest departure time t_e^P with perfect information are derived:

$$t_o^P = \mathbb{E}[t_o] = t^* - \bar{\tau} - \frac{\gamma}{\beta + \gamma} \frac{N}{s}, \quad (18)$$

$$t_e^P = \mathbb{E}[t_e] = t^* - \bar{\tau} + \frac{\beta}{\beta + \gamma} \frac{N}{s}. \quad (19)$$

The expected travel cost with perfect information is:

$$C^P = \mathbb{E}[C] = \alpha\bar{\tau} + \frac{\beta\gamma}{\beta + \gamma} \frac{N}{s}. \quad (20)$$

which is equal to the travel cost for risk-neutral travelers. Then, the benefit of perfect information can be derived as:

$$C^R - C^P = (\alpha^2 + \beta\gamma)\lambda \frac{\theta^2}{3}. \quad (21)$$

It implies that when free-flow travel time is uncertain, the expected travel cost with perfect information is lower than that with no information, and that perfect information can indeed improve travelers' welfare.

Imperfect Information

Although the traffic information can be collected and disseminated via GNSS, it may be difficult to receive high-quality information and travelers maybe miss information dispatches. Consequently, in practice, the provided information is imperfect (*14*). In this subsection, a case is considered in which commuters receive imperfect travel time information. It is assumed that commuters are informed that free-flow travel time today τ follows a uniform distribution on an interval $[\hat{\tau} - \omega, \hat{\tau} + \omega] \subseteq [\bar{\tau} - \theta, \bar{\tau} + \theta]$, in which ω represents the degree of information imperfection and $0 \le \omega \le \theta$. When $\omega = \theta$ and $\hat{\tau} = \bar{\tau}$, it becomes the case with no information. When $\omega = 0$, it becomes the case with perfect information, and the information of τ is accurate.

Therefore, the expected earliest departure time t_o^I and the expected latest departure time t_e^I with imperfect information are derived as:

$$t_o^I = t^* - \bar{\tau} - \frac{\gamma}{\beta + \gamma}\frac{N}{s} - (2\alpha + \gamma - \beta)\lambda\frac{\omega^2}{3}, \quad (22)$$

$$t_e^I = t^* - \bar{\tau} + \frac{\beta}{\beta + \gamma}\frac{N}{s} - (2\alpha + \gamma - \beta)\lambda\frac{\omega^2}{3}. \quad (23)$$

In addition, the expected travel cost at equilibrium with imperfect information is:

$$C^I = \alpha\bar{\tau} + \frac{\beta\gamma}{\beta + \gamma}\frac{N}{s} + (\alpha^2 + \beta\gamma)\lambda\frac{\omega^2}{3}. \quad (24)$$

Therefore, the gap between the travel costs in the imperfect-information case and the perfect-information case is:

$$C^I - C^P = (\alpha^2 + \beta\gamma)\lambda\frac{\omega^2}{3}. \quad (25)$$

The efficiency of imperfect information in reducing travel cost relative to perfect information can be measured using the index:

$$E^I = \frac{C^R - C^I}{C^R - C^P} = 1 - (\frac{\omega}{\theta})^2. \quad (26)$$

As ω ranges from 0 to θ and $C^I(\omega)$ ranges from C^P to C^R, the efficiency index E^I ranges from 0 to 1. $E^I = 0$ indicates that there is no value of information and the cost associated with risk cannot be reduced owing to the poor quality of information. $E^I = 1$ indicates that the cost associated with risk has been eliminated by perfect information.

Information Cost and Optimal Quality

This section examines the trade-off between the travel cost and the information cost. The cost is incurred for the collection and the dissemination of traffic information. The cost of information provision increases with information quality. We will derive the optimal level of information quality. Specifically, it is therefore assumed that:

(**A5**) The information cost decreases with the information imperfection parameter ω.

According to the assumption A5, compared to the total system travel cost with no information, that is, $SC^R = C^R N$, the total system cost with imperfect information is formulated as:

$$SC^I = (C^I + \delta \ln\frac{\theta}{\omega})N, \quad (27)$$

in which δ is unit cost of information quality improvement for each traveler.

Substituting Equation 24 into Equation 27, the derivative of SC^I with respect to ω can be given by:

$$\frac{dSC^I}{d\omega} = (\frac{2}{3}(\alpha^2 + \beta\gamma)\lambda\omega - \frac{\delta}{\omega})N. \quad (28)$$

Let $\frac{dSC^I}{d\omega^*} = 0$, the optimal quality level ω^* which minimizes the total system cost is derived as:

$$\omega^* = \sqrt{\frac{3}{2}\frac{\delta}{(\alpha^2 + \beta\gamma)\lambda}}. \quad (29)$$

Equation 29 shows that ω^* is determined by the travel preference parameters, that is, the value of travel time α, the product of unit schedule delay costs $\beta\gamma$, and the level of risk-aversion λ. The higher the preference parameter, the higher the optimal information quality is desired. Consistent with the intuition, when the information becomes more costly (the larger δ), the system will desire a lower quality of information at the system optimum. There are two system performance measures, that is, the total system cost SC and the individual travel cost C. As shown above, the travelers always benefit from better information quality which reduces the travel cost. However, the system cost SC^I decreases with ω when $\omega \in [0, \omega^*)$ and increases with ω when $\omega \in (\omega^*, +\infty)$. Therefore, it indicates two different regimes for information provision. When the uncertainty with no information is small enough, that is, $\theta < \omega^*$, improving the information quality will actually increase the total system cost even though it can reduce the individual travel cost. On the other hand, when the uncertainty with no information is larger than ω^*, that is, $\theta > \omega^*$, improving information quality will reduce both the total system cost and the individual travel cost, simultaneously. This finding may support us to guide the information system on a congested highway during the peak hour.

Numerical Example

This section illustrates the equilibrium patterns and the effects of information provision. Unless otherwise specified, following Small, the three parameter values of $\alpha = 6.4$ dollar/hour, $\beta = 3.9$ dollar/hour and $\gamma = 15.21$ dollar/hour were adopted (*10*). In addition, the desired arrival time is 9 am, the total number of travelers $N = 6000$, bottleneck capacity $s = 4000$ vehicle/hour, and mean free-flow travel time $\bar{\tau} = 0.5$ hours. Finally,

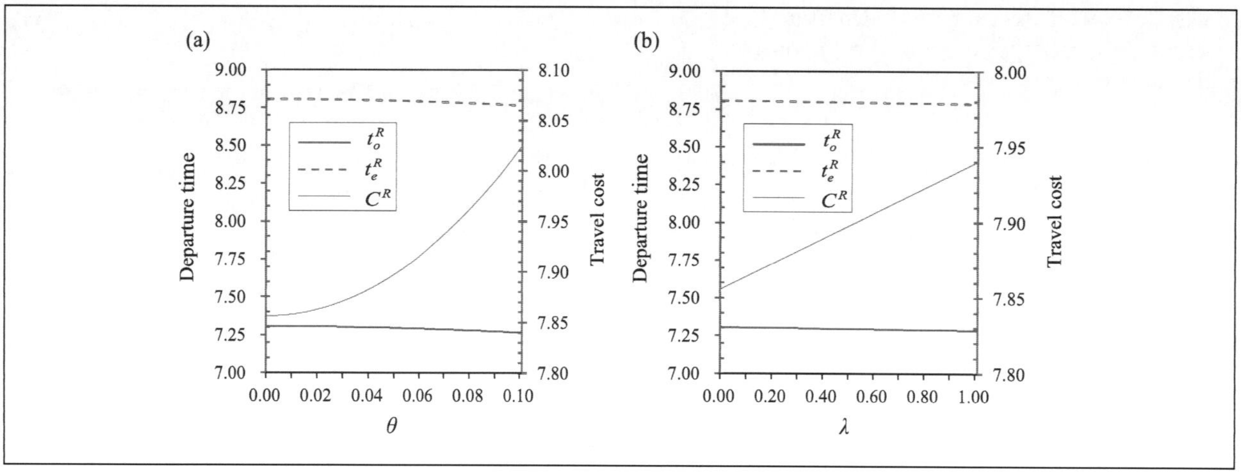

Figure 1. Effects of (*a*) θ and (*b*) λ on departure times and travel cost.

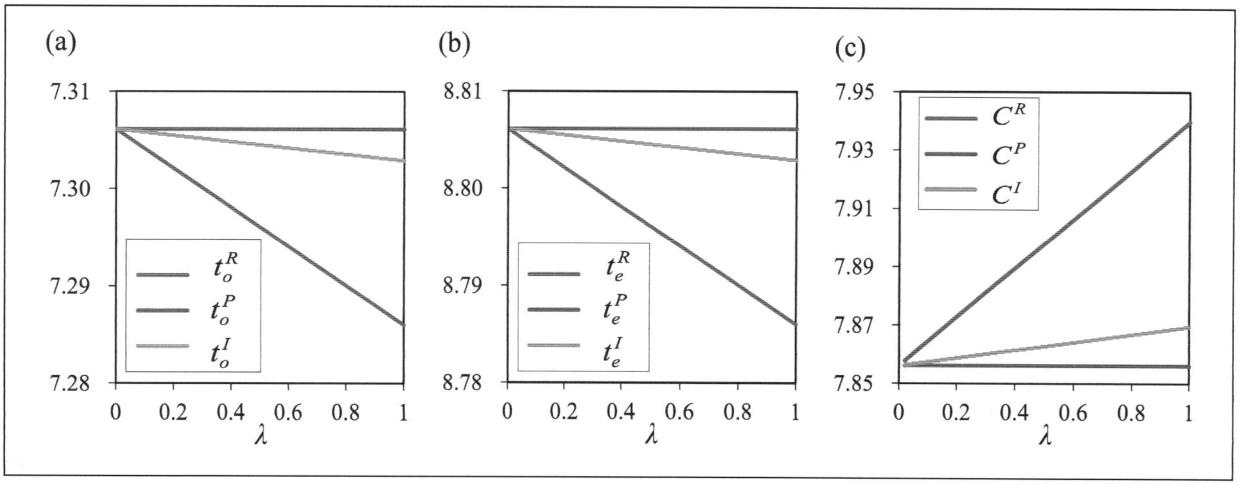

Figure 2. Equilibrium results of (a) earliest departure time, (b) latest departure time, and (c) travel cost over λ with three information types.

the risk parameter is $\theta = 0.05$ hours and the risk-aversion parameter is $\lambda = 0.5$.

Equilibrium Departure Times and Risk-Aversion Travel Cost

The equilibrium solutions to the bottleneck model with risk-averse travelers are provided. The earliest and latest departure times are 7:18 am and 8:21 am respectively, and thus the length of the morning peak is 1.5 hours. The risk-aversion travel cost is 7.9 dollars. It is also interesting to investigate the effect of the uncertainty θ and the risk-aversion parameter λ on the equilibrium results of the proposed risk-aversion bottleneck model. First, the value of θ was varied from 0 to 0.1 hours, and the λ value was varied from 0 to 1. It should be noted that $\theta = 0$ represents the deterministic case and $\lambda = 0$ represents the risk-neutral case. The results are shown in Figure 1.

Figure 1*a* depicts that the earliest and latest departure times of risk-averse commuters, that is, t_o^R and t_e^R, both decrease monotonically with the stochastic parameter θ. However, the length of the peak hour remains unchanged, and it is independent of θ. The risk-aversion travel cost C^R increases quadratically with θ. Figure 1*b* depicts that both t_o^R and t_e^R decrease monotonically with the degree of risk-aversion λ. However, the length of the peak hour also remains unchanged, which is independent of λ. The risk-aversion travel cost C^R increases linearly with λ.

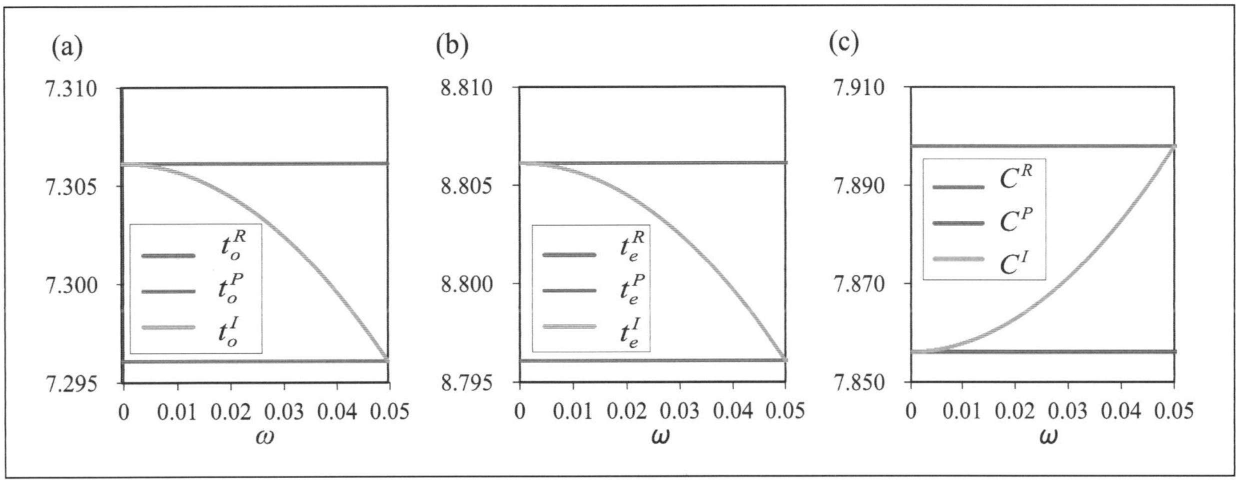

Figure 3. Equilibrium results of (a) earliest departure time, (b) latest departure time, and (c) travel cost over ω with three information types.

Effects of Information on Departure Times and Individual Travel Cost

Next, the effects of traffic information on departure times and individual travel cost was investigated. The information is first divided into three types, these are, no information, imperfect information, and perfect information. The λ value was changed from 0 to 1 and the information quality parameter ω was changed from 0 to θ ($\theta = 0.05$). Figure 2 illustrates the effects of the risk-aversion parameter λ. Figure 3 illustrates the effects of the information quality parameter ω.

Figure 2*a–c* shows that the earliest and latest departure times as well as the travel cost with perfect information, that is, t_o^P, t_e^P and C^P, are independent of the risk-aversion parameter λ. On the other hand, in the imperfect information case, both the earliest and latest departure times t_o^I and t_e^I monotonically decrease with λ, and the risk-aversion travel cost C^I increases with λ. Compared with the no-information case, the congestion time window is shifted later and the travel cost is reduced in the imperfect-information case.

Figure 3*a–c* shows that, in the imperfect information case, both the earliest and latest departure times, that is, t_o^I and t_e^I, quadratically decrease with the imperfect information parameter ω. The risk-aversion travel cost C^I monotonically increases with ω. This implies that the higher information quality will shift the peak hour later and result in a lower individual travel cost. As ω varies from 0 to 0.05, t_o^I changes from t_o^R to t_o^P, t_e^I changes from t_e^R to t_e^P, and C^I changes from C^P to C^R.

Information Efficiency and Provision Strategy

Although the traffic information can be collected and disseminated, a cost for the information provision will be incurred. In this subsection, the unit cost of the information quality was set as $\delta = 0.03$ for each traveler, and the degree of information imperfection ω range was from 0 to $\theta = 0.05$. The effects of the information quality ω on the information efficiency index and the total system cost are shown in Figure 4.

In Figure 4*a*, the imperfect information efficiency index E^I decreases quadratically with the degree of information imperfection ω. As ω ranges from 0 to θ, the E^I ranges from 0 to 1. Consequently, the higher the information quality, the lower the individual travel cost. In Figure 4*b*, the total system cost SC^I decreases within $\omega \in (0, \omega^*]$ and increases within $\omega \in (\omega^*, 0.05]$. The minimized total system cost is found at $\omega^* = 0.03$. Therefore, two information regimes are obtained. If the risk with no information is smaller than $\omega^* = 0.03$, reducing uncertainty by information provision may increase the system cost. If the risk with no information is larger than $\omega^* = 0.03$, by trading off the cost of information and the commuting cost, we should consider the optimal quality of information provision $\omega^* = 0.03$, which will reduce the system cost and the cost of commute simultaneously.

Conclusions and Future Works

This paper examined a stochastic bottleneck model with risk-averse travelers, in which the free-flow travel time follows a uniform distribution. The mean-variance approach to measure the travel cost under risk is adopted. We prove that the individual travel cost at a bottleneck equilibrium monotonically increases with the level of risk-aversion. With a higher level of risk-aversion, the morning peak hour starts earlier, but the duration of the peak hour remains constant regardless of the risk-aversion levels. When information provision can

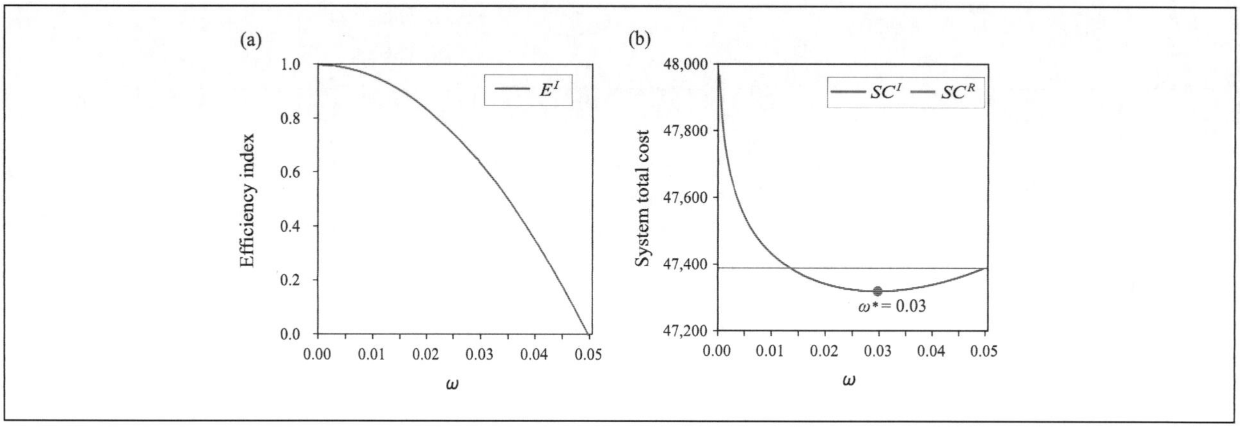

Figure 4. Effects of ω on (a) efficiency index and (b) total system costs.

reduce the travel time uncertainty, the risk-averse commuters will always benefit from the higher quality of information. Nevertheless, if the cost of information provision is considered, improving information quality will not always reduce the total system cost. The trade-off between the travel cost and information cost is analyzed and the optimal level of information quality which minimizes the total system cost was derived. This reveals that there are two different regimes for information provision. When the travel time uncertainty is small enough, higher information quality will actually increase the total system cost even though it can reduce the travel cost. On the other hand, when the travel time uncertainty is larger than the optimal level, higher information quality will improve both the total system cost and the travel cost simultaneously. Finally, the numerical examples demonstrate the information efficiency and optimal provision strategy. The findings reveal the characteristics of stochastic bottleneck congestion with risk-averse travelers and guide us to properly provide traffic information for a congested highway during the peak hours.

Some crucial components have not been considered in our study. First, public transport plays a critical role in reducing road traffic congestion because it offers an alternative travel mode in peak hours. This paper focuses on the departure time choices but ignores the mode choices. Second, travelers are assumed to be homogeneous. In fact, travelers may have heterogeneous attitudes toward the risk. Therefore, the heterogeneous risk-aversion preferences should be taken into account. Third, congestion pricing will affect the departure time choices of risk-averse commuters, but congestion pricing is not investigated in this paper. Fourth, this paper implies that all travelers have the traffic information. In fact, some of the travelers may not have access to information. Thus, the travelers can be categorized into user classes with and without information. These possible extensions may be discussed in future research.

Appendix

Proof of All Three Considered Intervals Occurring If $N \geq (\frac{\beta+\gamma}{\beta}\theta + \frac{\lambda}{\beta}\frac{\theta^2}{3})s$

To ensure the existence of all three considered intervals, that is, always arrive early, arrive early or late and always arrive late, it is necessary to consider that:

$$t_o^R + \bar{\tau} + \theta \leq t^* \leq t_o^R + \bar{\tau} + \theta + \frac{N}{s},$$
$$t_o^R + \bar{\tau} - \theta \leq t^* \leq t_o^R + \bar{\tau} - \theta + \frac{N}{s}.$$

According to equilibrium conditions, it can easily be proved that $t_o^R + \bar{\tau} - \theta \leq t^* \leq t_o^R + \bar{\tau} + \theta + \frac{N}{s}$, which implies that there are travelers who always arrive early under the best situation $\tau = \bar{\tau} - \theta$ and there are also traveler who always arrive late under the worst situation $\tau = \bar{\tau} + \theta$. In addition, it should also be supposed that $t_o^R + \bar{\tau} + \theta \leq t^* \leq t_o^R + \bar{\tau} - \theta + \frac{N}{s}$, which is equal to:

$$2\theta \leq t^* - t_o^R - \bar{\tau} \leq \frac{N}{s}.$$

Substituting the above equation into Equation 13, gives:

$$\frac{\gamma}{\beta+\gamma}\frac{N}{s} + \frac{\lambda}{\beta+\gamma}\frac{\theta^2}{3} \geq \theta,$$
$$\frac{\beta}{\beta+\gamma}\frac{N}{s} - \frac{\lambda}{\beta+\gamma}\frac{\theta^2}{3} \geq \theta.$$

From the above two equations, it is known that $\frac{\gamma}{\beta+\gamma}\frac{N}{s} + \frac{\lambda}{\beta+\gamma}\frac{\theta^2}{3}$ is always larger than $\frac{\beta}{\beta+\gamma}\frac{N}{s} - \frac{\lambda}{\beta+\gamma}\frac{\theta^2}{3}$ on account of $\gamma > \beta$. Therefore, all three considered intervals occur if $N \geq (\frac{\beta+\gamma}{\beta}\theta + \frac{\lambda}{\beta}\frac{\theta^2}{3})s$.

Acknowledgments

The work was supported by Singapore Ministry of Education Academic Research Fund Tier 1 (WBS No.R-266-000-094-112).

Author Contributions

The authors confirm contribution to the paper as follows: study conception and design: PL, YL; data collection: PL, YL; analysis and interpretation of results: PL, YL; draft manuscript preparation: PL, YL. All authors reviewed the results and approved the final version of the manuscript.

References

1. de Palma, A., M. Ben-Akiva, C. Lefevre, and N. Litinas. Stochastic Equilibrium Model of Peak Period Traffic Congestion. *Transportation Science*, Vol. 17, No. 4, 1983, pp. 430–453.
2. Miller-Hooks, E. D., and H. S. Mahmassani. Least Expected Time Paths in Stochastic, Time-Varying Transportation Networks. *Transportation Science*, Vol. 34, No. 2, 2000, pp. 198–215.
3. Jackson, W. B., and J. V. Jucker. An Empirical Study of Travel Time Variability and Travel Choice Behavior. *Transportation Science*, Vol. 16, No. 4, 1982, pp. 460–475.
4. de Palma, A., and N. Picard. Route Choice Decision under Travel Time Uncertainty. *Transportation Research Part A*, Vol. 39, No. 4, 2005, pp. 295–324.
5. Uchida, T., and Y. Iida. Risk Assignment: A New Traffic Assignment Model Considering Risk of Travel Time Variation. In *12th International Symposium Transportation Traffic Theory*, Elsevier, Amsterdam, the Netherlands, 1993, pp. 89–105.
6. de Palma, A., and N. Picard. Equilibria and Information Provision in Risky Networks with Risk-Averse Drivers. *Transportation Science*, Vol. 40, No. 4, 2006, pp. 393–408.
7. Chen, A., and Z. Zhou. The α-Reliable Mean-Excess Traffic Equilibrium Model with Stochastic Travel Times. *Transportation Research Part B*, Vol. 44, No. 4, 2010, pp. 493–513.
8. de Palma, A., R. Lindsey, and N. Picard. Risk Aversion, the Value of Information, and Traffic Equilibrium. *Transportation Science*, Vol. 46, No. 1, 2012, pp. 1–26.
9. Vickrey, W. S. Congestion Theory and Transport Investment. *The American Economic Review*, Vol. 59, No. 2, 1969, pp. 251–261.
10. Small, K. A. The Scheduling of Consumer Activities: Work Trips. *The American Economic Review*, Vol. 72, No. 3, 1982, pp. 467–479.
11. Noland, R. B., and K. A. Small. Travel-Time Uncertainty, Departure Time Choice, and the Cost of Morning Commutes. *Transportation Research Record: Journal of the Transportation Research Board*, 1995. 1493: 150–158.
12. Xiao, L. L., H. J. Huang, and R. H. Liu. Congestion Behavior and Tolls in a Bottleneck Model with Stochastic Capacity. *Transportation Science*, Vol. 49, No. 1, 2015, pp. 46–65.
13. Liu, Y., Y. Li, and L. Hu. Departure Time and Route Choices in Bottleneck Equilibrium under Risk and Ambiguity. *Transportation Research Part B Methodological*, 2017, In press.
14. Arnott, R., A. de Palma, and R. Lindsey. Does Providing Information to Drivers Reduce Traffic Congestion? *Transportation Research Part A*, Vol. 25, No. 5, 1991, pp. 309–318.
15. Arnott, R., A. de Palma, and R. Lindsey. Information and Time-of-Usage Decisions in the Bottleneck Model with Stochastic Capacity and Demand. *European Economic Review*, Vol. 43, No. 3, 1999, pp. 525–548.
16. Bates, J., J. Polak, P. Jones, and A. Cook. The Valuation of Reliability for Personal Travel. *Transportation Research Part E*, Vol. 37, No. 2, 2001, pp. 191–229.
17. Fosgerau, M., and L. Engelson. The Value of Travel Time Variance. *Transportation Research Part B*, Vol. 45, No. 1, 2010, pp. 1–8.
18. Batley, R. Marginal Valuations of Travel Time and Scheduling, and the Reliability Premium. *Transportation Research Part E*, Vol. 43, No. 4, 2007, pp. 387–408.
19. Börjesson, M., J. Eliasson, and J. P. Franklin. Valuations of Travel Time Variability in Scheduling versus Mean–Variance Models. *Transportation Research Part B*, Vol. 46, No. 7, 2012, pp. 855–873.
20. Xiao, Y., N. Coulombel, and A. de Palma. The Valuation of Travel Time Reliability: Does Congestion Matter? *Transportation Research Part B*, Vol. 97, 2017, pp. 113–141.
21. de Palma, A., and M. Fosgerau. Random Queues and Risk Averse Users. *European Journal of Operational Research*, Vol. 230, No. 2, 2013, pp. 313–320.
22. Siu, B. W., and H. K. Lo. Equilibrium Trip Scheduling in Congested Traffic under Uncertainty. In *Transportation and Traffic Theory 2009* (Lam, W., S. Wong, and H. Lo, eds.), Golden Jubilee, Springer, Boston, MA, 2009, pp. 147–167.
23. Siu, B. W., and H. K. Lo. Punctuality-Based Departure Time Scheduling under Stochastic Bottleneck Capacity: Formulation and Equilibrium. *Transportmetrica B: Transport Dynamics*, Vol. 1, No. 3, 2013, pp. 195–225.
24. Siu, B. W., and H. K. Lo. Punctuality-Based Route and Departure Time Choice. *Transportmetrica A: Transport Science*, Vol. 10, No. 7, 2014, pp. 585–621.
25. Friesz, T., R. Mookherjee, and T. Yao. Securitizing Congestion: The Congestion Call Option. *Transportation Research Part B*, Vol. 42, No. 5, 2007, pp. 407–437.
26. Yao, T., T. L. Friesz, M. M. Wei, and Y. Yin. Congestion Derivatives for a Traffic Bottleneck. *Transportation Research Part B*, Vol. 44, No. 10, 2010, pp. 1149–1165.
27. Yao, T., M. M. Wei, B. Zhang, and T. Friesz. Congestion Derivatives for a Traffic Bottleneck with Heterogeneous Commuters. *Transportation Research Part B*, Vol. 46, No. 10, 2012, pp. 1454–1473.
28. Li, H., P. H. Bovy, and M. C. Bliemer. Departure time distribution in the stochastic bottleneck model. *International Journal of ITS Research*, Vol. 6, No. 2, 2008, pp. 79–86.
29. Li, H., M. C. Bliemer, and P. H. Bovy. Reliability-Based Dynamic Discrete Network Design with Stochastic Networks. In *Transportation and Traffic Theory 2009* (Lam,

W., S. Wong, and H. Lo, eds.), Golden Jubilee, Springer, Boston, MA, 2009, pp. 651–673.
30. Li, H., H. Tu, and D. A. Hensher. Integrating the Mean-variance and Scheduling Approaches to Allow for Schedule Delay and Trip Time Variability under Uncertainty. *Transportation Research Part A*, Vol. 89, 2016, pp. 151–163.
31. Lo, H. K., X. W. Luo, and B. W. Siu. Degradable Transport Network: Travel Time Budget of Travelers with Heterogeneous Risk Aversion. *Transportation Research Part B*, Vol. 40, No. 9, 2006, pp. 792–806.
32. Coulombel, N., and A. de Palma. The Marginal Social Cost of Travel Time Variability. *Transportation Research Part C*, Vol. 47, 2014, pp. 47–60.

The Standing Committee on Transportation Network Modeling (ADB30) peer-reviewed this paper (18-04280).

Article

TRR
JOURNAL OF THE TRANSPORTATION RESEARCH BOARD

Transportation Research Record
2018, Vol. 2672(48) 79–92

DOI: 10.1177/0361198118791360
journals.sagepub.com/home/trr
SAGE

Improving Scalability of Generic Online Calibration for Real-Time Dynamic Traffic Assignment Systems

A. Arun Prakash[1], Ravi Seshadri[2], Constantinos Antoniou[3], Francisco C. Pereira[4], and Moshe Ben-Akiva[1]

Abstract

Flexible calibration of dynamic traffic assignment (DTA) systems in real time has important applications in effective traffic management. However, the existing approaches are either limited to small networks or to a specific class of parameters. In this light, this study presents a framework to systematically reduce the dimension of the generic online calibration problem, making it more scalable. Specifically, a state–space formulation of the problem in the reduced dimension space is proposed. Following this the problem is solved using the constrained extended Kalman filter, which is made tractable because of the low dimensionality of the formulated problem. The effectiveness of the proposed approach is demonstrated using a real-world network leading to better state estimation by 13% and better state predictions by 11%—with a 50 fold dimensionality reduction. Insights into choosing the right degree of dimensionality reduction are also discussed. This work has the potential for a more widespread application of real-time DTA systems in practice.

The pervasive problem of traffic congestion in urban transportation networks worldwide has led to an increasing emphasis on developing tools for real-time traffic management and operations. In this context, simulation-based dynamic traffic assignment (DTA) model systems (e.g., DynaSMART [*1*], DynaMIT [*2*]) have proven effective solutions for traffic management, which includes providing effective traveler guidance and optimizing network control strategies. However, the efficacy of these DTA systems depends on their ability to accurately estimate and predict traffic conditions. A crucial component in traffic state estimation and prediction is online calibration, which involves adjusting the demand and supply parameters of a DTA system in real time—using observed traffic measurements—so as to accurately estimate and predict traffic states. One widely used approach for solving the calibration problem is the extended Kalman filter (EKF) (*3*), which is appealing as it can incorporate both demand and supply-side parameters and any type of measurement data. However, a key challenge in adopting the EKF for online applications is its scalability to large networks—an issue that this work attempts to address.

The motivation for this study is twofold. First, the computationally intensive nature of the EKF poses challenges for online deployment. This is in large part a result of the need for computing numerical derivatives in real time, which involves expensive simulation runs—an issue that is further exacerbated as the number of calibration parameters increases. The second motivating factor is applications—such as traveler guidance generation and control strategy optimization—which require real-time performance of the DTA system.

In view of these motivating considerations, this study pursues the following objectives: (1) To develop a framework to reduce the dimensionality of the generic online calibration problem; (2) to formulate the state–space model and the EKF solution approach for online calibration in the reduced parameter space; and (3) to examine the performance of the principal component-based EKF on the Singapore expressway network using real data and perform a sensitivity analysis on the degree of dimensionality reduction.

This paper contributes to the existing literature in the following respects. First, the state–space model and EKF

[1]Massachusetts Institute of Technology, Cambridge, CA
[2]Singapore-MIT Alliance for Research and Technology (SMART), Singapore
[3]Technical University of Munich, Munich, Germany
[4]Technical University of Denmark, Denmark

Corresponding Author:
Address correspondence to A. Arun Prakash: arunakkin@gmail.com

solution approach for the online calibration problem is formulated in a reduced parameter space. Specifically, the approach is demonstrated using the principal component analysis for dimensionality reduction. Second, a case study conducted on the Singapore expressway network based on real data demonstrates that using principal components which explain 95% of the variance achieves a 52-fold dimensionality reduction while improving state estimation and prediction accuracy. The dimensionality reduction makes the EKF tractable and feasible for real-time application. Third, the numerical experiments and sensitivity analysis yield several insights regarding the effect of using principal components within the EKF-based formulation.

The following salient assumptions are made in applications discussed in the paper. First, it is assumed that the principal component directions are time-invariant. Second, a linear transition equation is assumed. Third, a sequential online calibration approach is adapted, where the estimated parameters of the previous interval are assumed to be fixed. Each of these assumptions are discussed in greater detail in the subsequent sections.

Literature Review

The calibration of simulation-based DTA models is critical in obtaining accurate traffic state estimates and predictions, and has traditionally been studied in two contexts: offline and online. The offline calibration problem involves estimating historical values for simulation parameters so as to best replicate average or historical traffic conditions for a given network. Solution methodologies for the dynamic origin–destination (OD) estimation problem (which focuses on demand-side parameters) typically involve optimization formulations including generalized least squares (GLS) approaches (*4–6*) and state–space formulations (*7, 8*). The GLS approach, however, does not lend itself well to the combined calibration of demand and supply parameters, and to address this, Balakrishna et al. proposed a more generic solution method for the offline problem based on the simultaneous perturbation stochastic approximation algorithm that can incorporate any type of measurement data (*9*). A more detailed discussion of offline calibration can be found in Balakrishna (*10*); the focus here is online calibration.

The online calibration problem involves updating the historical simulation parameters (such as OD-flows on the demand side and segment capacities/traffic dynamics parameters on the supply side) in real time to match simulated and prevailing traffic conditions as closely as possible (*11*). The existing literature can broadly be categorized into studies that consider demand-side parameters, supply-side parameters, and the combination of demand and supply-side parameters.

Several studies on the online calibration problem have focused solely on the calibration of OD demands using sensor count data. Ashok and Ben-Akiva proposed a Kalman filtering approach formulated in terms of deviations of OD-flows from historical values (as opposed to the OD-flows themselves) (*12*). This enables incorporation of a priori information regarding underlying structural relationships between the OD-flow variables. Ashok and Ben-Akiva extended this approach to explicitly account for stochasticity in the assignment matrix (which maps OD-flows to traffic counts) (*13*). Zhou and Mahmassani also use the Kalman filter for the real-time OD estimation problem where the transition equation is a polynomial trend filter that captures historical trends and structural deviations (*14*). Zhang et al. explicitly modeled constraints (such as non-negativity of OD-flows) in the Kalman filter-based approach and demonstrated significant improvements in accuracy for large networks where the constraints are often violated (*15*).

With regard to the calibration of supply-side parameters, Zhou and Mahmassani proposed a dynamic programming approach that uses an approximation of the simulator based on macroscopic models to adjust flow propagation (*16*). Antoniou et al. proposed the use of the extended, limiting, and unscented Kalman filters for the calibration of speed–density parameters (*17*).

A key challenge that arises in using the EKF for online applications is the linearization of the measurement equation which necessitates the computation of numerical derivatives that is computationally intensive, particularly when the number of calibration parameters is large. To address this issue, Antoniou et al. proposed limiting EKF as a practical alternative (*3*); the limiting EKF utilizes offline computed derivatives and drastically reduces the computational complexity of the EKF, without sacrificing significantly on accuracy. The authors simultaneously calibrate supply and demand parameters. Hashemi and Abdelghany also address the simultaneous calibration problem and propose an approach where the OD demands are calibrated using a least squares approach and supply-side parameters are calibrated using a feedback controller (*18*).

The scalability of online calibration approaches has received some attention in the recent literature. Frederix et al. proposed a hierarchical approach that decomposes the OD estimation problem for the entire network and enables the application of different OD estimation methods to different subnetworks (*19*). Djukic et al. apply the concept of principal components analysis to the dynamic OD estimation problem using a Kalman filtering approach (*20, 21*). Another study that addresses dimensionality reduction is that of Prakash et al., who propose a principal components-based GLS approach that significantly reduces dimensionality of the online calibration problem (*22*). However, the current work differs from that of earlier studies in three aspects. First, the proposed framework is generic and can incorporate (i) any technique for dimensionality reduction (like autoencoders and kernel principal component analysis), (ii) any set of parameters (pertaining to both supply models and demand models), and (iii) a wide variety of measurements (like sensor counts, automated vehicle identification [AVI], and Global Positioning System [GPS]). Second, the work examines the performance of dimensionality reduction in the context of online calibration (with a focus on both traffic state estimation and prediction) as opposed to offline calibration. Third, real-world data are used to test the performance of the principal component-based formulation and perform a detailed sensitivity analysis to gain insights on the effect of the number of principal components on estimation and prediction accuracy.

In summary, a large body of work on the calibration of DTA systems utilizes extended Kalman filtering-based approaches which are plagued by issues of scalability that have not been systematically addressed in an online context. This paper attempts to address this gap.

Online Calibration: State–Space Problem Formulation

The online calibration problem is briefly formulated in the following paragraphs. The reader is referred elsewhere for a more detailed discussion (*3*, *13*). The calibration problem is presented here as a standard state–space formulation, which is a classical approach to model dynamical systems.

Consider an analysis period T which is divided into equal intervals $h = 1, 2, 3 \ldots n$ of size t. The transportation network is represented by $G(N, L, S)$, where N represents the set of nodes, L represents the set of links, and S represents the set of segments. The network has n_N nodes, n_L links, and n_S segments. The segments are sections of road with homogeneous geometry; a link comprises one or more segments. The set of OD pairs are represented by K and are n_K in number. Further, n_s of the n_S segments are assumed to be equipped with surveillance sensors.

The state–space formulation consists of three main components: (i) a state-vector that succinctly characterizes the system; (ii) a transition equation that captures the evolution of the system—through the state-vector—in time; and (iii) a measurement equation that captures the relationship between the state-vector and the measurements or observations of the system. Following the proposal in Ashok and Ben-Akiva (*13*), the state–space formulation is expressed in terms of the *deviations* of the relevant variables from their historical values. As mentioned in Antoniou et al. (*3*), by modeling in terms of the deviations in parameters, the structural information of the trip and network patterns present in the historical values can be captured.

In the current context, let $\boldsymbol{\pi}_h$ represent vector containing the parameters of the DTA system in the time interval h; it can contain the OD-flow variables along with behavioral and supply parameters. Let $\boldsymbol{\pi}_h^H$ represent the historical values of the parameters in interval h. The historical values $\boldsymbol{\pi}_h^H$ are generally obtained through offline calibration (*10*). The state-vector is then denoted by $\Delta\boldsymbol{\pi}_h = \boldsymbol{\pi}_h - \boldsymbol{\pi}_h^H$, whose evolution can be represented through the following generic transition equation:

$$\Delta\boldsymbol{\pi}_h = f(\Delta\boldsymbol{\pi}_{h-1}, \ldots, \Delta\boldsymbol{\pi}_{h-q}) + \eta_h \tag{1}$$

where q denotes the number of previous interval's states that influence the current interval's state; and, η_h is the vector random errors in the transition equation.

Further, let $\mathbf{M}_h$ denote the vector of measurements in interval h; these can be both point-based measurements (like flows and speeds) or spatial measurements (like GPS or AVI). As before, let $\mathbf{M}_h^H$ represent the historical values of the measurements in interval h. The historical values of measurements are generally obtained by running the DTA system with the historical parameters or they can also be actual measurements used for offline calibration. The measurement-vector in deviations is then denoted by $\Delta\mathbf{M}_h = \mathbf{M}_h - \mathbf{M}_h^H$, which is related to the state-vector through the following generic measurement equation:

$$\Delta\mathbf{M}_h = \boldsymbol{g}(\Delta\boldsymbol{\pi}_h, \ldots, \Delta\boldsymbol{\pi}_{h-p}) + \varsigma_h \tag{2}$$

where p denotes the number of previous interval's states that influence the current interval's measurements; and ς_h is the vector random errors in the measurement equation.

Equations 1 and 2 together form the state–space formulation of the generic online calibration problem. In practice, two assumptions are made on the generic formulation: (i) the transition equation is approximated through a linear autoregressive process, and (ii) the measurement equation is expressed in terms of the simulator. After applying the above assumptions, the resulting transition and measurement equations are as follows:

$$\Delta\boldsymbol{\pi}_h = \sum_{i=h-q}^{h-1} \mathcal{F}_i^h \Delta\boldsymbol{\pi}_i + \eta_h \tag{3a}$$

$$\begin{aligned}\Delta\mathbf{M}_h = {} & \boldsymbol{S}(\Delta\boldsymbol{\pi}_h + \boldsymbol{\pi}_h^H, \ldots, \Delta\boldsymbol{\pi}_{h-p} + \boldsymbol{\pi}_{h-p}^H) \\ & - \boldsymbol{S}(\boldsymbol{\pi}_h^H, \ldots, \boldsymbol{\pi}_{h-p}^H) + \boldsymbol{\varsigma}_h\end{aligned} \tag{3b}$$

In Equation 3a, $\mathcal{F}_i^h$ represents a matrix relating the parameter estimates of interval i to the estimates of interval h, and q denotes the degree of the autoregressive process in the *deviations*. In Equation 3b, $S(.)$ represents the simulation model whose inputs are the parameters and outputs are the simulated measurements in the current interval, and p represents the the maximum number of previous interval's parameters that influence the measurements in the current interval.

Online OD-Flow Estimation Problem

The scalable version of the EKF proposed in this study can be applied to any set of parameters and measurements. To demonstrate the proposed approach the OD-flow estimation problem is chosen, which is a special case of the online calibration problem in Antoniou et al. (3). The parameters of interest are the time-dependent OD-flows and the measurements are the sensor-flow counts.

Let $\mathbf{x}_h$ represent vector of OD-flows in interval h, whose length is n_K. Let $\mathbf{x}_h^H$ represent the historical OD-flow values in interval h. Then the state-vector in the OD-flow estimation problem is given by $\Delta\mathbf{x}_h = \mathbf{x}_h - \mathbf{x}_h^H$. Similarly, let $\mathbf{y}_h$ represent vector of sensor-flow counts in interval h whose length is n_s; and let $\mathbf{y}_h^H$ denote the historical sensor-flow counts in interval h. The measurement-vector in deviations is then denoted by $\Delta\mathbf{y}_h = \mathbf{y}_h - \mathbf{y}_h^H$.

The transition and measurement equations for the state–space formulation of the OD-flow estimation problem can be given by:

$$\Delta\mathbf{x}_h = \sum_{i=h-q}^{h-1} \mathbf{F}_i^h \Delta\mathbf{x}_i + \mathbf{u}_h \tag{4a}$$

$$\begin{aligned}\Delta\mathbf{y}_h = {} & S(\Delta\mathbf{x}_h + \mathbf{x}_h^H, \ldots, \Delta\mathbf{x}_{h-p} + \mathbf{x}_{h-p}^H) \\ & - S(\mathbf{x}_h^H, \ldots, \mathbf{x}_{h-p}^H) + \mathbf{v}_h\end{aligned} \tag{4b}$$

where $\mathbf{F}_i^h$ is an $n_K \times n_K$ matrix relating the OD estimates of interval i to the OD estimates of interval h; $\mathbf{u}_h$ is the vector of random errors in the transition equation of OD-flows with covariance matrix $\boldsymbol{\Omega}_{u_h}$; $\mathbf{v}_h$ is the vector of random errors in the measurement equation of sensor-flow counts with covariance matrix $\boldsymbol{\Omega}_{v_h}$.

In standard OD-flow estimation problem the measurement Equation 4b is linearized and written as

$$\Delta\mathbf{y}_h = \sum_{j=h-p}^{h} \mathbf{A}_j^h \Delta\mathbf{x}_j + \mathbf{v}_h$$

using an assignment matrix $\mathbf{A}_j^h$ which assigns the proportion of OD-flows from interval j to measurements in interval h. The authors refrain from making this assumption to demonstrate that the proposed procedure is applicable even when seemingly linear relationships between measurements and parameters do not exist.

In the context of online systems, the problem in (4) is solved to obtain an estimate of *only* the OD-flow deviations in interval h, $\Delta\mathbf{x}_h$. The OD-flow estimates of the earlier intervals $h-1, h-2, \ldots$ are not re-estimated. In effect, the OD-flow estimates of the previous intervals are treated as constants in the current interval; this is inherently assumed in the subsequent discussions.

The sensor-flow counts in the current time interval might contain information about OD-flow parameters from the previous time intervals. Therefore, ideally, the OD-flow estimates from the previous time intervals need to be corrected based on the sensor-flow counts in the current interval. In the context of state–space formulation, this correction of the parameters estimated in previous time intervals is formulated using state augmentation approach (7). The authors would like to emphasize that the proposed approach can be extended to the state augmentation formulation where the state comprises the principal components corresponding to multiple time intervals. Although state augmentation results in better estimates of the parameters, it can be computationally intensive as it requires "rolling back" the simulator in real time. In other words, the simulator needs to re-run with the new parameter estimates (from the previous time intervals) to update the simulated measurements. Ashok and Ben-Akiva (13) found that the state augmentation can be reasonably approximated using the sequential estimation procedure described in the paper.

To predict the OD-flow vectors in the subsequent intervals $h+1, h+2, \ldots$, the estimates calculated using Equation (4a) are used. For example, the prediction for the time interval $h+1$, if the current interval is h, is calculated as

$$\mathbf{x}_{h+1} = \mathbf{x}_{h+1}^H + \sum_{i=h-q+1}^{h} \mathbf{F}_i^h \Delta\hat{\mathbf{x}}_i \tag{5}$$

where $\Delta\hat{\mathbf{x}}_i$ represent the a posteriori estimates of OD-flow deviations in interval i.

Extended Kalman Filter

The Kalman filter-based approaches are a natural and efficient way to recursively solve the system of equations in (4). The Kalman filter efficiently determines the estimate of the current interval using the estimate from the previous interval and the measurements in the current interval. The classical Kalman filter—which is a minimum mean square estimator—is applicable to linear state–space models. For the non-linear state–space models—as in the problem in (4)—theEKF was introduced. The EKF linearizes the non-linear transition or measurement equations around the a priori estimates and adopts the procedure for linear state–space models to estimate the state-vector. The main conceptual drawback of the EKF is that it is not an optimal estimator;

however, its practical application has been demonstrated in various studies (*11*). Note that for the application of the Kalman filter-based methods, an assumption is imposed that the error terms $\mathbf{u}_h$ and $\mathbf{v}_h$—from Equations 4a and b—are zero mean Gaussian variables and that they are independent over time.

There are three main steps in the traditional EKF algorithm when applied to the OD-flow estimation problem: (i) time-update, (ii) linearization, and (iii) measurement update. In the time-update step, the prediction for OD-flow deviation vector $\Delta\mathbf{x}_{h|h-1}$ and its covariance matrix $\mathbf{P}_{h|h-1}$ are made using the transition equation and the optimal estimates from the previous interval h, $\Delta\mathbf{x}_{h-1|h-1}$ and $\mathbf{P}_{h-1|h-1}$. These estimates—$\Delta\mathbf{x}_{h|h-1}$ and $\mathbf{P}_{h|h-1}$—are termed a priori estimates as they use data only until the previous time interval $h-1$. In the second step, the measurement equation is linearized around the a priori estimates of the current interval—$\Delta\mathbf{x}_{h|h-1}$—resulting in

$$\begin{aligned}\Delta\mathbf{y}_h = &\ S(\Delta\hat{\mathbf{x}}_{h|h-1} + \mathbf{x}_h^H, \Delta\mathbf{x}_{h-1} + \mathbf{x}_{h-1}^H \ldots \Delta\mathbf{x}_{h-p} + \mathbf{x}_{h-p}^H) \\ &-S(\mathbf{x}_h^H, \ldots, \mathbf{x}_{hp}^H) + \mathbf{H}_h\Delta\mathbf{x}_h - \mathbf{H}_h\Delta\hat{\mathbf{x}}_{h|h-1} + \mathbf{v}_h\end{aligned} \tag{6}$$

where

$$\mathbf{H}_h = \nabla S(\Delta\mathbf{x}_h + \mathbf{x}_h^H \ldots \Delta\mathbf{x}_{h-p} + \mathbf{x}_{h-p}^H)|_{\Delta\mathbf{x}_h = \Delta\mathbf{x}_{h|h-1}} \tag{7}$$

represents the $n_s \times n_K$ Jacobian of the simulator with respect to $\Delta\mathbf{x}_h$ at $\Delta\mathbf{x}_{h|h-1}$. As discussed, $\Delta\mathbf{x}_{h-1} \ldots \Delta\mathbf{x}_{h-p}$ are held fixed in the current estimation interval h. In the third step, the a priori estimates are updated using the linearized measurement equation and the matrix $\mathbf{H}_h$ to obtain the a posteriori estimates for OD-flow deviation vector $\Delta\mathbf{x}_{h|h}$ and its covariance matrix $\mathbf{P}_{h|h}$. Please refer to Antoniou (*11*) for a more detailed discussion of the procedure.

The main bottleneck to apply the EKF algorithm to the online calibration problem is the calculation of the Jacobian of the measurement equation, that is, the matrix $\mathbf{H}_h$. As the measurement equation represents the simulator, it has no closed-form expression. Therefore, $\mathbf{H}_h$ is calculated through the numerical derivatives of $S(.)$. Determining the centered numerical derivate involves perturbing each of the parameters and running the simulator $2n_K$ times, which can be computationally intractable. The procedure outlined in the next section alleviates this problem.

State–Space Formulation in the Reduced Parameter Space

This section discusses the approach to reduce the dimensionality of online calibration problem and solve it using the state–space formulation in the reduced parameter space. The dimensionality-reducing technique discussed is the principal component analysis; however, the approach is generic and can incorporate other techniques. Subsequently, the online calibration problem in formulated as a state–space formulation with the principal components as the state-vector. Finally, the estimates of parameter vector are constructed back from the estimates of the principal components.

Principal Components

The principal components are the linear combinations of the parameters vector that explain most of the variability of the parameters. The coefficients of the linear combinations are the principal component directions. The principal components also have the following desirable properties: (i) they are orthogonal to each other in the sample data, and (ii) the first principal component explains the most variance, followed by the second principal component and so on. Thus, the principal components not only approximate the parameter vector, they also decouple them. How such decoupling of parameters is beneficial is discussed in the next section where the problem is formulated. Please refer to Prakash et al. (*22*) for a detailed discussion on the construction of the principal components and their interpretation in the context of online calibration.

In the subsequent discussion, let $\mathbf{V}$ represent a $n_K \times n_d$ the matrix whose columns are the first n_d chosen principal components of the OD-flow vector:

$$\mathbf{V} = [\mathbf{v}_1 \quad \mathbf{v}_2 \quad \mathbf{v}_3 \quad \ldots \quad \mathbf{v}_{n_d}] \tag{8}$$

Then the $n_d \times 1$ principal component vector $\mathbf{z} = [z_1, z_2 \ldots z_{n_d}]^T$ of the OD-flow vector $\mathbf{x}$ can be written as

$$\mathbf{z} = \mathbf{V}^T\mathbf{x} \tag{9}$$

and the OD-flow vector $\mathbf{x}$ can be approximately reconstructed as

$$\mathbf{x} \approx \mathbf{V}\mathbf{z} \tag{10}$$

In the following discussion, it is assumed that the principal components directions $\mathbf{V}$ is constant across time intervals. This implies that the statistical correlations between the OD demands is assumed to be constant across time intervals h. But note that the structural differences and evolution in OD demands (and their principal components) across time intervals are captured using the transition equation (Equations 4a and 12). Indeed, the statistical correlations between the OD demands can change with the time of day. In this case, principal components directions specific to the time intervals $\mathbf{V}_h$ should be used. However, calculating the time-interval or time-period-specific principal components, though desirable, can have high data requirements.

Problem Formulation

Similar to the state–space formulation in terms of the OD-flow deviations, the online calibration problem is formulated in terms of the deviations of the principal components. The principal component deviations can be written as

$$\Delta \mathbf{z}_h = \mathbf{z}_h - \mathbf{z}_h^H \tag{11}$$

where $\mathbf{z}_h^H = \mathbf{V}^T \mathbf{x}_h^H$. The transition equation expressing the evolution of the principal component state-vector is given by

$$\Delta \mathbf{z}_h = \sum_{i=h-q}^{h-1} \mathbf{G}_i^h \Delta \mathbf{z}_i + \mathbf{w}_h \tag{12}$$

where $\mathbf{G}_i^h$ is an $n_d \times n_d$ matrix relating the principal component estimates of interval i to the principal component estimates of interval h; and $\mathbf{w}_h$ represents the error vector of size $n_d \times 1$ in the transition equation of principal components with covariance matrix $\mathbf{\Omega}_{w_h}$. As before, it is assumed that $\mathbf{w}_h$ are zero mean normal random variables that are independent over time intervals.

Note that the transition equation in (12) assumes that only the considered n_d principal components are related through the autoregressive process. The other $n_K - n_d$ principal components are assumed not to represent the structural variation in the data and are assumed to be inherently random.

In practice, three assumptions are made in the transition equations—both in (4a) and (12): (i) the matrices in the transition equations are assumed to be diagonal; (ii) the error components in the error vector are assumed to be independent; and (iii) the errors are assumed to be normally distributed. These three assumptions are more justifiable for the principal component-based formulation as discussed below.

The matrices in the transition equations are assumed to be diagonal as it is generally difficult to estimate the relationships between different parameters across time intervals when the state-vector is large. Fortunately, assuming that $\mathbf{G}_i^h$ in Equation 12 is diagonal is more reasonable compared with assuming that $\mathbf{F}_i^h$ in Equation 4a is diagonal. This follows from the property that the principal components are orthogonal to each other: a principal component does not explain—at least linearly—another principal component. Therefore, the non-diagonal terms in matrix $\mathbf{G}_i^h$ can be ignored, which simplifies its estimation. Note that the dependence between the OD-flows is implicitly captured in (12) through the principal components. By following the same reasoning—that the principal components are orthogonal to each other—the random vector $\mathbf{w}_h$ can reasonably be assumed to consist of n_d independent random variables. Finally, as the principal components are a linear combination of many OD-flows they can be treated as normal random variables by invoking the central limit theorem.

Substituting (10) in Equation 4b, the measurement equation in terms of the principal components can be written as

$$\begin{aligned} \Delta \mathbf{y}_h = {} & S(\mathbf{V}\Delta \mathbf{z}_h + \mathbf{V}\mathbf{z}_h^H, \ldots, \mathbf{V}\Delta \mathbf{z}_{h-p} + \mathbf{V}\mathbf{z}_{h-p}^H) \\ & - S(\mathbf{V}\mathbf{z}_h^H, \ldots, \mathbf{V}\mathbf{z}_{h-p}^H) + \mathbf{v}_h \end{aligned} \tag{13}$$

The set of Equations 12 and 13 can be solved recursively using the EKF algorithm. The EKF procedure is discussed in the following section.

Constrained EKF Algorithm for the Principal Component Formulation

This section presents the steps of the EKF algorithm to determine the estimate of the principal components. It also incorporates the enhancement to the EKF algorithm—which was proposed in Zhang et al. (*15*)—to impose constraints on the principal components. These constraints ensure that the OD-flows constructed from the estimated principal components are non-negative. After determining the estimates of the principal components in the current time interval, the OD-flows which are assigned to the network are reconstructed.

The EKF algorithm presented above has five main steps. First, the time-update step performed through Equations 16 and 17 obtains a priori estimates of the state-vector $\Delta \mathbf{z}_{h|h-1}$ and its covariance matrix $\mathbf{P}_{h|h-1}$. The second step involves computing the Jacobian $\mathbf{J}_h$—which is a $n_s \times n_d$ matrix—of measurement equation with respect to the state-vector of the current interval at its a priori estimate, that is, at $\Delta \mathbf{z}_h = \Delta \mathbf{z}_{h|h-1}$. In the third step, which is the measurement update, the posteriori estimates of state-vector $\Delta \mathbf{z}_{h|h}$ and its covariance matrix $\mathbf{P}_{h|h}$ are determined as per the traditional Kalman filter equations (refer Equations 19–21). In the fourth step, the constraint is imposed that the estimated state-vector must result in non-negative OD-flows. This is done by solving an optimization problem to determine the most likely value in the a posteriori distribution obtained from the measurement update subject to the constraints. Finally, in the fifth step, the OD-flows estimate are determined by constructing them from the principal component estimates and the historical OD-flows as given in Equation 23. The linearization and constraint imposition steps are discussed in the subsequent paragraphs.

Algorithm 1 Constrained EKF in Principal Components

Initialization

$$\Delta \mathbf{z}_{0|0} = \Delta \mathbf{z}_0 \tag{14}$$

$$\mathbf{P}_{0|0} = \mathbf{P}_0 \tag{15}$$

for $h = 1$ to N **do**
Time Update

$$\Delta \mathbf{z}_{h|h-1} = \sum_{i=h-q}^{h-1} \mathbf{G}_i^h \Delta \mathbf{z}_{i|i} \tag{16}$$

$$\mathbf{P}_{h|h-1} = \sum_{i=h-q}^{h-1} \mathbf{G}_i^h \mathbf{P}_{i|i} \mathbf{G}_i^{hT} + \Omega_{w_h} \tag{17}$$

Linearization

$$\mathbf{J}_h = \nabla_{\Delta \mathbf{z}_h} \mathbf{S}(\mathbf{V}\Delta \mathbf{z}_h + \mathbf{V}\mathbf{z}_h^H, \ldots, \mathbf{V}\Delta \mathbf{z}_{h-p} + \mathbf{V}\mathbf{z}_{h-p}^H)|_{\Delta \mathbf{z}_h = \Delta \mathbf{z}_{h|h-1}} \tag{18}$$

Measurement Update

$$\mathbf{K}_h = \mathbf{P}_{h|h-1}\mathbf{J}_h^T\left(\mathbf{J}_h \mathbf{P}_{h|h-1}\mathbf{J}_h^T + \Omega_{v_h}\right)^{-1} \tag{19}$$

$$\Delta \mathbf{z}_{h|h} = \Delta \mathbf{z}_{h|h-1} + \mathbf{K}_h \left[\Delta \mathbf{y}_h - \left(\mathbf{S}(\mathbf{V}\Delta \mathbf{z}_{h|h-1} + \mathbf{V}\mathbf{z}_h^H, \ldots) - \mathbf{S}(\mathbf{V}\mathbf{z}_h^H, \ldots, \mathbf{V}\mathbf{z}_{h-p}^H)\right)\right] \tag{20}$$

$$\mathbf{P}_{h|h} = \mathbf{P}_{h|h-1} - \mathbf{K}_h \mathbf{J}_h \mathbf{P}_{h|h-1} \tag{21}$$

Imposing Constraints

$$\Delta \hat{\mathbf{z}}_{h|h} = \text{argmax} \left\{(\Delta \mathbf{z} - \Delta \mathbf{z}_{h|h})^T \mathbf{P}_{h|h}^{-1}(\Delta \mathbf{z} - \Delta \mathbf{z}_{h|h}) | \mathbf{V}\Delta \mathbf{z} + \mathbf{x}_h^H \geq 0\right\} \tag{22}$$

Constructing parameter estimates

$$\hat{\mathbf{x}}_{h|h} = \mathbf{V}\Delta \hat{\mathbf{z}}_{h|h} + \mathbf{x}_h^H \tag{23}$$

end for

In the linearization step in Equation 18, the Jacobian is computed for the function $S(.)$ which represents the simulator. Therefore, the Jacobian cannot be computed analytically. The Jacobian is computed numerically using a centered finite-difference method as follows. Let $\delta \mathbf{z}_i$ represent $n_d \times 1$ the perturbation vector, which represents the perturbation of the ith principal component, that is, $\delta \mathbf{z}_i = [0, 0 \ldots c_i \ldots 0]^T$. In other words, $\Delta \mathbf{z}_{h|h-1} + \delta \mathbf{z}_i$ represents the perturbation of only the ith principal component by an amount c_i while holding all the other principal components the same. Let $\mathbf{j}_{ih}$ denote the ith column in the Jacobian $\mathbf{J}_h$—it represents the gradient of function $\mathbf{S}$ with respect to the ith principal component. The vector $\mathbf{j}_{ih}$ is computed numerically as

$$\mathbf{j}_{ih} = \frac{S(\mathbf{V}(\Delta \mathbf{z}_{h|h-1} - \delta \mathbf{z}_i) + \mathbf{V}\mathbf{z}_h^H, \ldots) - S(\mathbf{V}(\Delta \mathbf{z}_{h|h-1} + \delta \mathbf{z}_i) + \mathbf{V}\mathbf{z}_h^H, \ldots)}{2c_i} \tag{24}$$

The Jacobian matrix $\mathbf{J}_h$ can be constructed by concatenating the vectors $\mathbf{j}_{ih}$, $\forall i = 1, 2, \ldots n_d$ into matrix as $\mathbf{J}_h = [\mathbf{j}_{1h}, \mathbf{j}_{2h}, \ldots \mathbf{j}_{n_d h}]$. Thus, to determine $\mathbf{J}_h$ the simulator should be run—for the duration of the current time interval h—$2n_d$ times. Note that, the number of runs required has been reduced to compute the Jacobian from $2n_K$ in OD-flow state–space formulation to $2n_d$ in the current principal component formulation. Further, as each of these runs of the simulator is independent, they can be run in parallel. Further, as the number of parameters is reduced and only the most important ones are selected in the principal component-based formulation, the error in the numerical approximation of the Jacobian might also be reduced.

In step four of the algorithm, constraints are imposed on the final estimate of the state-vector. These constraints ensure that the OD-flow vector constructed from the estimated principal component vector is non-negative. In absence of these constraints, the procedure might result in negative OD-flows. In practice, the estimated negative OD-flows are truncated at zero. However, such truncation might result in erroneous and diverging results, as discussed in Zhang et al. (*15*). Moreover, as the estimated principal components do not ensure that such truncation is minimal, it is necessary to impose the non-negative OD-flow constraints. Note that constraints on any other kinds of parameters can be similarly imposed.

The constraints are imposed by correcting the posteriori estimates from the measurement update step. Specifically, the measurement update step determines the posteriori estimates of state $\Delta \mathbf{z}_{h|h}$ and its covariance matrix $\mathbf{P}_{h|h}$. As the state-vector is assumed to be normal random vector in the EKF, the measurement update step essentially determines the posteriori distribution of the state-vector $\Delta \mathbf{z}_h$ with mean $\Delta \mathbf{z}_{h|h}$ and covariance $\mathbf{P}_{h|h}$. Constraints are imposed on the final state estimate by determining the value with the maximal probability in the posteriori distribution subject to the constraints. Such estimate is determined by solving the optimality problem defined in Equation 22. The constraints in Equation 22 ensure that the OD-flows derived from the final state estimate $\Delta \hat{\mathbf{z}}_{h|h}$ are non-negative. Please refer to

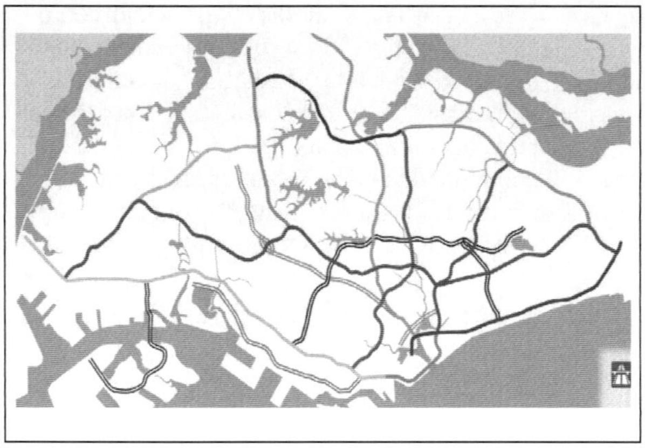

Figure 1. Singapore expressway network.

Zhang et al. (*15*) for more details regarding the procedure.

Case Study on Singapore Expressway Network

In this section, we apply the procedure presented above to the Singapore Expressway network. First, the simulation setup and data are explained, followed by the estimation of the inputs to the online calibration. Finally, the results are discussed.

Overview

The case study is conducted on the Singapore Expressway network, and the real-time DTA system used is DynaMIT-R (*2*). The road network is depicted in Figure 1, which has 939 nodes, 1157 links, and 3906 segments. The network specification also contains information about segment lengths, segment curvatures, speed limits, lanes specifications, lane connections, and dynamic tolling gantries which are replicated from the real world.

The network has 4121 OD pairs, whose locations and historical values were determined by an earlier work through offline calibration (*23*). The work also determined the supply-side parameters, which include the modified Greenshield's speed–density equation parameters and the segment capacities. The network has 357 sensors, each of which is associated with a segment; these video camera-based sensors count the vehicular flow for a period of 5 min. The Land Transport Authority of Singapore provided the measured sensor counts after preprocessing the raw data.

For the current case study, the simulation time-period was taken from 06:00 hrs to 12:00 hrs which includes the morning peak period and also the peak to off-peak transition. The estimation interval was 5 min and the prediction interval—to estimate future traffic states and provide guidance—was 15 min. Thus, there are 72 estimation intervals with a total of $72 \times 4121 = 102,312$ variables.

For this study, sensor count data of 30 weekdays in August–September 2015 were used. The 30-day data were divided into a training set of the first 25 days, and the calibration procedures were tested on the last 5 days.

Estimating the Inputs to the Calibration

To solve the online OD-flow estimation problem in (4), estimates of the following input variables are required: covariance matrix of the errors in the transition equation $\mathbf{\Omega}_u$, covariance matrix of the errors in the measurement equation $\mathbf{\Omega}_v$, and the autoregressive matrices $\mathbf{F}_i^h$ for $i = h, h-1, \ldots h-q$. In addition, to solve the principal component-based estimation problem in Equations 12 and 13 the following variables are needed: the principal component directions of the OD-flow vector $\mathbf{V}$, covariance matrix of the errors in the transition equation $\mathbf{\Omega}_w$, and the autoregressive matrices $\mathbf{G}_i^h$ for $i = h, h-1, \ldots h-q$.

The covariance matrix $\mathbf{\Omega}_v$, which measures the variance in sensor measurements, was determined based on sensor credibility. The covariance matrix $\mathbf{\Omega}_u$ and the autoregressive matrices $\mathbf{F}_i^h$ for $i = h, h-1, \ldots h-q$ were determined as follows. First, the online OD-flow estimation problem was solved for the days 11–25 using the GLS approach. The initial 10 days—from day 1 to day 10—were used to determine the inputs to the formulated GLS problem. Then, using the determined OD-flow estimates, the autogressive equation in (4a) was estimated for each OD pair assuming that the OD pairs are independent of each other; thus estimating the values of $\mathbf{F}_i^h$ for $i = h, h-1, \ldots h-q$ and $\mathbf{\Omega}_u$. Note that a random walk was assumed when no autoregressive relation could be established. For more details regarding the procedure please refer elsewhere (*11*, *22*).

The OD-flow estimates from days 11 to 25 were used to determine the principal component directions of the OD-flow vector. As discussed in Prakash et al. (*22*), a data matrix was constructed—with $n_p = 72 \times 15 = 1080$ data points—on which the principal component analysis was carried out. The results of the principal component analysis are presented in Figure 2.

From Figure 2, it can be seen that very few principal components explain the majority of the variance in the OD-flow vector. Specifically, 80 principal components explain 95% of the variance and only 150 principal components explain around 99% of the variance. Further, after 80 principal components, the percentage variance explained is minimal.

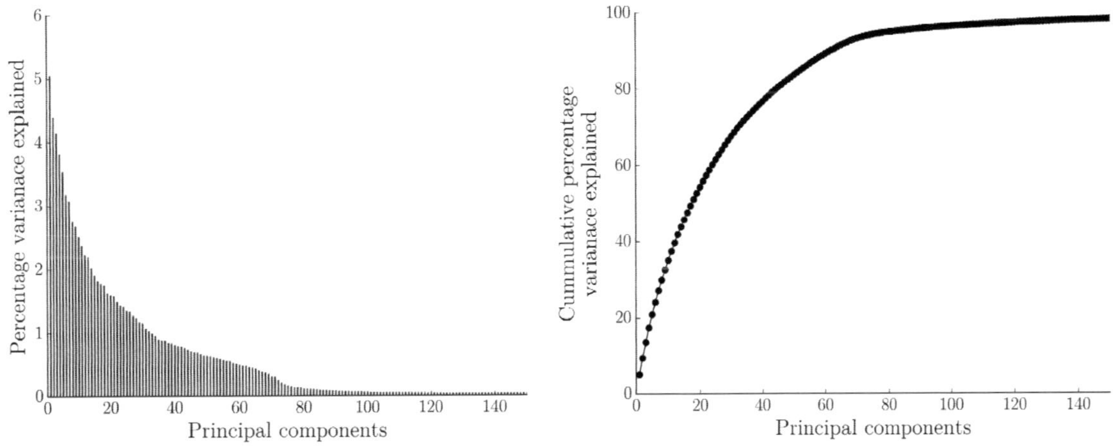

(a) Variance explained

(b) Cumulative variance explained

Figure 2. Principal component analysis of OD-flow vector using the estimated values from day 11 to day 25.

For the principal component-based calibration, covariance matrix of the errors in the transition equation $\mathbf{\Omega}_w$, and the autoregressive matrices $\mathbf{G}_i^h$ for $i = h, h-1, \ldots h-q$ were estimated as follows. First, the OD-flows estimated for days 11–25 were transformed into principal components. Then, an autoregressive equation was fit to each principal component independently. As before, if no autoregressive relation could be established for a principal component a random walk was assumed.

Calibration Results

The results from the online calibration on the 5 tests days for the Singapore Expressway network are discussed in this section. The performance measures adopted were the normalized root mean squared (RMSN) errors and mean absolute percentage errors (MAPE), which are defined as

$$RMSN = \frac{\sqrt{n \sum_{i=1}^{n} (\hat{y}_i - y_i)^2}}{\sum_{i=1}^{n} y_i} \tag{25a}$$

$$MAPE = \frac{100}{n} \sum_{i=1}^{n} \frac{|\hat{y}_i - y_i|}{y_i} \tag{25b}$$

where y_i represents actual measurement and $\hat{y}_i$ represents simulated measurement. Note that as MAPE is not normalized to consider the size of measurement, it can be high when the sensor flows are small even though the deviation from the observations is acceptable.

The performance of the OD-based calibration was compared with EKF algorithm (OD-EKF), principal component-based calibration with EKF algorithm (PC-EKF), and no online calibration. The no online calibration alternative represents the errors in the historical values which were obtained from the offline calibration. Further, the PC-EKF procedure was run by varying the number of chosen principal components. Specifically, there were four cases where the first 20, 40, 60, and 80 principal components were chosen; these components cumulatively explained 55%, 75%, 90%, and 95% of the variance in the data, respectively. The aggregate RMSNs and MAPEs of the procedures for estimation and prediction across the five test days are presented in Table 1.

In the context of estimation, the OD-EKF on average exhibits an RMSN of 0.287 and MAPE of 28.21%. It improves over historical by 32% in RMSN and 30% in MAPE. The PC-EKF with 80 principal components exhibited an RMSN of 0.25 and MAPE of 18.46%; its performance is better compared with OD-EKF by 13% in RMSN and 7% in MAPE. It appears that PC-EKF with sufficient principal components can perform well in estimation of the current state. In addition, the time-dependent RMSNs of the estimated sensor-flow counts for historical, OD-EKF, and PC-EKF are presented in Figure 3*a*. It can be seen that the performance of PC-EKF is better than OD-EKF for all the time intervals considered.

Comparing the estimation performance of PC-EKF by varying the number of chosen principal components, it can be seen that, curiously, the decreasing the number of principal components decreases the error in estimation. Specifically, the RMSN when 20 principal components were used is 0.24 and the RMSN when 80

Table 1. Aggregate Values of RMSNs and MAPEs of Sensor-Flow Counts (5 min) for Historical, OD-EKF, and PC-EKF Calibration Methodologies Over the 5 Test Days

Method	RMSN Est	RMSN Pred			MAPE Est	MAPE Pred		
		Step1	Step2	Step3		Step1	Step2	Step3
Hist	0.423	0.422	0.420	0.419	28.21	28.18	28.19	28.10
OD-EKF	0.287	0.290	0.295	0.305	19.77	24.60	26.44	27.71
PC-EKF 20	0.240	0.255	0.287	0.318	20.44	27.34	30.16	32.99
PC-EKF 40	0.241	0.255	0.284	0.310	20.34	26.95	28.16	29.24
PC-EKF 60	0.250	0.253	0.276	0.295	19.93	25.60	27.60	28.85
PC-EKF 80	0.250	0.252	0.265	0.275	18.46	22.13	24.31	25.66

Note: The number of principal components used is depicted in the method column. Method "Hist" refers to the offline calibrated parameter values.

principal components were used is 0.25. This seemingly counter-intuitive finding can be explained by noting that as the number of the principal components decreases, the weight given to the a priori estimates in the objective function—of the Kalman filter algorithm—also decreases. Therefore, decreasing the number of principal components essentially increases the weight given to the current measurements, resulting in the observed trend. It should be noted that this can happen only if the fewer number of chosen parameters can influence most of the observations. If this conjecture were true, poor predictions should be observed when the chosen principal components are too low in number even though the estimations are good—which is confirmed in the following paragraphs.

The results from the predictions are represented in three steps, which represent the three 5-min intervals in the complete 15-min prediction interval. In the context of prediction, from Table 1, the OD-EKF on average exhibits an RMSN of 0.290, 0.295, and 0.3 for the three steps and MAPE of 24.60%, 26.44%, and 27.71% for the three steps. By comparing the RMSNs and MAPEs after calibration with those of the historical values, it can be seen that RMSNs after calibration are significantly better than the historical, but the MAPEs after calibration are only marginally better than the historical. This implies that calibration improves on the predictions of sensor-flow counts of the sensors with higher flows than those with lower flows. The PC-EKF with 80 principal components on average exhibited RMSNs of 0.252, 0.265, and 0.275 for the three steps and MAPEs of 22.13%, 24.31%, and 25.66% for the three steps. PC-EKF outperforms OD-EKF in prediction in all the three steps of prediction. Specifically, PC-EKF with 80 principal components does better than OD-EKF by 11% in terms of RMSNs and by 8.5% in terms of MAPEs. The better predictive performance of PC-EKF is for the following three reasons. First, the PC-EKF formulation has lower "variance" compared with OD-EKF formulation, thus reducing the effect of overfitting. Second, the transition equation in PC-EKF is estimated taking into consideration the structural dependencies between OD-flows, which are ignored in OD-EKF. In addition, such an estimation also results in a better estimation of the error covariance structure Ω_w as the principal components are orthogonal. Third, the numerical approximation of the Jacobian in PC-EKF is better than that in OD-EKF. The time-dependent sensor count RMSNs in prediction are presented in Figure 3, *b, c*, and *d*. These trends show that PC-EKF with 80 principal components almost consistently performs better than the EKF in predicting the sensor counts.

Comparing the predictive performance of PC-EKF by varying the number of chosen principal components, it can be seen that decreasing the number of principal components worsens predictive performance. Specifically, the average predictive performance in RMSNs of PC-EKF with 20, 40, 60, and 80 principal components is 0.286, 0.283, 0.274, and 0.263, respectively. This is in line with the findings discussed in the earlier paragraphs. Decreasing the principal components introduces bias and gives more importance to the measurements in the current interval, leading to poorer predictions. This trend is also observed in time-dependent error plots in Figure 3, *c* and *d*, where the prediction errors of PC-EKF get worse as the chosen number of principal components decreases.

Scatter plots representing the estimated and predicted sensor-flow counts from PC-EKF versus actual sensor-flow counts are presented in Figures 4 and 5. Each figure contains plots corresponding to the four levels of principal components chosen in PC-EKF. The complete 5 days' sensor-flow count values are represented in a single plot as a heat map. A line was fit between the estimated/predicted and the actual values in each of the plots, and its equation is presented. From Figure 4, we see that the estimation accuracy does not depend much on the number of principal components chosen. This corroborates the earlier finding that the even a low number of principal components can influence many of the measurements. However, from Figure 5, it can be seen that PC-EKF with the lower number of principal components

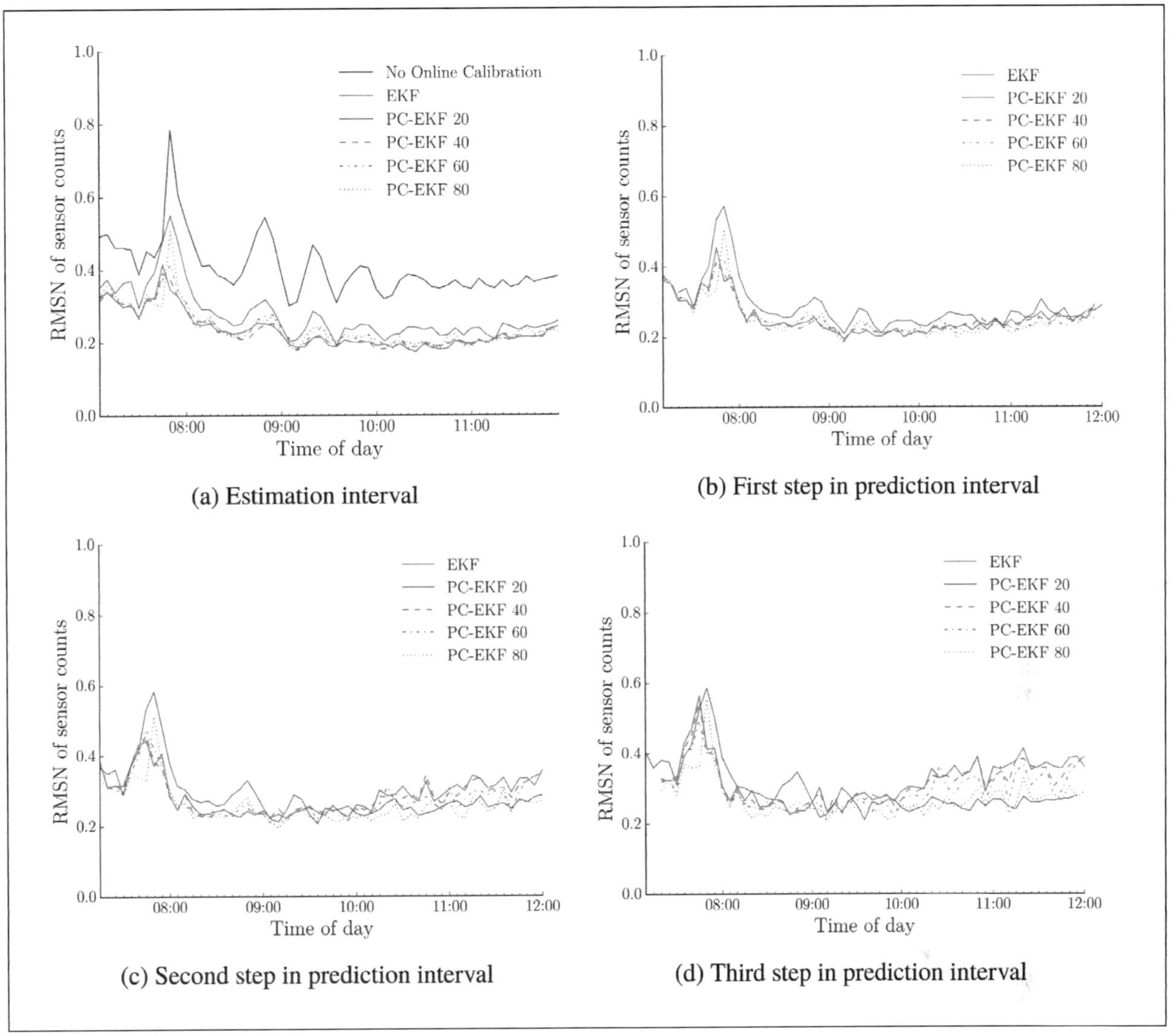

Figure 3. Plots of sensor count RMSNs for OD-EKF and PC-EKF with respect to time of day. The figure includes separate plots for estimation interval and each of the three steps in the prediction interval.

has poor predictive performance. Specifically, using PC-EKF introduces a bias in the formulation even though the variance is reduced—as in the typical bias–variance trade-off of the models. This bias can be quantified by the slope of the line-fit between the predicted values and the actual values; as the chosen number of principal decreases the bias increases. Specifically, the slope of the line with 20 principal components is 0.950 compared with the slope of the line with 80 principal components, which is 0.977. Thus choosing the right number of principal components while applying PC-EKF is crucial so that the net benefit—from reducing variance but introducing bias—compared with OD-EKF is positive.

Conclusions

This study presented a generic approach to make the complete online calibration of DTA systems scalable. Specifically, a scalable version of the problem formulation which incorporated parameters from *both* supply models and demand models was presented. To demonstrate the approach, the study adopted the calibration of OD-flows as the problem of interest and principal component analysis as the method to reduce the dimensionality of the parameter vector. The proposed state–space formulation was solved using the constrained EKF, which can incorporate the non-linear relationships between parameters and the corresponding

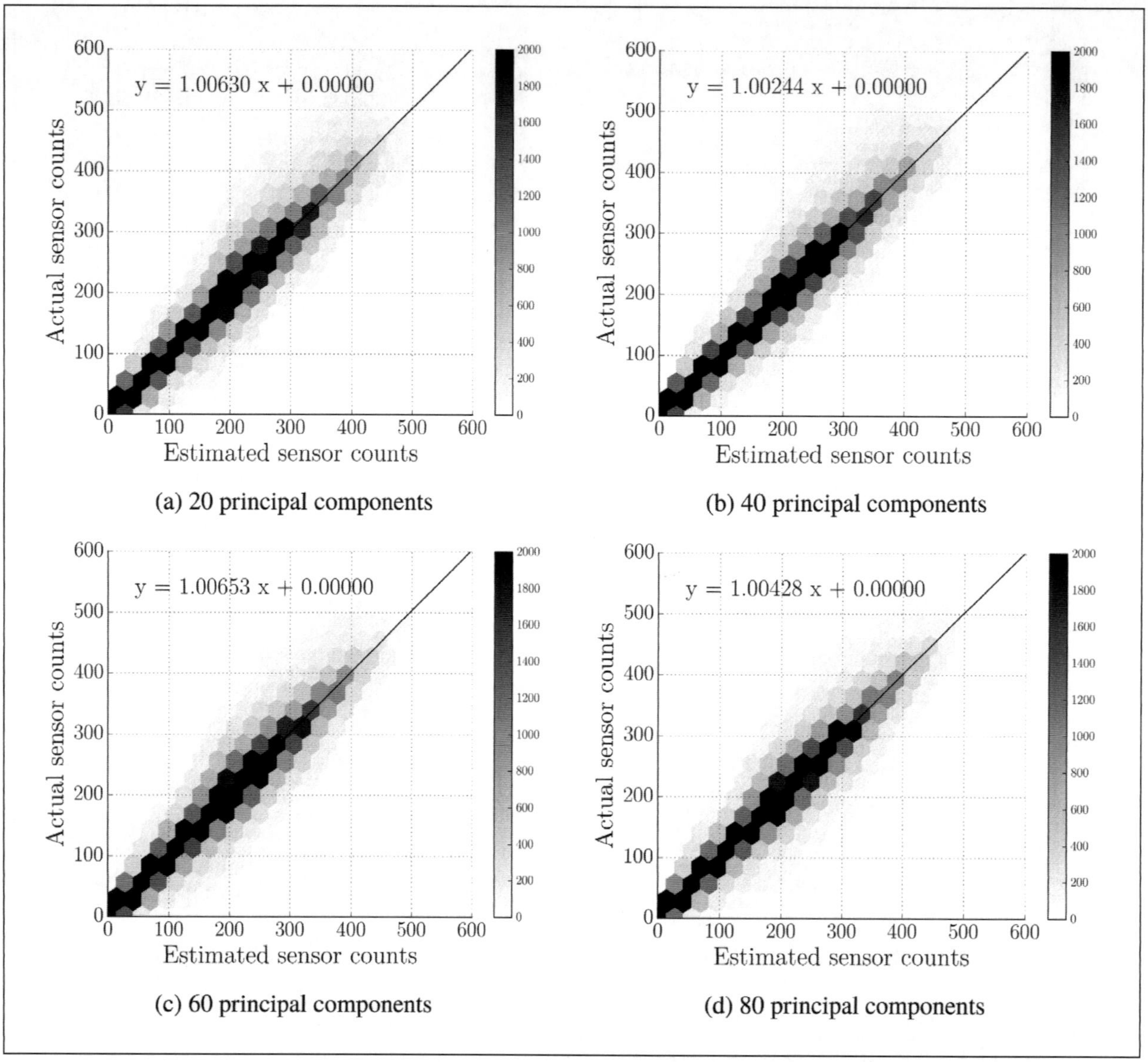

Figure 4. Comparison of estimation PC-EKF procedure with different principal components through the scatter plots of 5 min estimated vs. actual sensor counts on all 5 test days. The equations of the line-fit are depicted on the plots. The darker the cell, the higher the number of points in it.

measurements. Finally, the approach was demonstrated on a real-world transportation network in Singapore. The performance of the proposed approach was rigorously assessed, which led to following insights:

- Few principal components explain very a high percentage of the variance. Specifically, 80 principal components explain 95% of the variance (dimensionality reduction by 52 times) and only 150 principal components explain around 99% of the variance (dimensionality reduction by 27 times).
- The proposed approach (PC-EKF) makes the application of EKF feasible for application in real time. The traditional approach (OD-EKF) is intractable for real-time applications even with present computational resources.
- In state estimation, PC-EKF—with 80 principal components—outperformed OD-EKF by 13% in RMSNs and 7% in MAPEs. In state predictions, PC-EKF—with 80 principal components—outperformed OD-EKF across the three steps on average by 11% in RMSNs and 8.5% in MAPEs. The better overall performance of PC-EKF can be

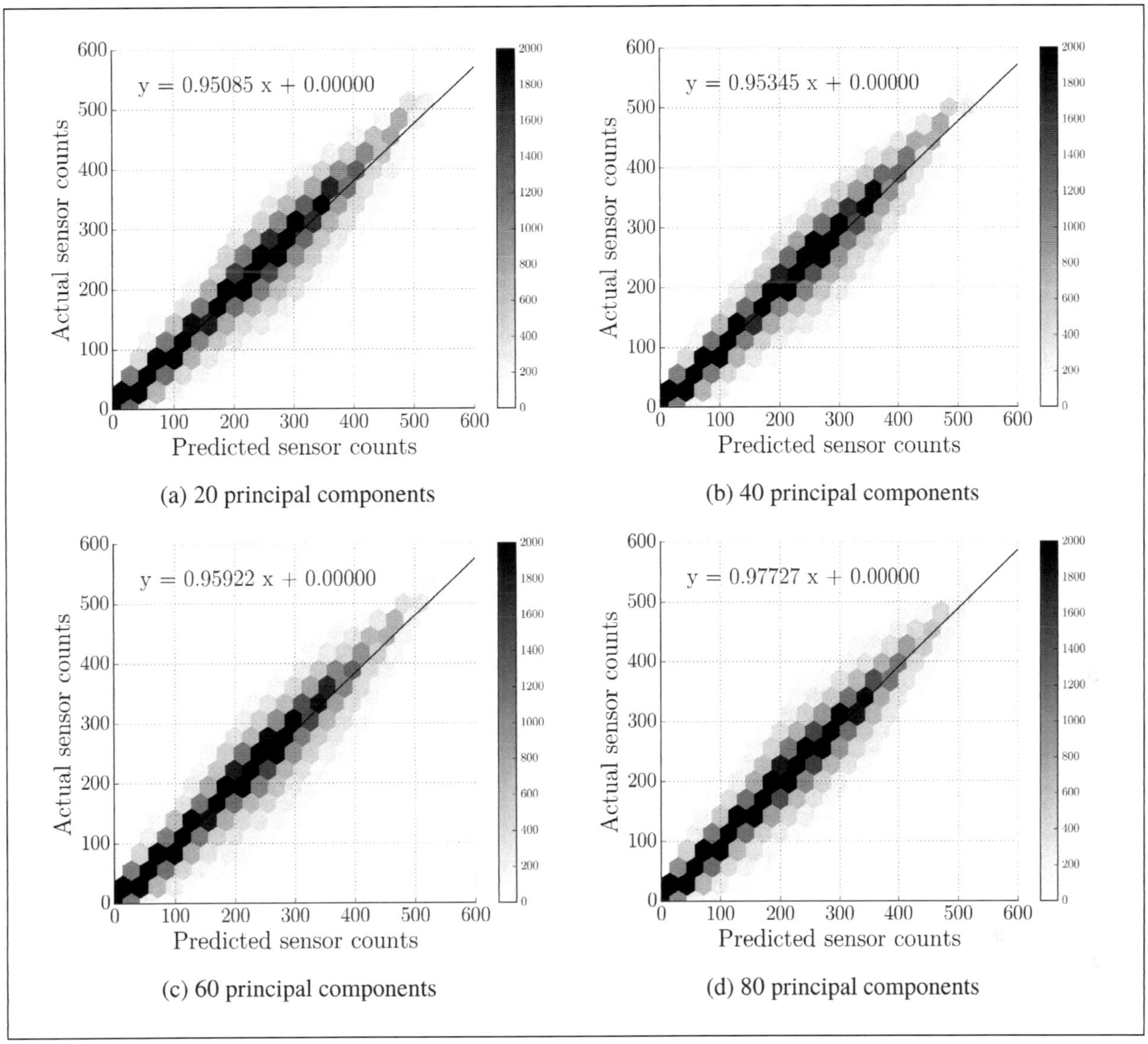

Figure 5. Comparison of step 3 predictions of PC-EKF procedure with different principal components through the scatter plots of 5 min predicted vs. actual sensor counts on all 5 test days. The equations of the line-fit are depicted on the plots. The darker the cell, higher the number of points in it.

explained as follows: (i) it lends itself to better formulation through more accurate transition equations and error covariances; (ii) it reduces the errors in the numerical approximation of Jacobian in EKF, leading to better performance of the algorithm; and, (iii) it tends to reduce the "variance" of the model, leading to better overall performance.

- Sensitivity analysis of principal components led to following insights. First, dimensionality reduction introduces bias into the model, which can be beneficial if the variance of the base model is too high. However, choosing very few principal components can bias the model too much, leading to poor predictive performance.

Acknowledgments

This research is supported by the National Research Foundation, Prime Minister's Office, Singapore, under its CREATE program, Singapore-MIT Alliance for Research and Technology (SMART) Future Urban Mobility (FM) IRG.

Author Contributions

The authors confirm contribution to the paper as follows: study conception and design: A. A. Prakash, R. Seshadri, C.

Antoniou, F. C. Pereira, M. E. Ben-Akiva ; data collection: A. A. Prakash, R. Seshadri ; analysis and interpretation of results: A. A. Prakash, R. Seshadri, C. Antoniou, F. C. Pereira; draft manuscript preparation: A. A. Prakash, R. Seshadri. All authors reviewed the results and approved the final version of the manuscript.

References

1. Mahmassani, H., T. Hu, and R. Jayakrishnan. Dynamic Traffic Assignment and Simulation for Advanced Network Informatics (DynaSMART). In *Urban Traffic Networks: Dynamic Flow Modeling and Control*. Springer, Berlin/New York, 1995.
2. Ben-Akiva, M., H. N. Koutsopoulos, C. Antoniou, and R. Balakrishna. Traffic Simulation with DynaMIT. In *Fundamentals of Traffic Simulation*, Springer, 2010, New York, pp. 363–398.
3. Antoniou, C., M. Ben-Akiva, and H. N. Koutsopoulos. Nonlinear Kalman Filtering Algorithms for On-Line Calibration of Dynamic Traffic Assignment Models. *IEEE Transactions on Intelligent Transportation Systems*, Vol. 8, No. 4, 2007, pp. 661–670.
4. Cascetta, E., D. Inaudi, and G. Marquis. Dynamic Estimators of Origin-Destination Matrices Using Traffic Counts. *Transportation Science*, Vol. 27, No. 4, 1993, pp. 363–373.
5. Park, E. S., L. R. Rilett, and C. H. Spiegelman. A Markov Chain Monte Carlo-Based Origin Destination Matrix Estimator that is Robust to Imperfect Intelligent Transportation Systems Data. *Journal of Intelligent Transportation Systems*, Vol. 12, No. 3, 2008, pp. 139–155.
6. Zhang, H., Y. Nie, and Z. Qian. Estimating Time-Dependent Freeway Origin-Destination Demands with Different Data Coverage: Sensitivity Analysis. *Transportation Research Record: Journal of the Transportation Research Board*, 2008. 2047: 91–99.
7. Ashok, K. *Estimation and Prediction of Time-Dependent Origin-Destination Flows*. Ph.D. thesis. Massachusetts Institute of Technology, Cambridge, MA, 1996.
8. Okutani, I. The Kalman Filtering Approaches in Some Transportation and Traffic Problems. *Transportation and Traffic Theory*, 1987, pp 397–416.
9. Balakrishna, R., M. Ben-Akiva, and H. Koutsopoulos. Offline Calibration of Dynamic Traffic Assignment: Simultaneous Demand-and-Supply Estimation. *Transportation Research Record: Journal of the Transportation Research Board*, Vol. 2003, Cambridge, MA, 2007, pp. 50–58.
10. Balakrishna, R. *Off-Line Calibration for Dynamic Traffic Assignment Models*. PhD thesis. Massachusetts Institute of Technology, Cambridge, MA, 2006.
11. Antoniou, C. *On-Line Calibration for Dynamic Traffic Assignment*. PhD thesis. Massachusetts Institute of Technology, Cambridge, MA, 2004.
12. Ashok, K., and M. E. Ben-Akiva. Alternative Approaches for Real-Time Estimation and Prediction of Time-Dependent Origin–Destination Flows. *Transportation Science*, Vol. 34, No. 1, 2000, pp. 21–36.
13. Ashok, K., and M. E. Ben-Akiva. Estimation and Prediction of Time-Dependent Origin-Destination Flows with a Stochastic Mapping to Path Flows and Link Flows. *Transportation Science*, Vol. 36, No. 2, 2002, pp. 184–198.
14. Zhou, X., and H. S. Mahmassani. A Structural State Space Model for Real-Time Traffic Origin–Destination Demand Estimation and Prediction in a Day-to-Day Learning Framework. *Transportation Research Part B: Methodological*, Vol. 41, No. 8, 2007, pp. 823–840.
15. Zhang, H., R. Seshadri, A. A. Prakash, C. Antoniou, F. C. Pereira, and M. Ben-Akiva. Improved Calibration Method for Dynamic Traffic Assignment Models: Constrained Extended Kalman Filter. *Transportation Research Record: Journal of the Transportation Research Board*, 2017. 2667: 142–153.
16. Zhou, X., and H. Mahmassani. Dynamic Programming Approach for Online Freeway Flow Propagation Adjustment. *Transportation Research Record: Journal of the Transportation Research Board*, 2002. 1802: 263–270.
17. Antoniou, C., M. Ben-Akiva, and H. Koutsopoulos. Online Calibration of Traffic Prediction Models. *Transportation Research Record: Journal of the Transportation Research Board*, 2005. 1934: 235–245.
18. Hashemi, H., and K. Abdelghany. Integrated Method for Online Calibration of Real-Time Traffic Network Management Systems. *Transportation Research Record: Journal of the Transportation Research Board*, 2015. 2528: 106–115.
19. Frederix, R., F. Viti, W. W. Himpe, and C. M. Tampère. Dynamic Origin–Destination Matrix Estimation on Large-Scale Congested Networks Using a Hierarchical Decomposition Scheme. *Journal of Intelligent Transportation Systems*, Vol. 18, No. 1, 2014, pp. 51–66.
20. Djukic, T., J. Van Lint, and S. Hoogendoorn. Application of Principal Component Analysis to Predict Dynamic Origin-Destination Matrices. *Transportation Research Record: Journal of the Transportation Research Board*, 2012. 2283: 81–89.
21. Djukic, T. *Dynamic OD Demand Estimation and Prediction for Dynamic Traffic Management*. PhD thesis. TU Delft, Delft University of Technology, Delft, Netherlands, 2014.
22. Prakash, A. A., R. Seshadri, C. Antoniou, F. C. Pereira, and M. Ben-Akiva. Reducing the Dimension of Online Calibration in Dynamic Traffic Assignment Systems. *Transportation Research Record: Journal of the Transportation Research Board*, 2017. 2667: 96–107.
23. Lu, L., Y. Xu, C. Antoniou, and M. Ben-Akiva. An Enhanced SPSA Algorithm for the Calibration of Dynamic Traffic Assignment Models. *Transportation Research Part C: Emerging Technologies*, Vol. 51, 2015, pp. 149–166.

The Standing Committee on Transportation Network Modeling (ADB30) peer-reviewed this paper (18-04982).

Article

TRR
JOURNAL OF THE TRANSPORTATION RESEARCH BOARD

Transportation Research Record
2018, Vol. 2672(48) 93–103

DOI: 10.1177/0361198118797486
journals.sagepub.com/home/trr
SAGE

Multimodal Transportation Network Modeling Based on the Generic Second Order Modeling Approach

Jean-Patrick Lebacque[1] and Megan M. Khoshyaran[2]

Abstract

The generic second order modeling (GSOM) approach to vehicular traffic flow combines the traditional kinematic description of traffic with dynamic equations for vehicle attributes. The GSOM model provides a unifying framework for many models in the literature. This framework includes tools such as local supply and demand, discretization, variational formulation, and generic intersection models.The object of the paper is to extend this approach to multimodal modeling of transportation networks. In multimodal networks two connected but distinct flows co-exist: vehicular and passenger flow. The main idea of the paper is to express the passenger flow as the flow of a specific attribute: the passenger load of vehicles, the unit of which is expressed in passengers per vehicle. Supplementary attributes pertaining to passengers can be added: destinations, activity plans, information availability, passenger category, and so forth. Intermodal poles, stations (bus, taxi, train), and intersections are modeled as nodes. It is shown that the stock variables associated to vehicles and passengers satisfy a system of conservation equations (on links) coupled with a system of ordinary differential equations (in nodes). Boundary conditions and numerical methods are proposed for the resolution of the model. Some examples are given, notably a method for managing a major incident on an automatic train line. The method uses most elements of the model in order to define the control strategy: passenger and train dynamics, station models, and local traffic supply and demand analysis.

The aim of this paper is to develop a generic macroscopic model for multimodal networks, analogous to macroscopic models for traffic flow. Expected gains of such an approach are: a simpler model, faster computations, and higher compliance with optimization methods for network management, routing, and dynamic assignment.

Two very different challenges must be met:

- Multimodal flows comprise two flows; vehicular and passengers. These flows are not independent; they are connected: passengers travel in vehicles.
- Passenger flows have a much higher complexity than vehicular flows: it is necessary to take into account activities, mode changes intermodality, and the related infrastructures.

A natural idea in order to model the innate complexity of multimodal transportation is to use a multi-agent approach. An early example of a generic model is provided by Urbansim (*1*), but this approach is still being developed, for instance with Matsim (*2*) but also other more specific models (*3–6*). Xu et al. propose to combine learning methods with the multi-agent approach (*7*). Some examples of application to optimization and management are provided by several authors (*8–10*), the latter proposing a dynamic model based on macroscopic differential equations.

Recently macroscopic traffic modeling tools have been introduced into multimodal problem solving. The macroscopic fundamental diagram (MFD) method, a concept from vehicular traffic modeling for global network analysis, has been applied to multimodal large-scale problems (*11–13*). In a similar direction, Sossoe proposes the application of bidimensional traffic flow modeling of very large networks to multimodal systems (*14*).

The main idea of the paper is to describe multimodal flows by extending macroscopic traffic flow models. The generic second order modeling (GSOM) approach to vehicular traffic was introduced in Lebacque et al. (*15*). It supplements the macroscopic kinematic description of traffic with a description of the dynamics of advected quantities called traffic attributes. Examples of attributes

[1]IFSTTAR, COSYS / GRETTIA, Cité Descartes, Champs-sur-Marne, Marne-la-Vallée, France
[2]ETC Economics Traffic Clinic, Paris, France

Corresponding Author:
Address correspondence to Jean-Patrick Lebacque:
jean-patrick.lebacque@ifsttar.fr

are: driver behavior or type, vehicle type and/or mode (in case of vehicular multimodality we might want to distinguish taxis, on-demand vehicles, etc), origin–destination (OD), destination, path, information and connectivity, and battery charge (electric vehicles). Refer to Lebacque and Khoshyaran (*16*) for a summary of attributes that have been used in GSOM models introduced in the literature. Refer to Khoshyaran and Lebacque (*17*) for an applicative example (pollutant emission) combining a non-trivial active attribute (engine state) with parking modeling.The GSOM model is thus by its very nature well adapted to multimodality wher two flows (passengers and vehicles) co-exist but one (the passenger flow) is advected by the other (the vehicular flow). Further, the GSOM models have many analytical properties which make their use convenient.

The proposal is thus to introduce a new attribute, the passenger load, which is a density (the number of passengers per vehicles). The dynamics of this attribute will provide a description of passenger flow advected by the vehicles of the transportation system. Change of mode, entry and exit of the system, occurs at nodes, thus most of the dynamics induced by activities are recaptured at the node level.

The outline of the paper is as follows. First some basic facts about GSOM models are discussed, as they will be used in the rest of the paper. Then the passenger load attribute is introduced, as well as other attributes pertaining to passengers (destinations), and the dynamics of these attributes on links and in nodes are described. A Eulerian discretization of the model is proposed. Finally, an application of the model to the management of an automated train system is presented.

GSOM Traffic Flow Models

Basic Concepts

In this section the basic elements of GSOM models for vehicular traffic flow are discussed. Details can be found elsewhere (*15*, *16*). GSOM models include many current macroscopic models, notably Lighthill–Whitham–Richards (LWR)-based multi-class (multi-commodity) models, some multi-lane models, and also stochastic models.

Basic traffic variables are: density $\rho(x,t)$, velocity $v(x,t)$, flow $q(x,t)$, and attributes $I(x,t)$. x,t denote the position and the time. Attributes constitute usually a vector: $= (I_\ell)_{\ell \in \mathcal{L}}$, as vehicles and drivers are liable to be endowed with many attributes of interest. The quantity ρI can be viewed as the density of attribute, whereas $\rho v I = qI$ is the flow of attribute. When necessary the position x must be specified in terms of link and local coordinate on the link: $x = \xi_a$ with ξ_a the local coordinate on link $a \epsilon A$ (A denotes the set of arcs).

The model equations are the following:

$$\begin{bmatrix} \partial_t \rho + \partial_x q = 0 & \text{(conservation of vehicular traffic)} \\ \partial_t \rho I + \partial_x q I = -\rho \Upsilon(I) & \text{(conservation of attribute vector)} \\ v = \mathfrak{I}(\rho, I) \text{i.e. } q = \mathfrak{R}(\rho, I) & \text{(attribute dependent fundamental diagram)} \end{bmatrix} \quad (1)$$

Equation (1) is a system of conservation equations with source terms and constitutes a higher order model in that sense. The terminology "second order" stems from the ARZ (Aw–Rascle–Zhang) model, which is an early example of GSOM model. In this model there was a single scalar attribute, expressing driver behavior, $I = v - V_e(\rho)$ (i.e., $\mathfrak{I}(\rho, I) = V_e(\rho) + I$), and the conservation of attribute equation was an acceleration equation.

By the definition of flow, $\mathfrak{R}(\rho, I) = \rho \mathfrak{I}(\rho, I)$. Attributes may include:

- Neutral attributes, which do not affect the fundamental diagram, such as ODs, destinations, passenger load, and so forth;
- Passive attributes, which are simply advected (i.e., $\Upsilon = 0$), such as vehicle type, passenger type, passenger load, and so forth.

The equilibrium supply and demand functions are deduced by the following expressions, which apply even in a context of piecewise continuous fundamental diagrams:

$$\begin{bmatrix} \Delta(\rho, I; x) = & \max_{r \leqslant \rho} \mathfrak{R}(\rho, I; x-) \\ \Sigma(\rho, I; x) = & \max_{r \geqslant \rho} \mathfrak{R}(\rho, I; x+) \end{bmatrix} \quad (2)$$

Local traffic supply and demands are given by:

$$\begin{bmatrix} \delta(x,t) = & \Delta(\rho(x-,t), I(x-,t); x) \\ \varsigma(x,t) = & \Sigma(\mathfrak{I}^{-1}[\mathfrak{I}(\rho(x+,t), I(x+,t); x+), I(x-,t; x+); x+], I(x-,t); x) \end{bmatrix} \quad (3)$$

(refer to [*18–20*]). The expression of the local supply $\varsigma(x,t)$ is complicated and depends both on upstream and downstream attribute values, because it concerns upstream vehicles, the velocity of which is imposed by

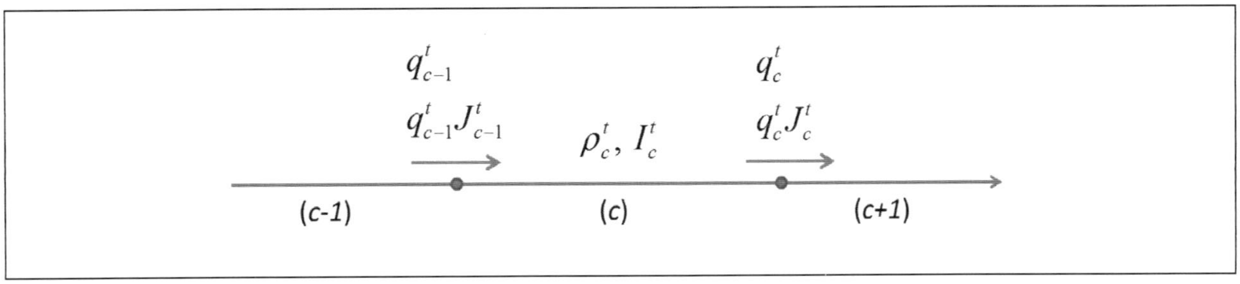

Figure 1. Eulerian discretization of the GSOM model on a link.

downstream conditions. The local supply $\varsigma(x,t)$ can be best understood by considering that a vehicle with attribute $(x-,t)$, crossing x, is subject to the supply conditions at $x+$, that is $\Sigma(.,I(x-,t);x)$. Its velocity is the velocity of vehicles at $x+$, that is $v_+ \underline{\underline{\text{def}}} \Im(\rho(x+,t),I(x+,t);x+)$ (indeed velocity is an invariant). The density to be taken into account for upstream vehicles in the calculation of the downstream supply is the density $\rho_m \underline{\underline{\text{def}}} \Im^{-1}[v_+,I(x-,t;x+);x+]$. This density satisfies: $v_+ = \Im(\rho_m,I(x-,t);x+)$ and corresponds to the upstream attribute and downstream speed. Thus finally the required expression is obtained for the supply: $\varsigma(x,t) = \Sigma(\rho_m,I(x-,t);x)$.

The flows at location x and time t are given by the Min formula:

$$\begin{bmatrix} q(x,t) = & \min[\delta(x,t),\varsigma(x,t)] \\ (qI)(x,t) = & q(x,t)I(x-,t) \end{bmatrix} \qquad (4)$$

All these expressions are valid even if the density, attributes, and fundamental diagrams are only piecewise continuous. Thus they can be used to solve Riemann problems, to formulate Godunov schemes, or to define boundary conditions (notably in nodes) (*16*, *18*).

Note that GSOM models admit a variational formulation: they can be recast as Hamilton–Jacobi equations with fundamental variable $X(N,t)$ the position of particle N at time t (*16*).

Eulerian Discretization of the GSOM Model

This paper considers only the discretization of a GSOM model on a link. The proposed scheme is an adaptation of the Godunov scheme, taking into account the formulas derived in the previous sub-section; refer to Mammar et al. (*18*), and also to Lebacque and colleagues (*15–17*). The link is divided into cells (c), time is discretized into time-steps $(t)\underline{\underline{\text{def}}}[t\Delta t,(t+1)\Delta t]$ and relevant quantities are (refer to Figure 1 below):

- q^t_c the flow from cell (c) into cell $(c+1)$ during time-step (t);
- ρ^t_c the mean density in cell (c) at time $t\Delta t$;
- I^t_c the mean attribute in cell (c) at time $t\Delta t$;
- Δ_c, Σ_c the equilibrium demand and supply functions in cell (c);
- Υ_c the mean source term function in cell (c);
- $\delta^t_c, \varsigma^t_c$ the demand and supply of cell (c);
- $\Im_c$ the speed-density fundamental diagram in cell (c);
- Δx_c the length of cell (c).

The conservation equation for density (vehicles) can be obtained as:

$$\begin{bmatrix} \rho^{t+1}_c = & \rho^t_c + {}^{\Delta t}\!/_{\Delta x_c}\left(q^t_{c-1} - q^t_c\right) \\ q^t_c = & \min\left[\delta^t_c, \varsigma^t_{c+1}\right] \end{bmatrix} \qquad (5)$$

with supply and demand given by:

$$\begin{bmatrix} \delta^t_c = & \Delta_c\left(\rho^t_c, I^t_c\right) \\ \varsigma^t_c = & \Sigma_c\left(\Im^{-1}_c\left[\Im_c\{\rho^t_c, I^t_c\}, I^t_{c-1}\right], I^t_{c-1}\right) \end{bmatrix} \qquad (6)$$

The dynamic attribute equation must be integrated (in the active attribute case); we consider a simple explicit Euler formula:

$$J^t_c = I^t_c - \Delta t \Upsilon_c\left(I^t_c\right) \qquad (7)$$

and the discretization is completed with the conservation equation for passengers, in terms of passenger density:

$$\rho^{t+1}_c I^{t+1}_c = \rho^t_c J^t_c + {}^{\Delta t}\!/_{\Delta x_c}\left(q^t_{c-1}J^t_{c-1} - q^t_c J^t_c\right) \qquad (8)$$

Equation (8) expresses that the mass of attribute is conserved, but that individual vehicle attributes change in time following (7).

A Courant–Friedrichs–Lewy (CFL) condition must be specified based on the consideration that waves generated at the extremities of each cell must not be able to interact within a time-step. Boundary data at the extremities of the link are the upstream demand (providing the upstream demand of the first cell) and the downstream supply, applied as supply to the last cell of the link. These quantities are provided by intersection models or by network boundary conditions (OD demand, traffic conditions, etc).

Multimodal GSOM and Network Modeling

Basic Concepts

The fundamental idea is to describe:

- The quantity and dynamics of vehicles by a density variable $\rho(x,t)$ which satisfies the traffic conservation and the fundamental diagram of Equation (1);
- The quantity of passengers and their location by a variable called the passenger load $\varpi(x,t)$, which is considerd as a passive and neutral attribute and which will satisfy the conservation of attribute in Equation (1). The physical meaning is that passengers are transported (advected) by vehicles.

The unit in which the passenger load $\varpi(x,t)$ is expressed is the number of passengers per vehicle. Thus $\rho\varpi(x,t)$ denotes the density of passengers (i.e., the number of passengers per unit length) and $\rho v\varpi(x,t) = q\varpi(x,t)$ denotes the flow of passengers at any given location and time. The rest of this paper will mainly consider the following neutral and passive attributes:

- $\chi^d(x,t)$ the fraction of vehicles with destination $d \epsilon D$;
- $\mu^d(x,t)$ the fraction of passengers with destination $d \epsilon D$;
- $L(x,t)$ other neutral and passive attributes (mode, vehicle type, etc). These should be treated in the same way as μ, χ and thus will be omitted in the rest of the paper.

Thus passenger and vehicle dynamics in a multimodal GSOM model an attribute of the following structure:

$$I = (\varpi, \mu, \chi, L).$$

In public modes (train, bus, tram...) the destination of passengers and of vehicles is not the same. Vehicles have their missions and passengers use vehicles to satisfy their own activity plans. Another important point: the time-scale of private vehicles is of the order of a second, whereas the time-scale of public transportation vehicles is greater (of the order of a minute), at least if the maximum frequencies of vehicles are considered (nevertheless in automated systems vehicle-to-vehicle interactions will take place at a much smaller time-scale).

From a physical point of view, the dynamics of vehicular traffic are described by the couple (ρ, χ) and the dynamics of passengers by the couple (ϖ, μ); nevertheless, from a mathematical point of view the system of Equation (1) is applied and the variable (ϖ, μ, χ) must be considered as attributes of the flow. Indeed, these variables are advected by the vehicle flow.

The network nodes will include intersections (vehicular traffic), stations, and intermodal poles (public transportation). Passengers are carried by vehicles on arcs, and possibly switch vehicles (and mode) at nodes. Parkings are easily accommodated in the model (*17*).

The GSOM multimodal GSOM model presented in this paper emphasizes the specific roles of vehicles and passengers and constitutes an instance of general GSOM models with a specific structure: vehicles (traffic density), passengers (passenger load attribute) carried by vehicles, and all other attributes, treated as regular GSOM attributes.

Notation

- The transportation system is described as a graph $G = (Z, A)$ with Z the set of nodes ($z \in Z$), and A the set of arcs (links): $a \in A$. Implicitly, a mode or set of modes is associated to an arc, which determines the type of vehicles allowed to travel on that arc;
- Time is t, position x, unless it is necessary to specify local coordinates on arc $a \in A$ and then we identify $x \cong \xi_a$;
- Destinations $d \epsilon D$, paths $p \in P$, ODs $w \in W$;
- Dynamic variables are as before: $(\rho, q, \varpi, \mu, \chi)$;
- Dynamic variables may be particularized to specify the link, thus we may note $\rho_a, q_a, \varpi_a, \mu_a^d, \chi_a^d$, (all of which are implicitly functions of (x,t) or (ξ_a, t)), in place of $(\rho, q, \varpi, \mu, \chi)$;
- $\varpi_{a,\max}$: the capacity of vehicles on arc $a \in A$, i.e. $\varpi_a \leqslant \varpi_{a,\max}\ \forall a \in A$.

Let us now introduce a few notations pertaining to nodes.

- A node (z) is endowed with entry points $i \in I(z)$ and exit points $j \in O(z)$;
- Each entry point $i \in I(z)$ respectively exit point $j \in O(z)$ of node (z) is associated to an entry arc $a = I(z,i)$ respectively exit arc $b = O(z,j)$;
- Any couple (i,j) with $i \in I(z), j \in O(z)$ is identified with a movement of node (z) (Figure 2);
- Dynamic variables associated with nodes and their exit points:
  - $\Pi_{z,j}^d(t)$ denotes the demand at node (z) for destination $d \epsilon D$ of those passengers who plan to use the exit arc $b = O(z,j)$;

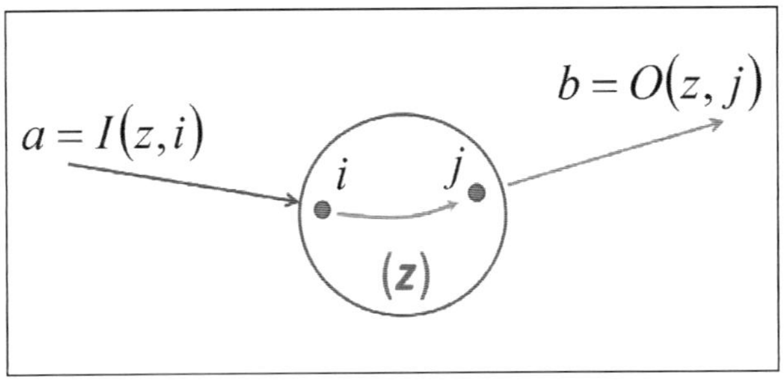

Figure 2. Representation of a movement inside a node.

- $N^d_{z,j}(t)$ denotes the stock at node (z) of those passengers who are waiting, whose destination is $d \epsilon D$ and who plan to use the exit arc $b = O(z,j)$. We denote by $N_{z,j}(t) \underline{\underline{\text{def}}} \sum_{d \in D} N^d_{z,j}(t)$ the total stock of passengers bound for arc $b = O(z,j)$;
- Recall that nodes are the locations of the network where passengers change mode, or change vehicle, or access or exit the network. Not all passengers may be processed at a given time, thus the stock of waiting passengers. This stock might also serve to model the transit of passengers in a station (pedestrian arc);

• Dynamic variables associated with nodes and movements:
  - $\gamma^d_{z,ij}(t)$: the fraction of vehicles entering node (z) through the entry point $i \in I(z)$, with destination $d \epsilon D$, who use the exit point $j \in O(z)$. These are the assignment coefficients for vehicles inside the node (z), and they are exogenous to the model. Indeed, they result from: vehicle mission (public transportation), control measures, user choice (private cars), and so forth;
  - $\theta^d_{z,ij}(t)$: the fraction of passengers entering node (z) through the entry point $i \in I(z)$, with destination $d \epsilon D$, who use the exit point $j \in O(z)$. These are the assignment coefficients for passengers inside the node (z), and they are exogenous to the model (they result from information or control strategies, user behavior, etc).

Arc Model

The arc model described by Equation (1) and its discretization given by Equations (5–8) carries over to multimodal systems with no modification. The attribute I is passive and neutral, and given by (as mentioned L is ignored for simplification):

$$I = (\varpi, \mu, \chi) \tag{9}$$

As the source term Υ is null and the fundamental diagram does not depend on the attribute, the Eulerian discretization simplifies.

The conservation equation for density (vehicles) is still obtained as:

$$\begin{bmatrix} \rho_c^{t+1} = & \rho_c^t + {}^{\Delta t}/_{\Delta x_c}\left(q_{c-1}^t - q_c^t\right) \\ q_c^t = & \min\left[\delta_c^t, \varsigma_{c+1}^t\right] \end{bmatrix} \tag{10}$$

with supply and demand now given by:

$$\begin{bmatrix} \delta_c^t = & \Delta_c\left(\rho_c^t\right) \\ \varsigma_c^t = & \Sigma_c\left(\rho_c^t\right) \end{bmatrix} \tag{11}$$

and the conservation equation for passengers, in terms of passenger density, is given by:

$$\rho_c^{t+1} I_c^{t+1} = \rho_c^t I_c^t + {}^{\Delta t}/_{\Delta x_c}\left(q_{c-1}^t I_{c-1}^t - q_c^t I_c^t\right) \tag{12}$$

It must be noted that (12) also includes the advection of the composition attributes with respect to destination.

Node Model

Many important features of the transportation network dynamics take place in nodes: change of mode and/or of vehicle, assignment and route choice, passenger arrival and departure, and so forth.

This paper considers the simplest node model possible: the point-wise model, thus neglecting such issues as:

- Physical extension of the node;
- Movement travel time (both for passengers and vehicles);
- Movement capacity, and conflicts between movements.

Vehicular Through-Flows. The flows through a node (z) will be essentially determined by the demand on the upstream links $a = I(z, i)$ for node entry points $i \in I(z)$, and on the supply of the downstream links $b = O(z,j)$ for node entry points $j \in O(z)$. The context is illustrated by the following Figure 3.

Let us define the supplies and demands that constitute the node boundary conditions and introduce some notations pertaining to node in- and out-flows for vehicles.

- $\delta_i = \Delta_a(\rho_a[z-, t])$, for $a = I(z, i)$;
- $\varsigma_j = \Sigma_b(\rho_b[z+, t])$, for $b = O(z,j)$;
- $q_i \underline{\underline{\text{def}}} q_a(z, t)$, for $a = I(z, i)$;
- $r_j \underline{\underline{\text{def}}} q_b(z, t)$, for $= O(z,j)$.

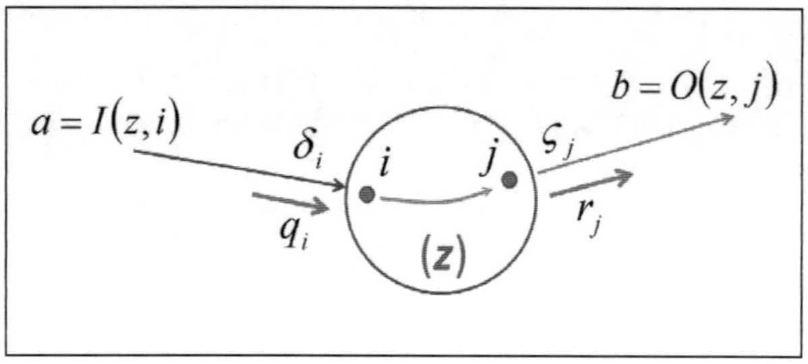

Figure 3. Calculation of node vehicular through-flows.

The flows must satisfy regular supply and demand constraints:

$$\begin{cases} 0 \leq q_i \leq \delta_i \\ 0 \leq r_j \leq \varsigma_j \end{cases} \tag{13}$$

They must also satisfy the conservation equation inside the node induced by the assignment coefficients and the composition of traffic in terms of destinations:

$$\sum_{i \in I(z), d \in D} \gamma^d_{z,ij}(t) \chi^d_{I(z,i)}(z-, t) q_i = r_j \tag{14}$$

(to simplify some expressions the dependency of various quantities on time may be omitted).

The flows are determined based on an optimization model (*21*, *22*):

$$\min \sum_{i \in I(z)} \mathcal{F}_i(q_i) + \sum_{j \in O(z)} \mathcal{H}_j(r_j)$$
$$\begin{cases} 0 \leq q_i \leq \delta_i & \forall i \in I(z) \\ 0 \leq r_j \leq \varsigma_j & \forall j \in O(z) \\ \sum_{i,d} \gamma^d_{z,ij} \chi^d_{I(z,i)} q_i = r_j & \forall j \in O(z) \end{cases} \tag{15}$$

The functions $\mathcal{F}_i(q_i)$, $\mathcal{H}_j(r_j)$ express how the conflicts between movements inside the node are resolved. Typically, these functions should be taken increasing positive and concave. For instance, quadratic functions provide expressions which are in a satisfactory agreement with observations in the case of vehicular traffic (*17*, *22*). Possible quadratic functions are:

$$\mathcal{F}_i(q_i) = -(q_i)^2/2 + q_i/q_{i,\max},\ \mathcal{H}_j(r_j) = (r_j)^2/2 + r_j/r_{j,\max}$$

These functions have been introduced in Lebacque and Khoshyaran (*22*), in the simple case of merges and diverges they yield classical models (first-in, first-out [FIFO] diverge and priority merge). The problem (15) is linear concave and therefore easily solvable, for instance by duality. In many cases the problem (15) can actually be solved in a finite number of steps.

Once problem (15) is solved the vehicular flow of the movement $(i) \to (j)$ is equal to:

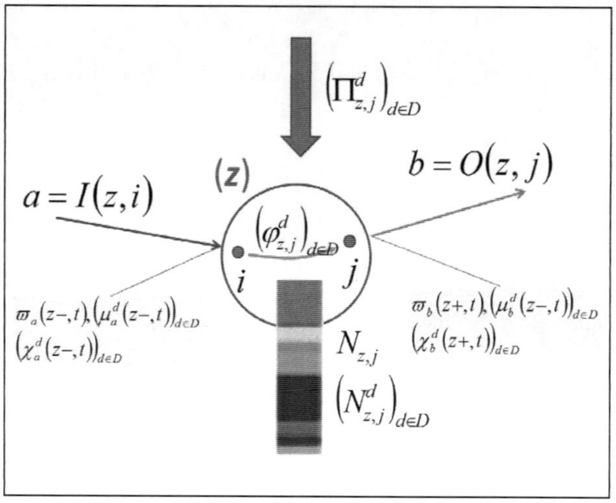

Figure 4. Calculation of node passenger through-flows. Color blocks in the bar representing $N_{z,j}$ depend on local destination and their size is proportional to $N^d_{z,j}$.

$$\sum_{d \in D} \gamma^d_{z,ij}(t) \chi^d_{I(z,i)}(z-, t) q_i \tag{16}$$

These through-flows could be affected by movement capacity and also by conflicting movements. Such constraints can be integrated into problem (15) if need be.

Finally, from the preceding analysis the composition in terms of destinations of the traffic entering the downstream link $= O(z,j)$ can be determined:

$$\chi^d_b(z+, t) = \left(\sum_{i \in I(z)} \gamma^d_{z,ij} \chi^d_{I(z,i)} q_i \right) / r_j \tag{17}$$

Passenger Through-Flows. Let us now consider passenger through-flows of the movement $(i) \to (j)$ of node (z). The calculation only applies to traffic with destination $d \neq z$ (otherwise passengers exit at the station $[z]$) (refer to Figure 4).

Analogous to Equation (16) the flow of passengers is deduced $(i) \to (j)$:

$$\sum_{d \in D} \theta^d_{z,ij}(t) \varpi_{I(z,i)} \mu^d_{I(z,i)}(z-, t) q_i \tag{18}$$

This flow must be added to the passenger arrival rate $\Pi^d_{z,j}(t)$ at node (z) in order to yield the total passenger demand with destination $d \in D$ requesting vehicles on link $= O(z,j)$:

$$\varphi^d_{z,j} \underline{\underline{\text{def}}} \Pi^d_{z,j} + \sum_{i \in I(z)} \theta^d_{z,ij} \varpi_{I(z,i)} \mu^d_{I(z,i)}(z-, t) q_i \tag{19}$$

The capacity of vehicles in terms of passenger transportation is $\varphi_{zj,\max} \underline{\underline{\text{def}}} r_b \varpi_{b,\max}$. This capacity may not be sufficient to accommodate the total passenger demand for transportation on link $b = O(z,j)$, that is

$$\varphi_{z,j} \underline{\underline{\text{def}}} \sum_{d \in D} \varphi^d_{z,j} \quad (20)$$

Further if the total stock of passengers $N_{z,j}(t)$ is positive then the demand for transportation is maximum and its composition in terms of destination is that of the stock $N_{z,j}$. Given these various elements, the following rules are proposed for the passenger dynamics exiting the node (z) on link $b = O(z,j)$:

- If $\varphi_{z,j} \geqslant \varphi_{zj,\max}$ or $N_{z,j} > 0$, the passenger flow is $\varphi_{zj,\max}$, meaning that the boundary value of the passenger load at the entrance of the link $b = O(z,j)$ is given by $\varpi_b(z+,t) = \varpi_{b,\max}$;
- Otherwise the total passenger flow is $\varphi_{z,j}$, thus the boundary value of the passenger load at the entrance of the link $b = O(z,j)$ is given by $\varpi_b(z+,t) = \varphi_{z,j}$;
- If $N_{z,j} > 0$ the composition of passenger traffic in terms of destination is given by $\mu^d_{z,j} \underline{\underline{\text{def}}} \mu^d_{O(z,j)}(z+,t) = N^d_{z,j}/N_{z,j}$
- Otherwise (i.e., $N_{z,j} = 0$) this composition is given by $\mu^d_{z,j} \underline{\underline{\text{def}}} s\varphi^d_{z,j}/\varphi_{z,j}$.

These results are summarized with $b = O(z,j)$:

$$\varpi_b(z+,t) = \begin{bmatrix} \varpi_{b,\max} & \text{if } \varphi_{z,j} \geqslant \varphi_{zj,\max} \text{or } N_{z,j} > \mu^d_{O(z,j)}(z+,t) = 0 \\ \varphi_{z,j} & \text{otherwise} \end{bmatrix} \quad (21)$$

$$\mu^d_b(z+,t) = \begin{bmatrix} N^d_{z,j}/N_{z,j} & \text{if } N_{z,j} > 0 \\ \varphi^d_{z,j}/\varphi_{z,j} & \text{if } N_{z,j} = 0 \end{bmatrix} \quad (22)$$

The dynamics of the stocks of passengers inside the node are deduced immediately from Equations (21) and (22):

$$\frac{d}{dt} N^d_{z,j} = \varphi^d_{z,j} - \begin{cases} \left(N^d_{z,j}/N_{z,j}\right)\varphi_{zj,\max} & \text{if } N_{z,j} > 0 \\ \min\left[\varphi^d_{zj,\max}, \varphi^d_{z,j}\right] & \text{if } N_{z,j} = 0 \end{cases} \quad (23)$$

Of course the total stock is given by

$$N_{z,j} = \sum_{d \in D} N^d_{z,j} \quad (24)$$

The intersection model is thus completed as we can calculate all through-flows (vehicles and passengers), the passenger stocks inside the node and also the boundary data and values for downstream links, such as $\chi^d_b(z+,t)$, $\varpi_b(z+,t)$ and $\mu^d_b(z+,t)$.

Remarks

1. The node model described by Equations (19), (20), (23), (24) is not FIFO, but can be made FIFO at small supplementary cost by keeping track of the detailed composition of the stock $N_{z,j}$ with respect to destination instead of simply keeping track of the global composition $N^d_{z,j}$. This can be achieved by indexing in some way passengers in the stock and noting their destination. The modification will be described in a further publication.
2. It should be noted that Equations (15) (calculation of node in- and out-flows), (17), (21), (22) (boundary data for exit link $\in O[z]$), and (19) (passenger demand) can be used with no or very little modification in a discretized setting in order to complement the link discretization (5–8). In a discretized context Equation (23) should be integrated over a time-step.

An Example of Application of Multimodal GSOM Network Modeling

The main benefits of this approach are:

- Its macroscopic character, making it suitable for large-scale problems (control, information, evaluation, dynamic assignment, etc.);
- The availability of GSOM tools, such as supply and demand (Equations 3, 4), discretization (Eulerian discretization as in Equations [9–12], but also Lagrangian discretizations as in references [*16, 23–25*]), and variational formulation (*16*);
- A general methodology for node modeling (*21, 22, 24, 26*);
- A unified framework for multimodal network modeling (very useful for instance for coding).

The aim of this section is to provide an example of the application of these concepts (especially the supply demand approach and the node model) to modeling and management of a major incident on an automated metro line.

The setting is described in Figure 5. The line is cut at peak hour, between 7.35 and 9.10 a.m., between stations (s) and $(s-1)$, in direction (E–W). The classical solution consists of cutting the line into two and operating each half-line separately.

The idea here is to use the possibility to switch tracks between stations $(s-1)$ and $(s-2)$ on one hand, (s) and

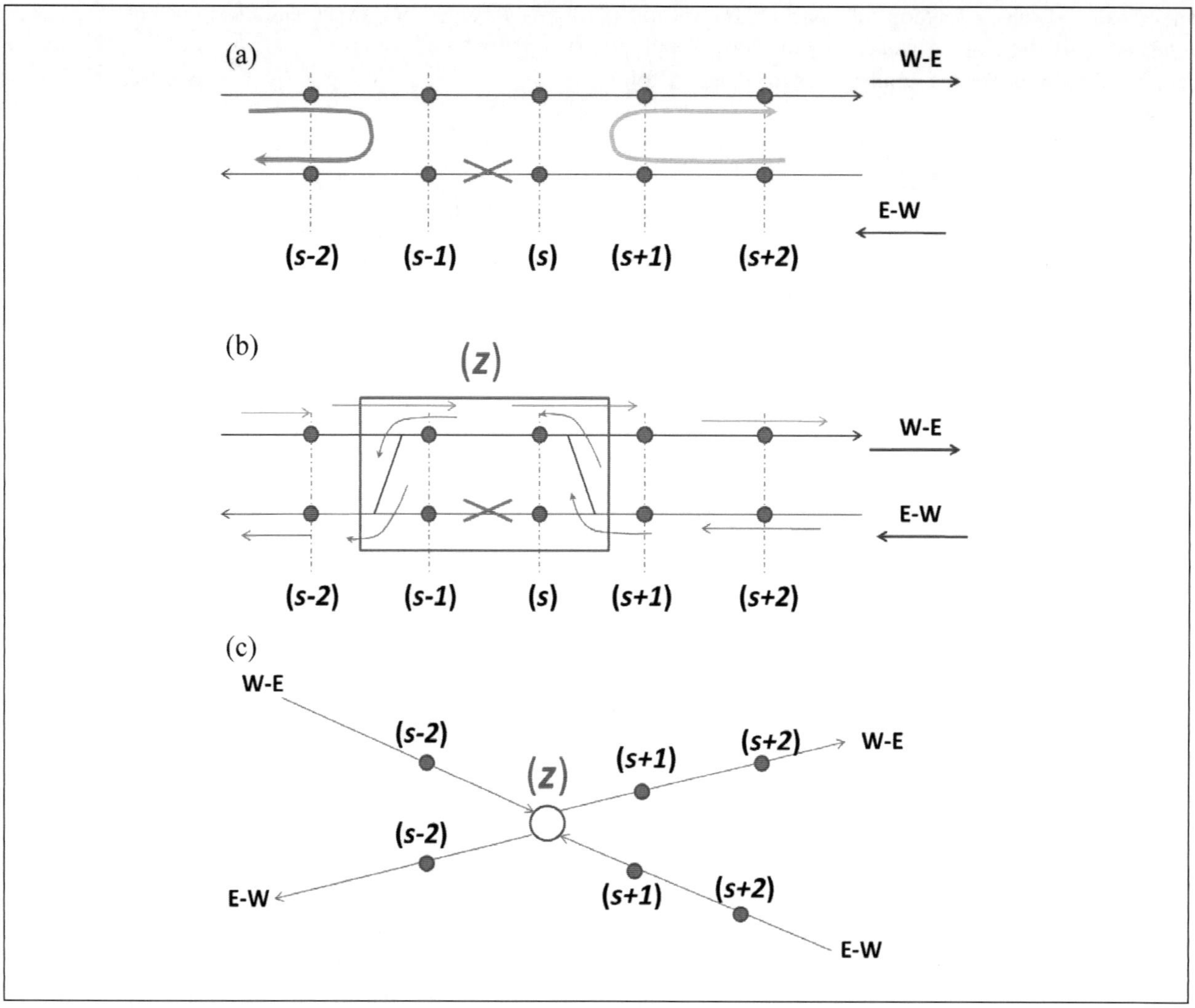

Figure 5. Incident management. (*a*) classical management (line cut between (s – 2) and (s + 1); (*b*) management as a controlled intersection; and (c) the intersection model.

$(s+1)$ on the other side, in order to manage the incident without loss of connectivity.

To manage conflicts on the remaining track between (s) and $(s-1)$ a traffic light controls the access to the remaining track at the exit of $(s+1)$ in the (E–W) direction and at the exit of $(s-2)$ in the (W–E) direction.

The system is modeled as follows:

- The incident zone is described, from a control point of view, as a single node (z) as depicted in Figure 5. Nevertheless stations (s) and $(s-1)$ are treated, from the point of view of traffic dynamics (vehicles or passengers) as distinct units.
- The trains are automated and thus they interact followind a law that induces a speed-density (or speed headway) fundamental diagram.
- The passengers in (z) are the passengers of stations (s) and $(s-1)$; the directions (j) in the node are the directions (E–W) and (W–E).
- The dynamics of the system are modeled as described in the previous section. Link dynamics are described by Equations (10–12) and node dynamics by Equations (15–23).
- To obtain a command of the system, the following demands are estimated:
  - Train demands (trains waiting when light is red) upstream of;
  - Passenger demands: these are evaluated in the node and downstream, that is, the passengers waiting in stations $(s-1)$, (s) and $(s+1)$ in direction (W–E), and in stations (s), $(s-1)$ and $(s-2)$ in direction (E–W);

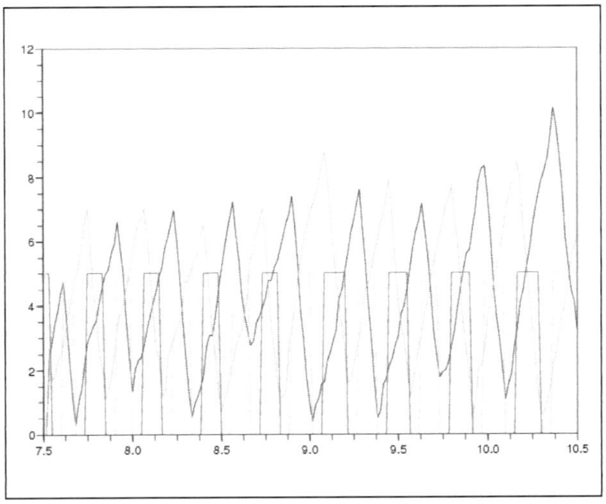

Figure 6. Total demand (blue: E–W, green: W–E) vs. time (horizontal axis), with arbitrary scale; commands (turquoise curve = E–W, red curve = W–E) with scale green light = 5, red light = 0.

- Passenger demand is converted into train demand (the number of trains required to statisfy it given the passenger load of available trains). The two demands are added for each direction.

- The traffic light becomes red on one direction when the demand in the opposite direction becomes higher than the demand on the current direction.

It must be noted that if the light becomes red in the direction say (E–W), the trains in direction (W–E) are not allowed green light until the node (z) is empty. Thus, when the light turns red in the direction (E–W), the train demand in direction (E–W) starts to increase while the passenger demand still decreases (passengers are still serviced, by the trains bound to exit node [z]). All demands on (W–E) increase until the light turns green in that direction, after which there is some delay before the passenger demand starts to decrease.

Figures 6 and 7 illustrate this process. Note that the units in these figures on the vertical axis are arbitrary (as only the relative value of the demands is relevant for system command).

In the considered case, the result of the proposed strategy was to form train platoons of 1–7 trains, resulting in waiting times of 10–15 min immediately downstream of the accident, but the system could still manage the passenger demand in both directions (the mean frequency of trains being about 1 train every 2 minutes).

As discussed above there is some time loss in order to empty the node each time the traffic light changes color. The impact of this time loss is reduced by platooning. The strategy is adaptative.

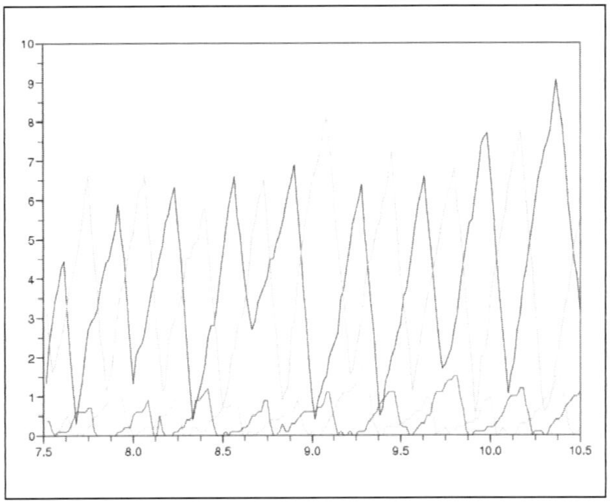

Figure 7. Queues (vertical axis), that is, vehicle queue (blue: E–W, green: W–E), passenger queues (red E–W, turquoise W–E) cumulated inside (z) and downstream of (z), vs. time in hours (horizontal axis). Queues unit is number of trains (passenger queues have been converted to trains via train capacities).

The model allows simulation of the system, evaluation of the strategy and fine-tuning of its parameters.

Conclusion

Multimodal traffic comprises two different flows: vehicular and passenger flow. These flows are not independent: passengers are transported by vehicles. The GSOM approach accounts for this fundamental fact by interpreting the passenger density as an attribute, the passenger load, and by equating the flow of passengers with the flow of vehicles times the passenger load.

The model presented in this paper provides a comprehensive framework for macroscopic multimodal network modeling. The paper has shown how to model vehicular and passenger traffic both on links and inside nodes, and how to carry out a Eulerian discretization of the model in order to make it numerically tractable.

Data requirements for the model depend essentially on the applicational context. At the vehicular level are those collected usually for road networks or public transportation operations. Data for passengers will become more readily available as the use of streaming data generalizes (portables, GPS, ticketing, etc.). It is also expected that these data will yield information pertaining to other attributes.

Current research concerns first of all improvements of the modeling aspects.

- The node model presented in this paper is basic: more realistic node models should take into account movement travel time, movement

capacity, and possible conflicts between movements. The internal state approach (*26*), also documented in the case of the simpler LWR (Lighthill–Whitham–Richards model) as the multi-buffer method (*27*, *28*), provides an efficient method to reach this goal.
- A Lagrangian discretization is more suitable for certain specific problems, in which it is necessary to individualize vehicles, and also for very large networks. In the case some attributes are not neutral, the supply function depends both on upstream and downstream attributes (Equation 3) and the resolution of problem (15) for node modeling becomes untractable. The recourse is then a lagrangian discretization of the node modeled with the internal state approach (*23*, *24*, *29*).

Ongoing applicative research on the model presented in this paper includes

- (A)MOD (mobility on demand) systems (possibly automated) in a multimodal context, an application to be introduced elsewhere (*30*);
- Management of automated guided transport systems;
- Large-scale multimodal simulation and management.

Author Contributions

The authors confirm contribution to the paper as follows: study conception and design: all authors; data collection: all authors; analysis and interpretation of results: all authors; draft manuscript preparation: all authors. All authors reviewed the results and approved the final version of the manuscript.

References

1. Waddell, P. UrbanSim: Modeling Urban Development for Land Use, Transportation, and Environmental Planning. *Journal of the American Planning Association*, Vol. 68, No. 3, 2002, pp. 297–314.
2. Horni, A., K. Nagel, and K. W. Axhausen, eds. *The Multi-agent Transport Simulation MATSim*. Ubiquity Press, London, 2016.
3. Monteiro, Nuno, R. Rossetti, P. Campos, and Z. Kokkinogenis. A Framework for a Multimodal Transportation Network: An Agent-based Model Approach. *Transportation Research Procedia*, Vol. 4, 2014, pp. 213–227.
4. Holmgren, J., P. Davidsson, J. A. Persson, and L. Ramstedt. TAPAS: A Multi-agent-based Model for Simulation of Transport Chains. *Simulation Modelling Practice and Theory*, Vol. 23, 2012, pp. 1–18.
5. Monteiro, N., R. Rossetti, P. Campos, and Z. Kokkinogenis. A Framework for a Multimodal Transportation Network: An Agent-Based Model Approach. *Transportation Research Procedia*, Vol. 4, 2014, pp. 213–227.
6. Ma, T.-Y., and J.-P. Lebacque. A Cross Entropy Based Multiagent Approach for Multiclass Activity Chain Modeling and Simulation. *Transportation Research Part C: Emerging Technologies*, Vol. 28, 2013, pp. 116–129.
7. Xu, J., M. Nourinejad, and J. Y. J . Chow. Network Learning via Multi-agent Inverse Transportation Problems. *arXiv preprint arXiv*: 1609.04117, 2016.
8. Moccia, L., J.-F. Cordeau, G. Laporte, S. Ropke, and M. P. Valentini. Modeling and Solving a Multimodal Transportation Problem with Flexible-time and Scheduled Services. *Networks*, Vol. 57, No. 1, 2011, pp. 53–68.
9. Dib, O., M.-A. Manier, and L. Moalic. Advanced Modeling Approach for Computing Multicriteria Shortest Paths in Multimodal Transportation Networks. *Proc., IEEE International Conference on Intelligent Transportation Engineering (ICITE)*, IEEE, 2016.
10. Yang, X., X. J. Ban, and J. Mitchell. Modeling Multimodal Transportation Network Emergency Evacuation Considering Evacuees' Cooperative Behavior. *Transportation Research Procedia*, Vol. 23, 2017, pp. 1033–1053.
11. Zheng, N., and N. Geroliminis. Modeling and Optimization of Multimodal Urban Networks with Limited Parking and Dynamic Pricing. *Transportation Research Part B: Methodological*, Vol. 83, 2016, pp. 36–58.
12. Zheng, N., G. Rérat, and N. Geroliminis. Time-dependent Area-based Pricing for Multimodal Systems with Heterogeneous Users in an Agent-based Environment. *Transportation Research Part C: Emerging Technologies*, Vol. 62, 2016, pp. 133–148.
13. Loder, A., L. Ambühl, M. Menendez, and K. W. Axhausen. Empirics of Multi-modal Traffic Networks–Using the 3D Macroscopic Fundamental Diagram. *Transportation Research Part C: Emerging Technologies*, Vol. 82, 2017, pp. 88–101.
14. Sossoe, K. *Modélisation des systèmes de transport multimodaux de grands réseaux*, 2017.
15. Lebacque, J.-P., S. Mammar, and H. H. Salem. Generic Second Order Traffic Flow Modeling. In *Transportation and Traffic Flow Theory 2007* (Allsop, R. E., M. G. H. Bell, and B. G. Heydecker eds.), London, 2007, pp. 755–776.
16. Lebacque, J.-P., and M. M. Khoshyaran. A Variational Formulation for Higher Order Macroscopic Traffic Flow Models of the GSOM Family. *Procedia-Social and Behavioral Sciences*, Vol. 80, 2013, pp. 370–394. *Transportation Research Part B: Methodological*, Vol. 57.C, 2013, pp. 245–265.
17. Khoshyaran, M. M., and J.-P. Lebacque. URBASIMCE: A Dynamic Model for Assignment and Cold Start Emission. *Procedia-Social and Behavioral Sciences*, Vol. 20, 2011, pp. 714–722.
18. Mammar, S., J.-P. Lebacque, and H. H. Salem. Riemann Problem Resolution and Godunov Scheme for the Aw-Rascle-Zhang Model. *Transportation Science*, Vol. 43.4 (2009): 531-545.
19. Lebacque, J-P., S. Mammar, and H. H. Salem. Second Order Traffic Flow Modeliing: Supply-demand Analysis of the Inhomogeneous Riemann Problem and of Boundary Conditions. *Proceedings of the Euro Working Group on*

Transportation (2005) and *Archives of Transport* Vol. 20, No. 1-2, 2008, pp. 47–67.
20. Lebacque, J-P., S. Mammar, and H. H. Salem. The Aw–Rascle and Zhang's Model: Vacuum Problems, Existence and Regularity of the Solutions of the Riemann Problem. *Transportation Research Part B: Methodological*, Vol. 41, No. 7, 2007, pp. 710–721.
21. Lebacque, J. P., and M. M. Khoshyaran. First Order Macroscopic Traffic Flow Models for Networks in the Context of Dynamic Assignment. *Transportation Planning*. 2002, pp. 119–140.
22. Lebacque, J.-P., and M. M. Khoshyaran. First-order Macroscopic Traffic Flow Models: Intersection Modeling, Network Modeling. *Transportation and Traffic Theory. Flow, Dynamics and Human Interaction. 16th International Symposium on Transportation and Traffic Theory*, 2005.
23. Khoshyaran, M. M., and J. P. Lebacque. Lagrangian Modelling of Intersections for the GSOM Generic Macroscopic Traffic Flow Model. *Proc., 10th International Conference on Application of Advanced Technologies in Transportation (AATT2008)*, Athens, Greece, 2008.
24. Costeseque, G., J.-P. Lebacque, and A. Khelifi. Lagrangian GSOM Traffic Flow Models on Junctions. *IFAC-PapersOnLine*, Vol. 48, No. 1, 2015, pp. 147–152.
25. Costeseque, G., and J.-P. Lebacque. A Variational Formulation for Higher Order Macroscopic Traffic Flow Models: Numerical Investigation. *Transportation Research Part B: Methodological*, Vol. 70, 2014, pp. 112–133.
26. Khoshyaran, M. M., and J. P. Lebacque. Internal State Models for Intersections in Macroscopic Traffic Flow Models. *Proceedings of Traffic and Granular Flow09*, Springer, Berlin, Heidelberg, 2009.
27. Garavello, M., and B. Piccoli. A Multibuffer Model for LWR Road Networks. *Advances in Dynamic Network Modeling in Complex Transportation Systems*. Springer, New York, 2013, pp. 143–161.
28. Delle, M., M. Laura, J. Reilly, S. Samaranayake, W. Krichene, P. Goatin, and A. M. Bayen. A PDE-ODE Model for a Junction with Ramp Buffer. *SIAM Journal on Applied Mathematics*, Vol. 74, No. 1, 2014, pp. 22–39.
29. Lebacque, J.-P., M. Salim, and H.-S. Habib. An Intersection Model Based on the GSOM Model. *IFAC Proceedings Volumes*, Vol. 41, No. 2, 2008, pp. 7148–7153.
30. Khoshyaran, M. M., and J. P. Lebacque. Extended GSOM Model for Multimodal Transportation Networks. In *Proceedings of Traffic Granular Flow TGF'2017*, Wahington, 2018.

The Standing Committee on Transportation Network Modeling (ADB30) peer-reviewed this paper (18-05152).

Article

TRR
JOURNAL OF THE TRANSPORTATION RESEARCH BOARD

Transportation Research Record
2018, Vol. 2672(48) 104–115

DOI: 10.1177/0361198118793329
journals.sagepub.com/home/trr
SAGE

Algorithm to Compute Urban Road Network Resilience

Bhattiyil Kuzhiyamkunnath Bhavathrathan[1] and Gopal R. Patil[1]

Abstract

Road network resilience is emerging as a vital planning criterion. Yet, unique and cross-comparable indices for road network resilience are scarce. One of the recent approaches determines resilience as a unique network attribute based on the system travel time at an upper envelope of operable disruptions. This upper envelope represents 'critical states' (or tipping points) of capacity disruptions. Critical state gives a bounding capacity degradation vector, beyond which the network cannot wholly cater to the origin–destination demand even under the best possible traffic assignment. However, solving the critical state identification problem (CSP) on real-scale networks has remained a challenge. This paper presents a weighted fictitious play algorithm to fill this gap. CSP has been previously envisaged as a two-player game between a network attacker and a network defender. Here, we make the players play iteratively, and make them learn from the competitor's past strategies so that they converge to an equilibrium. We illustrate the method on a simple toy network, and solve it on different real-life networks. Resilience of the Anaheim city network was computed in 42.8 min., considerably outperforming—both in problem-size and solution-time—the previous, two-space genetic algorithm.

The degradable nature of urban road networks makes resilience a vital design and planning criterion. Though various attempts to define urban transport network resilience have been attempted, a convergence to a definite methodology is yet to happen. Even in this emergent phase, resilience of transportation networks has had different interpretations. For example, resilience has been defined in the past as the weighted sum of the resilience of all nodes (*1*), while the resilience of a node is the weighted sum of the numbers of reliable passageways to all other nodes. The relationship of network resilience to the topological structure of the network has been systematically established (*2*), for example, the edge resilience and the node resilience of networks are in terms of the maximum number of edge or node failures that the network can tolerate (*3*). In yet other measures of network resilience, it is considered as the reroutability of a network under different failure scenarios (*4*), and also based on accessibility variations (*5*). Researchers have also systematically assessed the role of network topology in transportation network resilience (*2*) and have developed methods for optimally planning for pre-event mitigation and preparedness and post-event response (*6*). Transport network resilience has also been treated as a combination of many attributes (*7*). Previous resilience studies have focused on specific areas of transport network, such as freight networks (*8–12*), and also modeled its interactions with other lifelines, such as the power network (*13*).

Since the nature of degradation is uncertain, it is imperative that resilience be quantified as an inherent and unique attribute of the network, irrespective of the likeliness of disruption. Recently, the authors (*14*, *15*) have developed methodologies for such quantification, using which any network with a given origin–destination demand profile can be attributed with a unique resilience value. The authors have also analyzed the sensitivity of resilience to demand variability (*16*), thereby proposing a generalized index of resilience. The methodology first identifies a critical network state that marks the tipping point of disruptions, beyond which a part of the demand goes unmet at a given level of service. Subsequently, the operational cost at this tipping point is employed in the quantification of resilience.

The above mentioned problem of critical state identification (CSP) is formulated as a mathematical program involving minimax optimization. One method by which solution can be attempted is using a two-space genetic

[1]Transportation Systems Engineering, Department of Civil Engineering, Indian Institute of Technology Bombay, Powai, Mumbai, India

Corresponding Author:
Address correspondence to Gopal R. Patil: gpatil@iitb.ac.in

algorithm (2SGA). Herrmann (*17*) developed 2SGA using the general approach of coevolution, which is apt for minimax problems (*18*). This algorithm has been used to solve many minimax optimisation problems in engineering, including parallel machine scheduling problem (*17*), demand stochastic transportation network design (*19*), and damping controller design (*20*). Even though instances of the problem have been hitherto solved using 2SGA, they only involved test networks with limited size (*14–16*), and also proved to be computationally costly. As an alternative, this paper presents a weighted fictitious play algorithm to solve the CSP.

This paper is organised as follows. First, a brief reproduction of the methodology to quantify resilience is made. Here, the CSP is presented and the relation of a network's resilience to its critical state is established. Next, the basic fictitious play algorithm, and a modified version of it, known as the weighted fictitious play is described. Convergence of the two-space genetic algorithm as compared with the weighted fictitious play algorithm, and illustrations of solubility on real-life networks are then presented. Subsequently, the computational time required for the algorithm is discussed, and compared with state-of-the-art complexity.

Deducing Resilience from the Critical Network State

Network resilience can be defined as "the ability of the network to provide and maintain an acceptable level of service in the face of various faults and challenges to normal operation" (*21*). Resilience can thus be quantified by assessing the magnitude of 'faults' that a network can survive before collapsing from a state of 'normal operation'. This means that understanding the resilience of a system necessitates a unique definition of bounds on different states of operation. In the case of road networks, this is intricate, as here, 'faults and challenges to normal operation' causes queues and incremented travel times. This causes complexity in differentiating better and deteriorated levels of operation, at a network level. Under certain conventional assumptions, the authors had reported the existence of bounds on normal operability levels, and had related it to quantify resilience (*14–16*).

For a given demand, let STT_{Cr} indicate the unique operable bound on system travel time (STT), beyond which the network cannot meet the demand facilitating a defined level of service. If any of the links is disrupted a bit further from the levels at this critical state, the system would get displaced from one state (demand-meeting/higher-level-of-service) to another (demand-not-meeting/deteriorated-level-of-service). Hence, STT_{Cr} is the maximum STT that the system can suffer, before moving from a normal state to an abnormal state. It can thus be used to measure the resilience of the network. Network resilience can be interpreted as the difference between the best operational cost and the critical operational cost, as both these values are unique at a given demand for a network. It can further be normalized relative to the critical operational cost to limit the value between zero and one to impart comparability. The index for network resilience ρ is formally written as:

$$\rho = \frac{STT_{Cr} - STT_0}{STT_{Cr}} \tag{1}$$

Note that the best operational cost denoted as STT_0, is the system optimal travel time. It can be determined by a minimisation problem (*22*).

A network that takes in greater magnitudes of disruption and yet caters to the traffic demand without suffering a drop in level of service will have a higher value in the numerator of Equation 1. As the numerator increases, ρ tends to one. Conversely, if the demand and the network parameters are such that the network cannot take in any disruption at all without a drop in level of service, then STT_{Cr} will be equal to STT_0 and the value of ρ remains zero. A value of ρ close to zero indicates that the difference in STT_{Cr} and STT_0 is small, and the network cannot endure much disruption if it has to satisfy the demand. Such a network is less resilient. On the other hand, a value of ρ close to 1 indicates that the network can withstand significant disruption before becoming inoperable. A representation of the resilience contour is presented in Figure 1, where the variation of resilience is contoured on a STT_{Cr}–STT_0 plot. It should be noted that the resilience depends on the current demand between OD pairs, since both STT_{Cr} and STT_0 are functions of current demand. We direct interested readers to a deeper discussion into this measure of resilience, its sensitivity to demand, a resulting generalized indicator, its comparison to some previous measures etc., which are available in the literature (*14–16*).

As can be seen from the above discussion, the pivoting part of resilience quantification is the identification of STT_{Cr}. This CSP is presented in the next section.

The Critical State Identification Problem

Roadways are susceptible to a variety of disruptions such as traffic incidents, snowing, flooding, road space reallocations, road space infiltrations, infrastructural failures, etc. Disregarding the physical cause, they may be viewed as simultaneous decrement on links' capacities. The objective here is to establish a consequent critical state, in which the network is at a worst possible demand-meeting state. Such a state serves as a benchmark of the network, like the system-optimal state.

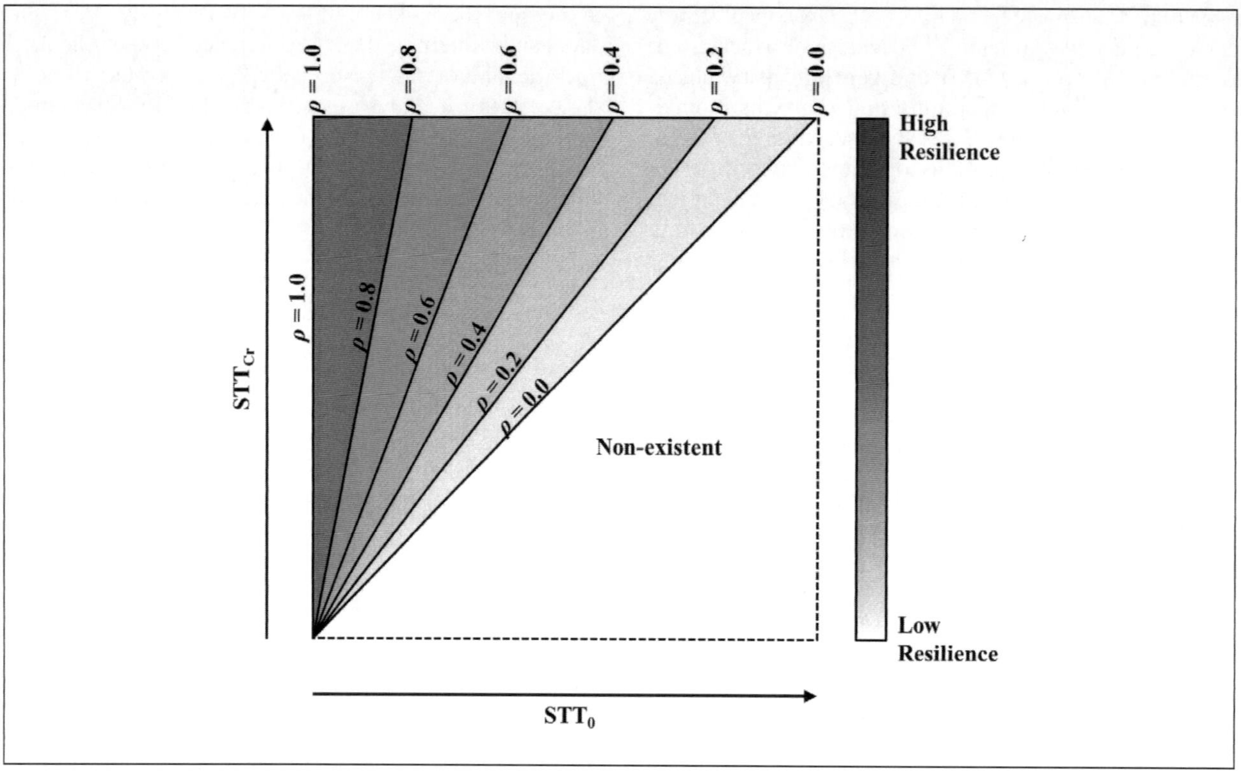

Figure 1. Resilience contour.

We briefly present a general version of the CSP (*14*) here for ease of reference. On a traffic network represented as a digraph $\mathcal{G}$, where $\mathcal{N}$ is the set of nodes and $\mathcal{A}$ is the set of links, $\mathcal{R}$ denotes the set of origins and $\mathcal{S}$ denotes the set of destinations, $\mathcal{D}$ is the degraded capacities, and $\mathcal{F}$ is the flows on each link of the set $\mathcal{A}$.

The critical system travel time ($\mathrm{STT_{Cr}}$) is defined as the maximum possible STT at any demand-meeting degraded state, and which cannot be minimized even by the best possible re-assignment. It can be found by solving the following program:

$$\mathrm{STT_{Cr}} = \min_{x_{ij}} \max_{y_{ij}} \sum_{ij} x_{ij}.c_{ij}(x_{ij}, y_{ij}) \tag{2}$$

Subject to:

$$\sum_{ij \in A_i^+} f_{ij}^s - \sum_{ji \in A_i^-} f_{ji}^s = q^{is}, \quad \forall i,\ i \neq s \tag{3}$$

$$\sum_s f_{ij}^s = x_{ij}, \quad \forall ij \tag{4}$$

$$0 \leq x_{ij} \leq (k_{ij}/y_{ij}), \quad \forall ij \tag{5}$$

$$1 \leq y_{ij} \leq m_{ij}, \quad \forall ij \tag{6}$$

where

q^{rs} is the demand between the origin r and destination s ($r \in \mathcal{R}, s \in \mathcal{S}$);

ij is a directed link between node i and j;

x_{ij} is the flow on link ij;

f_{ij}^s is the flow specific to destination s on link ij;

k_{ij} is the capacity on link ij before degradation;

y_{ij} represents the degradation such that $1/y_{ij}$ is the available proportion of capacity after degradation;

$\mathcal{A}_i^+$ and $\mathcal{A}_i^-$ denote the set of outgoing links and the set of incoming links, respectively, at node i;

m_{ij} is the upper bound on divisive factor that the analyst inputs; the link cost function such that: m_{ij} is the upper limit on y_{ij} such that $1/m_{ij}$ is a guaranteed minimum proportion of capacity after degradation. In other words, m_{ij} is a finitely large value such that $(100(m_{ij} - 1)/m_{ij})\%$ is the maximum allowable capacity degradation on any link ij.

For a network to remain operable after disruption, it must be ensured that no link shall be required to carry a flow which is more than its leftover capacity after disruption. This is confirmed using the constraint in Equation 5, where the right-hand-side (RHS) is the leftover capacity and x_{ij} is the flow that the link ij must carry in order to cater to the demand. $c_{ij}(x_{ij}, y_{ij})$ is the cost function as given in Equation 7:

$$c_{ij}(x_{ij}, y_{ij}) = T_{ij} + \beta \left(\frac{x_{ij}}{\left(\frac{k_{ij}}{y_{ij}}\right)} \right)^{\alpha} \quad (7)$$

The 2SGA has previously been found as a feasible choice to solve the critical state problem for small case studies (*14*, *15*). However, when the network size increases, the number of constraints increases, and as a result the 2SGA becomes inefficient due to the issues associated with constraint handling. Also, even though the genetic algorithm does not get trapped in local optimums, it also does not guarantee convergence to global optimum.

Fictitious Play Algorithm

Basic Fictitious Play

Introduced by Brown (*23*) and Robinson (*24*), fictitious play (FP) is an effective method to learn Nash equilibria in different types of games (*25–27*), but most importantly for two-player zero-sum games (*24*). The procedure involves repeated play of a game, thereby approximating its equilibrium value as required. FP algorithm is proven to follow best response trajectories of players (*28*), before eventually nullifying the mistakes of discretisation (*29*) and converging to equilibrium.

FP algorithm has been reported effective for large games, and has even been opined to be the only option for very large problems (*30*). The applicability of FP algorithm in solving problems related to road networks is demonstrated by Garcia et al. (*31*), where it is employed to find dynamic system optimal routing in large scale networks. In the implementation, authors report considerable decrease in computational expense without trading off solution accuracy.

An overview of the preliminary concepts on FP can be found in Fudenberg and Levine (*32*). Basic FP process involves a series of repeated plays, where a player responds in the best possible manner to an assumption on the opponent's mixed strategy profile. The general idea is that the players play the best response to all previous rounds, and not just the last round. It is proven that FP will necessarily converge for the type of game being discussed here, i.e. finite two-player zero-sum games (*24*).

In spite of being favourable theoretically, the practical significance of basic fictitious play is limited, for large scale implementation. One of the main hindrances is the computational complexity involved in calculating the mixed utility for finding the best response (*33*). An improved form of the basic FP called the weighted fictitious play (WFP), which is also shown to be computationally efficient for networks of bigger sizes, is presented in the next section.

Weighted Fictitious Play

In a WFP, players' beliefs are updated by depreciating or nullifying the strategies that the opposite player played before a few rounds. In effect, the players weigh the opponent's recent actions more (*32*), as opposed to the basic FP where all previous actions are equally weighed. In other words, an instance of the WFP with equal placements of weights on each of the previous steps throughout the history of the play, will result in the basic FP. In this sense, the WFP may be called a generalised version of the basic FP. The WFP is interpreted as a process in which the players have limited memory, and forget the opponent's actions beyond a certain number of plays. This interpretation intuitively explains how the WFP relaxes the issues pertaining to memory and computational complexity while dealing with large games. The players only need to keep track of the opponent's latest M actions, and can discard those before M steps. As the convergence to equilibrium would require a large number of iterations, the option of discarding historic information eases the computational requirements.

As discussed, the friend assigns flow to each of the links of the network, and has x_{ij} as the strategy for each link ij, such that $x_{ij} \in [0, K_{ij}]$, where K_{ij} is the maximum physically possible capacity on link ij. This constitutes the strategy space X of the friend and similarly, the foe too has an associated strategy space Y. At an instance, let $\mathbb{X}$ represent the strategy played by friend, and let $\mathbb{Y}$ represent that by foe. The resultant value of the game at this instance will be STT ($\mathbb{X}$, $\mathbb{Y}$), which is nothing but the system travel time, given a set of flows and disruptions. Now suppose that the friend observes the past moves and makes a learned guess that the foe will play a strategy $\tilde{\mathbb{Y}} \in Y$. Then, in the subsequent move, the friend will play the best response $\tilde{\mathbb{X}} \in X$ to this $\tilde{\mathbb{Y}}$. This action of finding the best response is denoted as $BR_X(\tilde{\mathbb{Y}})$. The pseudocode of the WFP algorithm is presented in Algorithm 1. lb and ub represent the lower and upper bounds respectively, of the variables.

Computing the Best Response

It can be seen from Algorithm 1 that computing the best response is its centre-stone. In typical finite games, computing best response is just a standard linear program (*34*), as in such games, there exists a cost-vector c for each player such that $c \times \mathbb{X}$ explicitly defines the payoff associated with a strategy x. However, the cost function of the friend–foe game is continuous in nature and the responses are bound by a set of constraints as described in Equations 2–6. Hence, computing the best response is not as straightforward as in typical matrix-form games.

In a routing perspective, the best response at any instant is a flow pattern that minimizes the cost. Since

Algorithm 1 WFP Algorithm

procedure WFP	
$\mathbb{X}^{cntrd} \leftarrow$ any strategy in X	
$\mathbb{Y}^{cntrd} \leftarrow$ any strategy in Y	
lb $\leftarrow -\infty$	ub $\leftarrow \infty$
$t \leftarrow 0$	
while Δ(ub − lb) $> \epsilon$ **do**	
$t \leftarrow t + 1$	
$\mathbb{X}_t^{srch} \leftarrow BR_X(\mathbb{Y}^{cntrd})$	$\mathbb{Y}_t^{srch} \leftarrow BR_Y(\mathbb{X}^{cntrd})$
$v_{\mathbb{X}} = STT(\mathbb{X}^{cntrd}, \mathbb{Y}^{srch})$	$v_{\mathbb{Y}} = STT(\mathbb{X}_t^{srch}, \mathbb{Y}^{cntrd})$
lb $\leftarrow$ max(lb, $v_{\mathbb{Y}}$)	ub $\leftarrow$ min(ub, $v_{\mathbb{X}}$)
if $t<M$ **then**	
$\mathbb{X}^{cntrd} \leftarrow \frac{1}{t}\sum_{i=1}^{t} \mathbb{X}_i^{srch}$	$\mathbb{Y}^{cntrd} \leftarrow \frac{1}{t}\sum_{i=1}^{t} \mathbb{Y}_i^{srch}$
else	
$\mathbb{X}^{cntrd} \leftarrow \frac{1}{M}\sum_{i=t-(M+1)}^{t} \mathbb{X}_i^{srch}$	$\mathbb{Y}^{cntrd} \leftarrow \frac{1}{M}\sum_{i=t-(M+1)}^{t} \mathbb{Y}_i^{srch}$
end	

the cost is modelled as STT, it follows that the best response is the system-optimal flow pattern. Thus, a traditional method—the Frank-Wolfe (FW) algorithm—is employed to obtain the system-optimal solution. It is known that the solution of any user-equilibrium program with marginal link travel times would produce a system-optimal flow pattern (*22*). Marginal travel time is the marginal contribution to link travel time from an additional infinitesimal flow. As a result, at an equilibrium of marginal travel times, no link would gain a reduction in travel time by reassigning some of its flows to another link. This property is harnessed in obtaining the system-optimal flow pattern using the process.

FW algorithm was originally developed to solve quadratic optimisation problems (*35*). It is a feasible direction method, which identifies a descent direction, and chooses a step size that would minimise the objective function (*22*). Determining the step size requires evaluating the objective function along the descent direction. In a traditional user equilibrium solution employing the FW algorithm, the initial feasible flows are determined using the all-or-nothing assignment. Then, the direction of descent is obtained by solving a linearised version of the objective function. For the user-equilibrium formulation, when the objective function is linearised based on Taylor series expansion, the resulting problem becomes the shortest path problem (*36*). Hence, in the traditional approach, all-or-nothing assignment is performed to find the auxiliary flows. Subsequently the step size is determined, which is used to find the convex combination of the initial feasible flow and the auxiliary flow. The steps of the FW algorithm are popular (*22*), and are not reproduced here.

As observed in the literature (*37–39*), step-size computation in the FW algorithm can become costly for large networks. With the advent in computing power, this is no longer a concern for conventional planning requirements. However, note that in most of the cases only one instance of equilibrium is sought, as opposed to the multiple instances required in the WFP algorithm. In this case, it would thus be interesting to involve methods that in some sense bypass step-size computation. Such methods exist, in which the step size is predetermined, without resorting to its exact computation in each iteration. These methods include approximating the linear sub-problem by optionally performing the line search (*40*), by depending on an approximate line search (*41*), employing simple averaging, or even using a constant step size (*42*). For instance, with any sequence of step sizes obeying $\lim_{k\to\infty} t^{(k)} = 0$ and $\sum_{k=0}^{\infty} t^{(k)} = \infty$, gradient descent with steps of sizes $t^{(k)}$ will converge to the optimum (*37*). An example is the popular $t^{(k)} = 2/(k+2)$ step-size sequence. The basic idea behind bypassing the line search is that it becomes undesirable to allocate significant resources to determining a value of α_n to precisely minimize the objective function, as the resources could instead be channelized to identify a better search direction. Then, the aim is just to find an α_n that offers a reasonable improvement in the objective function, rather than determining the actual minimizing value of α_n. In this study, we determine step sizes by truncating the operations of line search before an optimal α_n is found. This would yield step sizes that confirm to the regularity conditions. The steps of the modified FW algorithm are given below:

1. *Initialization*. Find a feasible initial solution $\mathbb{X}'$ by performing all or nothing based on $\tilde{t}_a = \tilde{t}_a(0)$ which yields x_a^1. $n \leftarrow 1$.
2. *Update*. Set $\tilde{t}_a^n = \tilde{t}_a(x_a^n) \forall a$.
3. *Direction finding*. Find the shortest paths for all OD pairs based on $\tilde{t}_a^n = \tilde{t}_a(x_a^n) \forall a$. This yields auxiliary flows $\tilde{x}_a^n$.
4. *Step size*. Find α_n that loosely solves:

$$\min_{0 \le \alpha \le 1} \sum_a \int_0^{x_a^n + \alpha(\tilde{x}_a^n - x_a^n)} t_a(\omega) d\omega \tag{8}$$

5. *Move*. $x_a^{n+1} \leftarrow x_a^n + \alpha_n(\tilde{x}^n - x_a^n), \forall a$.
6. *Convergence test*. Stop, if a convergence criterion (similarity of two successive solutions or the reduction of the objective function values between successive iterations) is met. $x_a^{n+1} \forall a$ is the solution. Else, $n \leftarrow n + 1$ and go to Step 2.

An Illustration

Consider the simple example shown in Figure 2. Assume a demand of five units from node 1 to node 2, and that the free travel times are 1.5 units and 1.4 units respectively on Links 1 and 2.

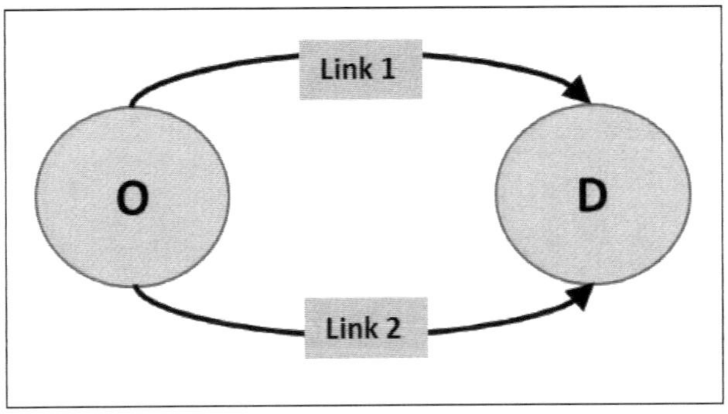

Figure 2. Simple network to illustrate FP process.

In the FP process, the zero-sum game (*14*) is simulated as being repeatedly played by the two competing players: a friend and a foe of the network. The foe is an imaginary player, though it is possible that malevolent agencies or even nature can don the role of a network's foe. The friend too is a virtual entity, though recognizable as a traffic manager/planner who beholds a bird's eye view of the network state. The foe tries to maximise the STT by deciding how much to disrupt the link capacities. The friend tries to minimise the STT by deciding the flow on the links. The critical state is a rarely realized demand-meeting worst case, where neither the foe, nor the friend can better their objectives by unilaterally changing their strategies, because this point will have the maximum possible irreducible STT. Whereas the friend cannot modify the supply side, the foe is constrained to keep the degradations within the demand-meeting regime. Note that these players are just instruments constituted with the sole purpose of identifying an abstract critical state, and hence, they need not be realistic players with realistic goals.

If both links have equal capacity of 10 units, then, in the first step, the friend assigns the flow so that the total network cost is minimized. Note that this best response from the friend would result in all flow allocated to the shorter link. In this simple case, a nonlinear programming method known as generalized reduced gradient algorithm is used to compute the best response. However, a standardized method like the FW algorithm would be more appealing,as it would ease the augmentation of this methodology to popular transportation planning packages. In the next step, the foe plays to maximize the cost, thereby degrading the capacity of the shorter link. Subsequently, the friend observes the past action of the foe, and plays by switching some part of the flow, such that the cost is minimized. This process continues, and as can be seen from Figure 3 attains the critical cost.

Computational Instances

Three road networks: (*a*) Sioux Falls, South Dakota, USA, (*b*) Anaheim, California, USA, and (*c*) Fort Area, Mumbai, India are chosen. Before presenting the real-life examples, we use a small test network to present the efficiency of the WFP algorithm by comparing its performance with the 2SGA method.

Comparison of Convergence: WFP and 2SGA

To compare the convergence of the WFP and the two-space genetic algorithm, a small test network as shown in Figure 4 is used. A demand of six units is assumed from node 1 to node 6, and equal capacity of 12 units is assumed on all the five links. By comparing the results from the 2SGA and the WFP, existence of equal-payoff multiple equilibria, i.e. multiple solutions with same value of objective function is ascertained. Note in Table 1 that both the solution algorithms converge to the same (in fact, comparable) value of STT, even though the corresponding critical states are different. 2SGA is run twice with two different initial points. Since genetic algorithm exhibits asymptotic convergence, the instance that is run longer converges further to the solution obtained using WFP. This sensitivity is unusual in such small toy problems, and indicates that increase in network size may render 2SGA inefficient, likely due the nonlinear constraints in Equation 5. Also, the two 2SGA instances indicate that the solution is sensitive to the initial points supplied.

On Windows 8 Pro 64-bit OS with dual core Intel Core i5-650 CPU, 3.20 GHz and 3.8 GB RAM., while the two-space genetic algorithm took around 1.3 CPU seconds to solve this example, the WFP algorithm did so in under a second. As reported by Garcia et al. (*31*), the benefit in computation time by FP does not trade off the accuracy, as the same value of objective function repeats in both the algorithms. Note that, in spite of different values on the flow and disruption, the value of the objective function (STT) is same in all the equilibria.

Sioux Falls Network

An aggregated road network of the city of Sioux Falls, South Dakota, USA is taken up in this example. A map of the city and the network topology are presented in Figure 5. The network includes the roads of the primary, secondary, and tertiary importance, inside the area cordoned by Interstate Highways 229, 90, and 29. The network has 76 links, 24 nodes and 552 OD pairs. The network topology and link attributes are adopted from LeBlanc et al. (*43*). As the network is popular in the literature, we do not reproduce its details here, and they are available online (http://www.bgu.ac.il/bargera/tntp/).

The convergence of the WFP algorithm on the Sioux Falls network is shown in Figure 6. The changes in utility increment for either player are plotted against the

Link	Flow (Friend's Decision)	Capacity	Degradation (Foe's Decision)	Link Cost	
Step 1, Friend plays the best response					
1	0.00	10	1.00	0.00	min
2	5.00	10	1.00	7.31	
			Cost	7.31	
Step 2, Foe plays the best response					
1	0.00	10	1.00	0.00	max
2	5.00	10	2.00	12.00	
			Cost	12.00	
Step 3, Friend					
1	2.96	10	1.00	4.46	min
2	2.04	10	2.00	2.91	
			Cost	7.38	
Step 4, Foe					
1	2.96	10	3.38	7.40	max
2	2.04	10	4.90	4.89	
			Cost	12.30	
Step 5, Friend					
1	2.96	10	3.38	7.40	min
2	2.04	10	4.90	4.89	
			Cost (Critical)	12.30	

Figure 3. FP process on the two-linked network.

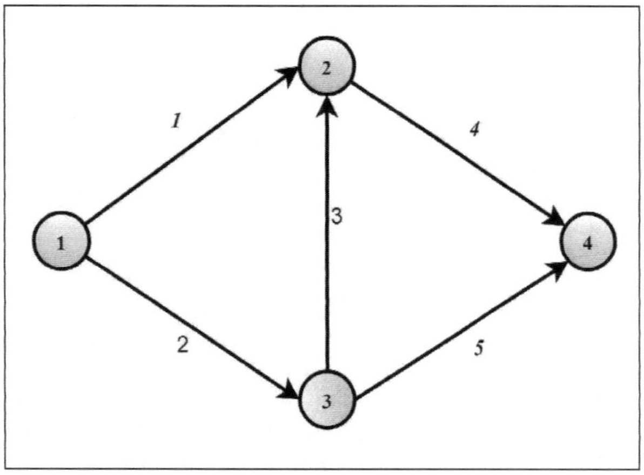

Figure 4. Test network to compare the WFP and the 2SGA.

iteration number. Note that the utility for friend is always negative as the STT is incremental throughout the process. They converge to zero at around 350 iterations when the accuracy of the step size is low. While the WFP algorithm solved this case with 552 OD pairs in under 4 CPU seconds, the 2SGA ran overnight to solve the same network with only two OD pairs, thus demonstrating the efficiency of the WFP algorithm.

Anaheim Network

This network is the road network of the city of Anaheim, outside Los Angeles, in Southern California, USA. The map of the city, and its topology (as presented by Oh and Jayakrishnan (*44*)) is given in Figure 7. The network, cordoned by California State Routes 55, 22, 91, and

Table 1. Comparison of Solutions from the WFP and the 2SGA

		Values of decision variables					
		At initial point		At equilibrium			
Method	Link	Flow (x_{ij})	Divisive factor (y_{ij})	Flow (x_{ij})	Divisive factor (y_{ij})	STT at equilibrium (z)	CPU Time
WFP	1	0.00	1.00	3.00	4.00	61.8	0.57 s
	2	0.00	1.00	3.00	4.00		
	3	0.00	1.00	0.00	1.00		
	4	0.00	1.00	3.00	4.00		
	5	0.00	1.00	3.00	4.00		
SGA	1	0.00	4.75	1.66	7.25	61.8	6.92 s
	2	6.00	2.00	4.34	2.76		
	3	0.00	9.82	0.00	9.82		
	4	0.00	5.63	1.66	7.25		
	5	6.00	2.00	4.34	2.76		
SGA	1	3.00	4.00	3.00	4.00	61.9	6.84 s
	2	3.00	4.00	3.00	4.00		
	3	0.00	4.00	1.00	12.0		
	4	3.00	4.00	4.00	3.00		
	5	3.00	4.00	2.00	6.00		

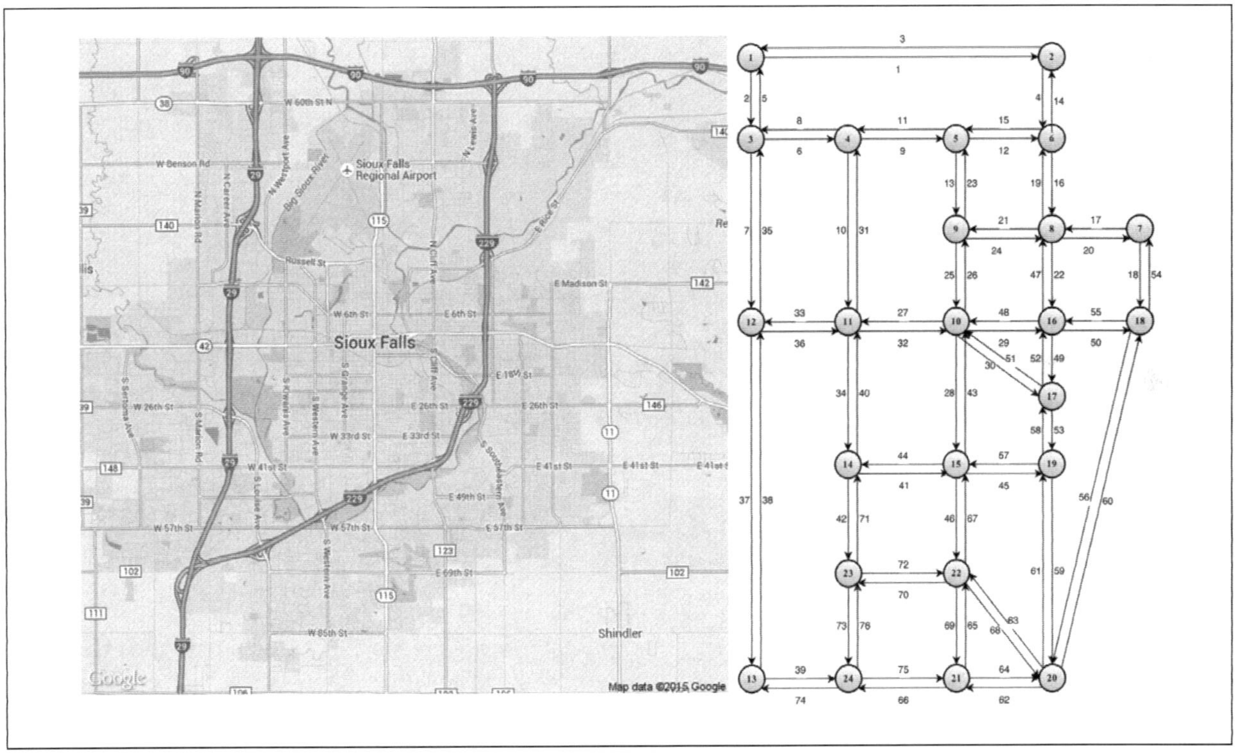

Figure 5. The map of the city of Sioux Falls, South Dakota, USA, and the topology of its road network.

Secondary Highway 39, has 416 nodes, 914 links and 1406 OD pairs. The link attributes for the network are obtained from http://www.bgu.ac.il/bargera/tntp/.

Figure 8 shows—with respect to the secondary (red) y axis—the fall in utility of both the players during the WFP process. With respect to the primary (black) y axis, the evolution of the STT value along the process is traced. The STT increases and converges simultaneous to the convergence of the players' utility increment. This value of the STT that both the players settle into is the critical-state-system-travel-time (STTCr). Therefore the y-intercept gives the value $\rho \times$ STTCr.

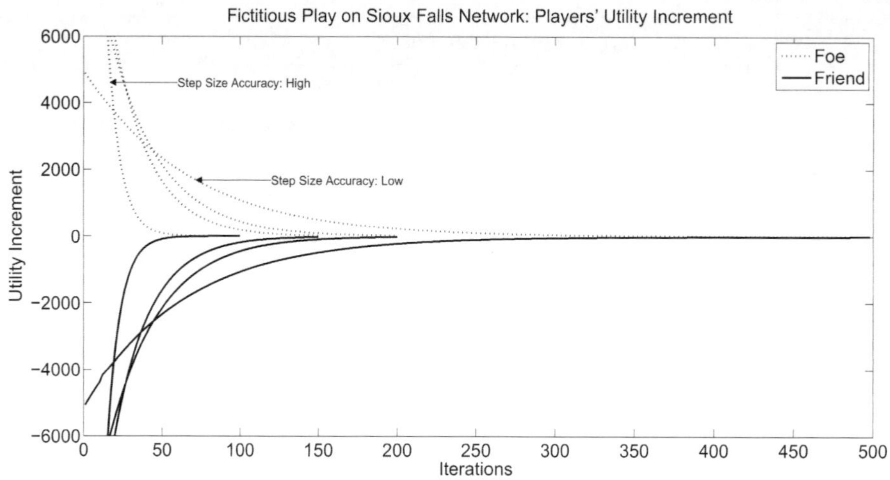

Figure 6. WFP instances with different step sizes on Sioux Falls network.

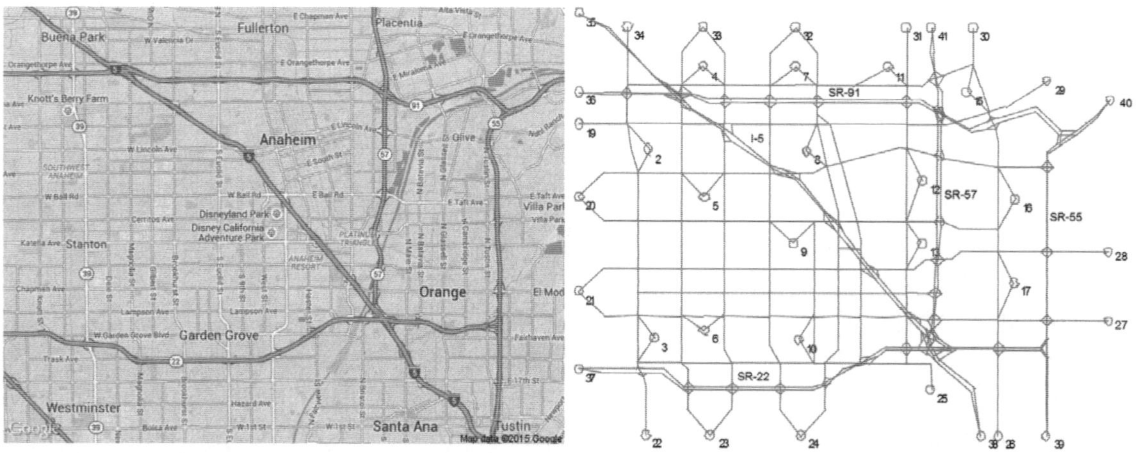

Figure 7. The map of the city of Anaheim, California, USA, and the topology of its road network.

Mumbai Fort Area Network

The Fort Area network in Mumbai is selected for this illustration. The map of the fort area is shown in Figure 9 and the network topology as presented by Sharma et al. (*45*) is also reproduced there. The network has 45 links, 17 nodes and 72 OD pairs.

Discussion of CPU Time

As mentioned, the WFP algorithm is executed under the following environment specifications: Windows 8 Pro 64-bit OS; Intel Core i5-650 CPU, 2 Cores, 3.20 GHz; 3.8 GB RAM. The CPU time required by the algorithm is shown in Table 2. The state-of-the-art CPU time required for one run of the user equilibrium traffic assignment using FW algorithm are 0.76 sec for the Sioux Falls network (*36*) and about 12 sec for the Anaheim network (*46*). Note that the times reported in Table 2 are considerably more than those required for one run of traffic assignment. However, considering the fact that the WFP algorithm requires multiple runs of system-optimal traffic assignment, the CPU times are promising.

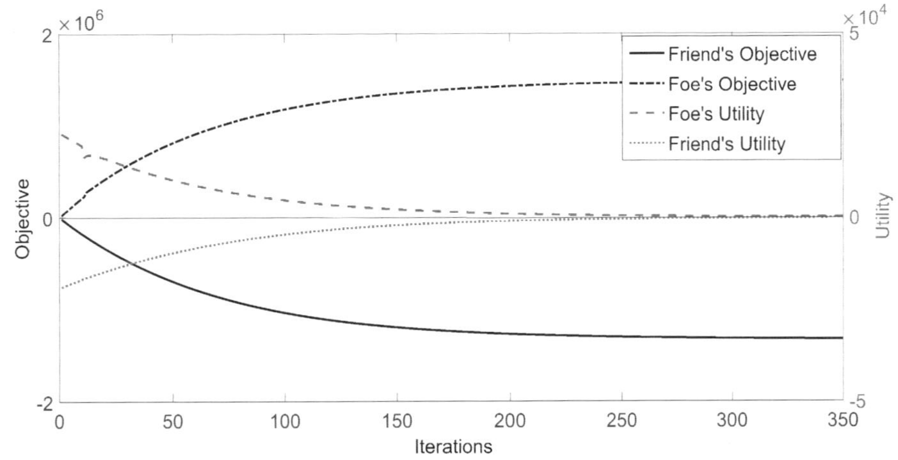

Figure 8. Performance of the FP algorithm on Anaheim network.

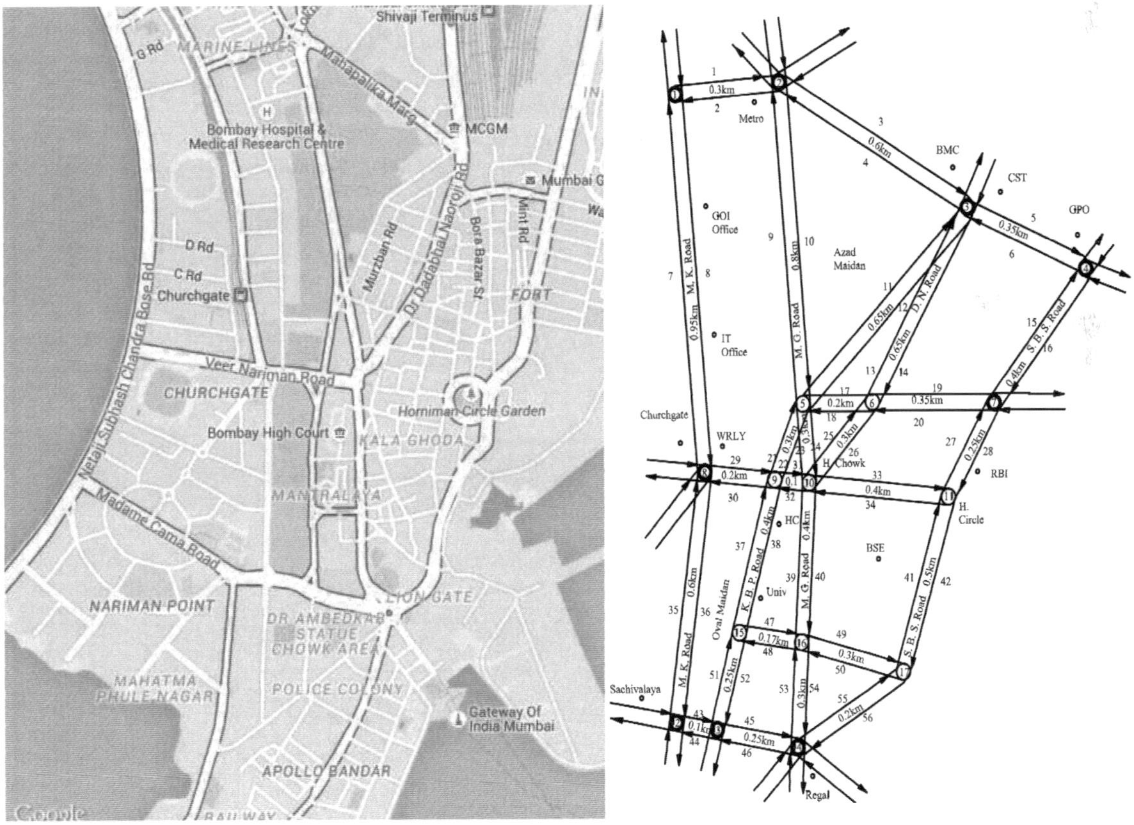

Figure 9. Map and road network topology of Fort Area, Mumbai, India.

Table 2. Convergence of the WFP Algorithm on Real-Life Networks

Network	Links	Nodes	OD pairs	Plays	CPU time	Resilience
Mumbai Fort Area	45	17	72	200	2.11 s	0.5
Sioux Falls	76	24	552	200	3.43 s	0.5
Anaheim	914	416	1406	350	486.23 s	0.7

Conclusion

Urban road networks are prone to disruptions due to a multitude of factors. This nature has effected the emergence of network resilience as a vital planning criterion. Recently, the solution of the CSP to identify a unique and cross-comparable measure has been proposed. However, the applicability of the method hitherto remained crippled in the absence of a specific algorithm that solves the CSP on bigger networks. This paper addresses this issue by presenting the WFP algorithm. Hence, the primary focus of this paper is to extend the practical applicabilities of the CSP to bigger networks.

The WFP harnesses the underlying game-nature of the CSP formulation by simulating repeated plays until equilibrium is attained. Even though the FW procedure is adopted for finding best response in this study, note that it can very well be substituted by other methods.

The results obtained here show that the method has good potential in being an augmentation to the existing planning packages. Such augmentation should be seamless, as the FW process—which is a corner stone of this algorithm—is already built into popular planning packages. Planners can thus have an understanding of the magnitude of shock the system can safely be subjected to, and thereby leverage the planning process.

Acknowledgments

The network dataset for Sioux Falls and Anaheim city are downloaded from Transportation Networks, a networks repository for transportation research (https://github.com/bstabler/TransportationNetworks). The network dataset for Mumbai Fort Area is obtained from the PhD thesis repository of Department of Civil Engineering, IIT Bombay, from the thesis titled *Transportation Network Design Considering Environmental Parameters and Demand Uncertainty* by Sushant Sharma. We express our thanks to these repositories and the respective contributors.

Author Contributions

The authors confirm contribution to the paper as follows: study conception and design: Bhavathrathan, Patil; data collection: Bhavathrathan; analysis and interpretation of results: Bhavathrathan, Patil; draft manuscript preparation: Bhavathrathan. All authors reviewed the results and approved the final version of the manuscript.

References

1. Ip, W., and Q. Wang. Resilience and Friability of Transportation Networks: Evaluation, Analysis and Optimization. *Systems Journal, IEEE*, Vol. 5, 2011, pp. 189–198.
2. Zhang, X., E. Miller-Hooks, and K. Denny. Assessing the Role of Network Topology in Transportation Network Resilience. *Journal of Transport Geography*, Vol. 46, 2015, pp. 35–45.
3. Rosenkrantz, D. J., S. Goel, S. S. Ravi, and J. Gangolly. *Structure-based Resilience Metrics for Service-Oriented Networks. In M. Dal Cin, M. Kaâniche, and A. Pataricza (eds.), Dependable Computing - EDCC 5.*, Springer, Berlin, Heidelberg, 2005.
4. Matisziw, T., and A. T. Murray. Modeling s–t Path Availability to Support Disaster Vulnerability Assessment of Network Infrastructure. *Computers and Operations Research*, Vol. 36, 2009, pp. 16–26.
5. Taylor, M. A., and S. Susilawati. Remoteness and Accessibility in the Vulnerability Analysis of Regional Road Networks. *Transportation Research Part A: Policy and Practice*, Vol. 46, 2012, pp. 761–771.
6. Faturechi, R., and E. Miller-Hooks. Travel Time Resilience of Roadway Networks under Disaster. *Transportation Research Part B: Methodological*, Vol. 70, 2014, pp. 47–64.
7. Murray-Tuite, P. A. *Comparison of Transportation Network Resilience under Simulated System Optimum and User Equilibrium Conditions.* Proc., 2006 Winter Simulation Conference, Monterey, CA, IEEE, 2006.
8. Zhang, D., D. Payam, and D. Ekwall. How Robustness and Resilience Support Security Business Against Antagonistic Threats in Transport Network. *Journal of Transportation Security*, Vol. 4, 2011, pp. 201–219.
9. Miller-Hooks, E., X. Zhang, and R. Faturechi. Measuring and Maximizing Resilience of Freight Transportation Networks. *Journal of Computers and Operations Research*, Vol. 39, 2012, pp. 1633–1643.
10. Lichun, C., and E. Miller-Hooks. Resilience: An Indicator of Recovery Capability in Intermodal Freight Transport. *Transportation Science*, Vol. 46, 2012, pp. 109–123.
11. Chen, C.-C. Modeling Resilience Enhancement Strategies for International Express Logistics. *Transportation Research Record: Journal of the Transportation Research Board*, 2013. 2378: 92–98.
12. Nair, R., H. Avetisyan, and E. Miller-Hooks. Resilience Framework for Ports and Other Intermodal Components. *Transportation Research Record: Journal of the Transportation Research Board*, 2010. 2166: 54–65.
13. Wilson, G., T. M. Wilson, N. I. Deligne, D. M. Blake, and J. W. Cole. Framework for Developing Volcanic Fragility

and Vulnerability Functions for Critical Infrastructure. *Journal of Applied Volcanology*, Vol. 6, No. 1, 2017, p. 14.
14. Bhavathrathan, B. K., and G. R. Patil. Capacity Uncertainty on Urban Road Networks: A Critical State and Its Applicability in Resilience Quantification. *Computers, Environment and Urban Systems*, Vol. 54, 2015, pp. 108–118.
15. Bhavathrathan, B. K., and G. R. Patil. Quantifying Resilience using a Unique Critical Cost on Road Networks Subject to Recurring Capacity Disruptions. *Transportmetrica A: Transport Science*, Vol. 11, No. 9, 2015, pp. 836–855.
16. Patil, G. R., and B. Bhavathrathan. Effect of Traffic Demand Variation on Road Network Resilience. *Advances in Complex Systems*, Vol. 19, 2016, p. 1650003.
17. Herrmann, J. W. A Genetic Algorithm for Minimax Optimization Problems. Proc., 1999 Congress on Evolutionary Computation-CEC99 (Cat. No. 99TH8406), Washington, D.C., 1999, pp. 1099–1103.
18. Jensen, M. T. A New Look at Solving Minimax Problems with Coevolutionary Genetic Algorithms. *Metaheuristics: Computer Decision-Making*, Vol. 86, 2004, pp. 369–384.
19. Herrmann, J. W. *A Genetic Algorithm for a Minimax Network Design Problem*. Institute for Systems Research, University of Maryland, College Park, MD, 1999.
20. Jiang, Q. Y., Y. J. Cao, and S. J. Cheng. A Genetic Approach to Design a HVDC Supplementary Subsynchronous Damping Controller. IEEE Transactions on Power Delivery, Vol. 20, No. 2, 2005, pp. 1059–1064.
21. Sterbenz, J. P., D. Hutchison, E. K. Çetinkaya, A. Jabbar, J. P. Rohrer, M. Schöller, and P. Smith. Resilience and Survivability in Communication Networks: Strategies, Principles, and Survey of Disciplines. *Computer Networks*, Vol. 54, 2010, pp. 1245–1265.
22. Sheffi, Y. *Urban Transportation Networks: Equilibrium Analysis with Mathematical Programming Models*. Prentice-Hall Inc., Englewood Cliffs, NJ, 1985.
23. Brown, G. W. Iterative Solution of Games by Fictitious Play. *Activity Analysis of Production and Allocation*, Vol. 13, No. 1, 1951, pp. 374–376.
24. Robinson, J. An Iterative Method of Solving a Game. *Annals of Mathematics*, Vol. 54, No. 2, 1951, pp. 296–301.
25. Berger, U. Fictitious Play in 2 × n Games. *Journal of Economic Theory*, Vol. 120, No. 2, 2005, pp. 139–154.
26. Sela, A., and D. Herreiner. Fictitious Play in Coordination Games. *International Journal of Game Theory*, Vol. 28, No. 2, 1999, pp. 189–197.
27. Milgrom, P., and J. Roberts. Rationalizability, Learning, and Equilibrium in Games with Strategic Complementarities. *Econometrica: Journal of the Econometric Society*, Vol. 58, No. 6, 1990, pp. 1255–1277.
28. Gilboa, I., and A. Matsui. Social Stability and Equilibrium. *Econometrica: Journal of the Econometric Society*, Vol. 59, No. 3, 1991, pp. 859–867.
29. Leslie, D. S., and E. J. Collins. Generalised Weakened Fictitious Play. *Games and Economic Behavior*, Vol. 56, No. 2, 2006, pp. 285–298.
30. Ferguson, T. S. Two-Person Zero-Sum Games. In *Game Theory* (2nd ed.), University of California, Los Angeles, CA, 2014.
31. Garcia, A., D. Reaume, and R. L. Smith. Fictitious Play for Finding System Optimal Routings in Dynamic Traffic Networks. *Transportation Research Part B: Methodological*, Vol. 34, No. 2, 2000, pp. 147–156.
32. Fudenberg, D., and D. K. Levine. *The Theory of Learning in Games*, Vol. 2, MIT Press, 1998.
33. Swenson, B., S. Kar, and J. Xavier. Empirical Centroid Fictitious Play: An Approach for Distributed Learning in Multi-Agent Games. *IEEE Transactions on Signal Processing*, Vol. 63, No. 15, 2015, pp. 3888 -3901.
34. McMahan, H. B., and G. J. Gordon. A Fast Bundle-Based Anytime Algorithm for Poker and Other Convex Games, *International Conference on Artificial Intelligence and Statistics*, Vol. 2, 2007, pp. 323–330.
35. Frank, M., and P. Wolfe. An Algorithm for Quadratic Programming. *Naval Research Logistics Quarterly*, Vol. 3, No. 1–2, 1956, pp. 95–110.
36. Perederieieva, O., M. Ehrgott, A. Raith, and J. Y. Wang. A Framework for and Empirical Study of Algorithms for Traffic Assignment. *Computers & Operations Research*, Vol. 54, 2015, pp. 90–107.
37. Powell, W. B., and Y. Sheffi. The Convergence of Equilibrium Algorithms with Predetermined Step Sizes. *Transportation Science*, Vol. 16, No. 1, 1982, pp. 45–55.
38. Fisk, C. Some Developments in Equilibrium Traffic Assignment. *Transportation Research Part B: Methodological*, Vol. 14, No. 3, 1980, pp. 243–255.
39. Sheffi, Y., and W. Powell. A Comparison of Stochastic and Deterministic Traffic Assignment over Congested Networks. *Transportation Research Part B: Methodological*, Vol. 15, No. 1, 1981, pp. 53–64.
40. Jaggi, M. Revisiting Frank-Wolfe: Projection-Free Sparse Convex Optimization. *Proc., 30th International Conference on Machine Learning (ICML-13)*, 2013, pp. 427–435.
41. Bach, F. Duality between Subgradient and Conditional Gradient Methods. *SIAM Journal on Optimization*, Vol. 25, No. 1, 2015, pp. 115–129.
42. Freund, R. M., and P. Grigas. New Analysis and Results for the Frank–Wolfe Method. *Mathematical Programming*, Vol. 155, No. 1–2, 2016, pp. 199–230.
43. LeBlanc, L. J., E. K. Morlok, and W. P. Pierskalla. An Efficient Approach to Solving the Road Network Equilibrium Traffic Assignment Problem. *Transportation Research*, Vol. 9, No. 5, 1975, pp. 309–318.
44. Oh, J.-S., and R. Jayakrishnan. Emergence of Private Advanced Traveler Information System Providers and Their Effect on Traffic Network Performance. *Transportation Research Record: Journal of the Transportation Research Board*, 2002. 1783: 167–177.
45. Sharma, S., S. Ukkusuri, and T. Mathew. Pareto Optimal Multiobjective Optimization for Robust Transportation Network Design Problem. *Transportation Research Record: Journal of the Transportation Research Board*, 2009. 2090: 95–104.
46. Mitradjieva, M., and P. O. Lindberg. The Stiff Is Moving-Conjugate Direction Frank-Wolfe Methods with Applications to Traffic Assignment. *Transportation Science*, Vol. 47, No. 2, 2013, pp. 280–293.

The Standing Committee on Transportation Network Modeling (ADB30) peer-reviewed this paper (18-06526).

Article

TRR
JOURNAL OF THE TRANSPORTATION RESEARCH BOARD

Transportation Research Record
2018, Vol. 2672(48) 116–126

DOI: 10.1177/0361198118799039
journals.sagepub.com/home/trr
SAGE

Network Partitioning Algorithms for Solving the Traffic Assignment Problem using a Decomposition Approach

Cesar N. Yahia[1], Venktesh Pandey[1], and Stephen D. Boyles[1]

Abstract

Recent methods in the literature to parallelize the traffic assignment problem consider partitioning a network into subnetworks to reduce the computation time. In this article, a partitioning method is sought that generates subnetworks minimizing the computation time of a decomposition approach for solving the traffic assignment (DSTAP). The aim is to minimize the number of boundary nodes, the interflow between subnetworks, and the computation time when the traffic assignment problem is solved in parallel on the subnetworks. Two different methods for partitioning are tested. The first is an agglomerative clustering heuristic that reduces the subnetwork boundary nodes. The second is a flow weighted spectral partitioning algorithm that uses the normalized graph Laplacian to partition the network. The performance of both algorithms is assessed on different test networks. The results indicate that the agglomerative heuristic generates subnetworks with a lower number of boundary nodes, which reduces the per iteration computation time of DSTAP. However, the partitions generated may be heavily imbalanced leading to a higher computation time when the subnetworks are solved in parallel separately at a particular DSTAP iteration. For the Austin network partitioned into four subnetworks, the agglomerative heuristic requires 3.5 times more computation time to solve the subnetworks in parallel. The results also show that the spectral partitioning method is superior for minimizing the interflow between subnetworks. This leads to a faster convergence rate of the DSTAP algorithm.

The traffic assignment problem (TAP) is used to predict route choice and link flows for a given travel demand. The static version of this problem can be formulated as a convex program and solved efficiently using modern specialized algorithms (*1–3*). However, there are computationally demanding problems that require solving TAP multiple times or solving TAP on a large network. Those problems include bi-level mathematical programs with equilibrium constraints, solving TAP on statewide or national network models, and Monte Carlo simulations (*4, 5*).

Methods for parallelizing the TAP to decrease computation time have been studied recently (*4, 6, 7*). The decomposition approach to the static TAP (DSTAP) was developed to decrease computation time by solving the TAP in parallel on partitions of the full network (*4*). This approach creates subproblems for each partition and a master problem that equilibrates traffic across subnetworks. The master problem also includes regional traffic that has an origin or a destination outside a certain subnetwork or in two different subnetworks. To find equilibrium in this master–subproblem framework, the DSTAP algorithm exploits the equilibrium sensitivity analysis method developed in Boyles (*8*) to generate artificial links that represent paths between network nodes. DSTAP is shown to converge to the global equilibrium solution for a general network and its computation time is stated to depend on the subnetwork partitions (*4*).

The objective of this study is to identify and test partitioning algorithms that can improve the performance of a DSTAP. It seeks to generate partitions that minimize the number of boundary nodes and the interflow between subnetworks. These requirements minimize the interactions between subnetworks, which influences the time needed to converge to a global equilibrium in a framework such as DSTAP. In addition, the study seeks partitions that minimize the computation time needed to solve the TAP in parallel for the subnetworks. This refers to the per iteration lower level subproblems in DSTAP. With this motivation and objective in place, two

[1]Department of Civil, Architectural and Environmental Engineering, The University of Texas at Austin, TX

Corresponding Author:
Address correspondence to Cesar N. Yahia: cesaryahia@utexas.edu

algorithms are tested in the analysis. The first algorithm is proposed by Johnson et al. (*9*) for objectives similar to those required in this paper. The second algorithm is based on flow weighted spectral partitioning. The performance of the algorithms on real-world networks is compared against the stated objectives.

The remainder of this paper describes methods to parallelize TAP and evaluates the performance of the partitioning algorithms. The next section reviews current methods for solving TAP and partitioning networks. The third section presents the algorithms evaluated and their use in the DSTAP framework. The fourth section presents demonstrations for different transportation networks. The fifth section concludes the paper.

Literature Review

This section summarizes existing literature in the following areas: the latest advancements in methods for solving the TAP, a parallelization approach to the TAP, and the need for efficient network partitioning algorithms.

Algorithms for solving TAP are generally classified as either link-based, path-based, or bush-based. Link-based methods work in the space of link flows and require less operational memory than path-based methods, but are much slower to converge (*10*, *11*). Bush-based methods exploit the acyclic nature of paths that are used by origins at equilibrium (*2*, *3*, *12*, *13*). Recent work also includes ϵ-optimal improved methods for solving TAP on large problems (*14*). Although recent advancements have improved the state-of-the-art for solving TAP efficiently, there is still a need for faster methods. Computationally demanding instances include solving TAP on large-scale statewide models and solving TAP iteratively in network design problems with equilibrium constraints (*5*, *15*).

To address the computational demands of large-scale or iterative traffic assignment problems, methods that aim to parallelize TAP have been developed. Bar-Gera describes a parallelization approach based on the paired-alternative segments (*6*). The algorithms proposed in Chen and Meyer (*16*) and Lotito (*17*) also parallelize TAP by decentralizing the computations for each origin–destination (OD) pair. A decomposition approach for static traffic assignment (DSTAP) developed by Jafari et al. (*4*) parallelizes TAP by network geography instead of the traditional decomposition approach by OD pairs. This article aims to identify partitioning algorithms that minimize the computation time per iteration of DSTAP and the total computation time required to reach convergence. Proofs of convergence and correctness of DSTAP are provided in Jafari et al. (*4*).

The literature on network partitioning algorithms is extensive. These algorithms can be broadly classified into agglomerative/divisive heuristics, integer programming based approaches, and spectral partitioning algorithms. Integer programming formulations for the partitioning problem are proven to be NP-hard (*18*) and approximation heuristics have been proposed (*18*, *19*).

Heuristics for generating partitions based on agglomerative and divisive clustering have been recently used in various transportation related applications. Saeedmanesh and Geroliminis used an agglomerative clustering heuristic for generating partitions based on "snake" similarities for applications of the macroscopic fundamental diagram (*20*). Etemadnia et al. developed similar heuristics for distributed traffic management (*18*). Johnson et al. developed another heuristic for decentralized traffic management (*9*). This heuristic aims to minimize boundary nodes in subnetworks and to create subnetworks of similar size. Their heuristic performed better than the METIS algorithm proposed in (*21*).

Spectral partitioning is an alternative approach for partitioning a graph (*22*–*26*). Bell applied a capacity-weighted form of the spectral partitioning methods to investigate network vulnerability (*27*). Other transportation applications include air traffic control and urban traffic signal control systems (*28*, *29*). The partitioning mechanism is based on the eigenvalues associated with the graph Laplacian. The partitions that result from spectral partitioning have low intercluster similarity (*23*). Additionally, using the normalized Laplacian generates graphs that are balanced by weight. This is an important feature because ignoring the balance requirement results in cuts that isolate a small number of peripheral nodes. For example, the minimum cut program that aims to minimize the weight between resulting partitions will often result in separating one node from the rest of the network (*25*). However, incorporating balance requirements causes cut problems to become NP-hard. Spectral partitioning is an approximate method for obtaining a cut with minimal cut cost while satisfying balance requirements (*22*, *25*, *26*).

Network Partitioning for Decentralized Traffic Assignment

Consider a directed network G defined by a set of nodes N and set of edges A. Let M be the node–node adjacency matrix for the network. M is an $|N| \times |N|$ matrix, with elements m_{ij} equal to 1 if there is a link connecting node i to j and zero otherwise. The weighted adjacency matrix M_G^D is also defined with elements $m_{(i,j)}^{G,D}$ equal to $w_{(i,j)}$ if $(i,j) \in A$ and zero otherwise, where $w_{(i,j)}$ is the weight assigned to link $(i,j) \in A$. In this article, it is assumed that $w_{(i,j)}$ is the flow on link (i,j). To construct a graph Laplacian, the method uses an undirected version of M_G^D, denoted by M_G, defined as the sum of M_G^D and its transpose. The elements of M_G are $m_{(i,j)}^G$. The graph diagonal matrix D_G is defined as a diagonal matrix with principal diagonal elements in row i as the sum of elements in row i of M_G: $d_{ii} = \sum m^G_{(i,j)}$. The graph Laplacian is defined as $L_G = D_G - M_G$.

Decomposition Approach to the Static TAP

This study aims to partition a large-scale network into subnetworks such that an algorithm based on the DSTAP is solved efficiently. To properly define the objectives of the partitioning algorithms, it is necessary to review the main elements of the DSTAP algorithm developed by Jafari et al. (*4*).

DSTAP is an iterative aggregation–disaggregation algorithm consisting of two levels, a master problem and a set of lower level subproblems corresponding to the respective subnetworks. A subproblem corresponds to solving the TAP for a specific subnetwork. The master problem is used to model interactions between the subproblems. In the master problem, the subnetworks are aggregated using first order approximation methods based on equilibrium sensitivity analysis (*8*, *30*). This results in artificial links representing the subnetworks in the master level problem. The algorithm proceeds by solving the subproblems in parallel, aggregating the subnetworks using artificial links, shifting flow towards equilibrium in the simplified master level network, obtaining subnetwork boundary flow from the master level iteration, and then proceeding to disaggregate the flow on subnetworks and solving the subproblems in parallel again. This procedure is repeated until convergence to a global equilibrium as shown in Figure 1.

The computational performance of DSTAP at each iteration depends on the number of artificial links. These links need to be updated at each iteration using equilibrium sensitivity analysis to incorporate the latest information on travel costs. To reduce the number of artificial links generated, the number of boundary nodes associated with the subnetworks needs to be minimized. In addition to the regional artificial links that approximate subnetworks at the master level, there are subnetwork artificial links generated for each subproblem to represent flow that originates from a subnetwork then traverses other subnetworks before returning to the subnetwork. To reduce subnetwork artificial links, the flow traversing multiple subnetworks needs to be minimized.

The computational performance at each iteration is also influenced by the time needed to solve the TAP in parallel for the subnetworks. This represents solving the K lower level subproblems in Figure 1. The computation time needed to solve the subproblems in parallel is dominated by the subproblem that requires the greatest computational cost. Therefore, to reduce this computation time, the subproblems need to be balanced in size. This can be achieved by balancing the flow distribution across subnetworks, as opposed to having few subnetworks containing the majority of travel demand.

Consider the maximum excess cost termination criteria defined as the greatest difference between the longest used path and the shortest path for each OD pair. It was shown in Jafari et al. (*4*) that the maximum excess cost for the full network ϵ_{OD} is bounded by the total number of boundary points across subnetworks $\tilde{B}$ multiplied by the sum of the maximum excess cost for the master level regional network ϵ^r_{OD} and the maximum excess cost for all subnetworks ϵ^s_{OD} as shown in Equation 1. Therefore, to reach convergence faster, the rate at which the bound in Equation 1 tightens must be increased. The subproblem maximum excess cost ϵ^s_{OD} can be reduced by solving the subproblems to a low gap level. After approximating the subnetworks with artificial links, the master level maximum excess cost ϵ^r_{OD} could be obtained. It is noted that if the interflow between subnetworks is minimized, then the artificial links representing the subnetworks will have a similar cost structure across successive iterations because the influence of external flows on subnetwork equilibrium is reduced. Therefore, the least cost path in the master level regional network would be relatively invariant across iterations. This implies that ϵ^r_{OD} could be reduced at a higher rate. In the extreme case in which the master level least cost path is completely dominated by constant costs on artificial links, the maximum excess cost could be reduced to zero by placing all the regional flow on the path with the least cost artificial links. Thus, faster convergence could be reached by minimizing the interflow between subnetworks. Convergence rate can also be increased by minimizing the number of boundary nodes $\tilde{B}$ as shown in:

$$\epsilon_{OD} \leqslant 2\tilde{B}\left(\epsilon^r_{OD} + \epsilon^s_{OD}\right) \quad (1)$$

Partitioning Algorithms

The study tests the performance of two algorithms that aim to partition the network such that the computation time for a decomposition approach to solve traffic assignment is minimized.

Domain Decomposition Algorithm. The first heuristic algorithm tested is the shortest domain decomposition algorithm (SDDA) proposed in Johnson et al. (*9*). This algorithm works in an agglomerative fashion and constructs a given number of partitions such that the number of boundary nodes between the subnetworks is minimized (primary objective) and the partitions are balanced in size (secondary objective). SDDA only depends on the topological properties of the graph. This feature is desirable when limited information is available

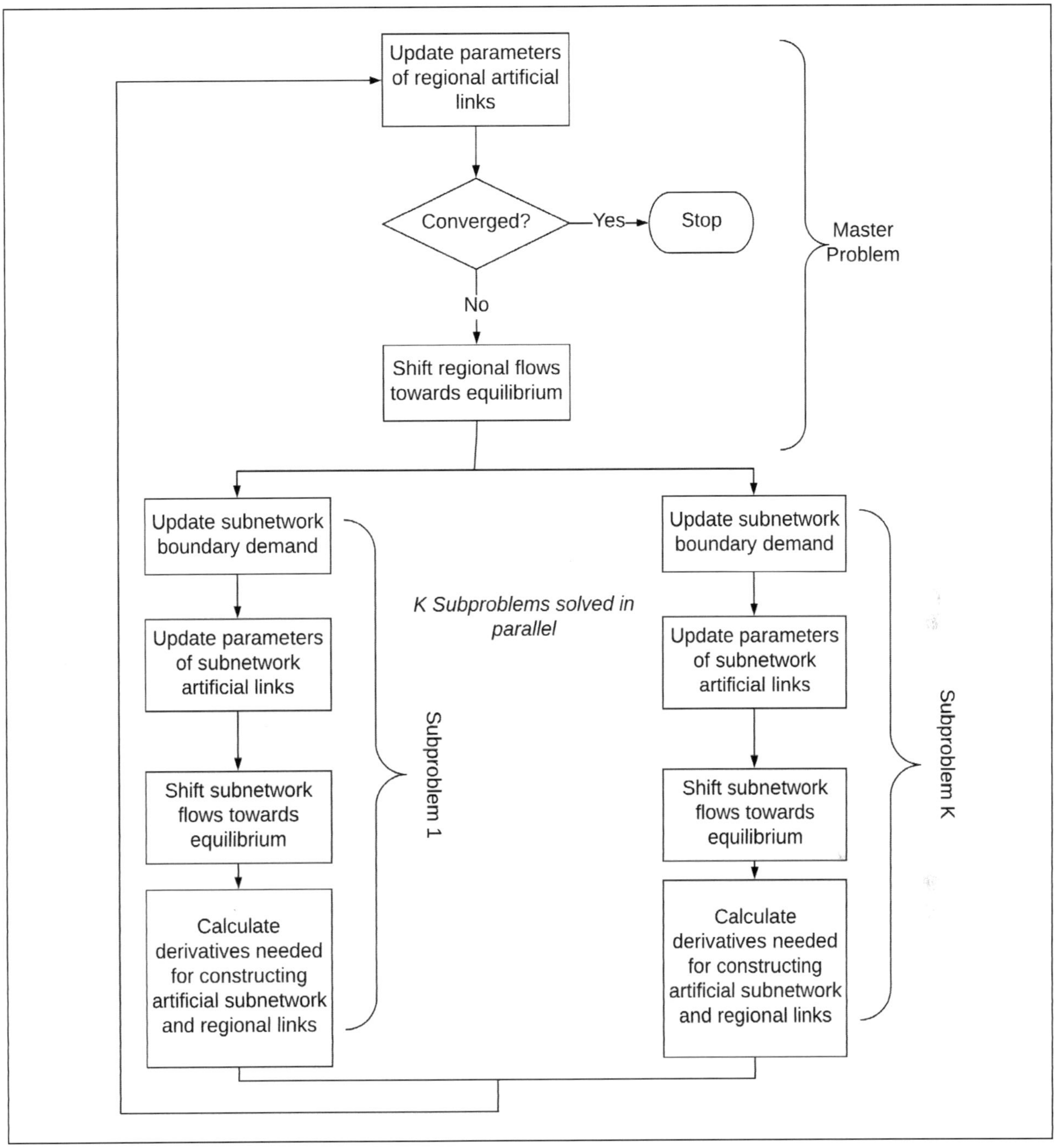

Figure 1. Algorithm for the DSTAP (*4*).

on link costs, flow between OD pairs, or other data that could form the basis of a partitioning algorithm. The computation time per DSTAP iteration is reduced by minimizing the boundary nodes and generating balanced subnetworks. Minimizing the number of boundary nodes would also improve the convergence rate.

The sequential steps of SDDA are shown in Algorithm 1. The algorithm constructs the partitions by identifying source nodes which are "far" from each other given a distance measure. The number of links on a breadth-first search tree between two nodes is used as the distance measure. This distance measure indicates the

Algorithm 1 Shortest domain decomposition algorithm (9)

Step 1: Initialize
Let n_s be the number of subnetworks/partitions to be generated.
Set $R_s^n := MAX$
Step 2: Determine first source node
Set the rank of each node as the sum of the number of incoming and outgoing links.
Choose the node with lowest rank s_1 as the first source node.
Step 3: Update the rank and determine other source nodes
For i in $2 : n_s$
  Perform breadth-first search from every source node, $s_j \quad \forall 1 \leq j < i$
  Determine the rank of node n as an $(i-1)$-dimensional vector of which the elements are the distance of node n from source nodes s_j where $1 \leq j < i$
  Choose the node which has the highest total rank (sum of all elements in the rank vector). Resolve ties in favor of nodes which have minimum value of the sum of pair-wise difference between each element of the rank vector.
  Assign the chosen node as the ith source node s_i
Step 4: Populate subdomain associated with each source node
For each node, assign it to the source node to which it has the minimum distance.
Step 5: Identify system boundary nodes and allocate the subnetworks
For $(i, j) \in A$ do
  if i and j are assigned to different source nodes then
    Add i and j to the set of boundary nodes.
Stop.

extent of separation of two nodes and is used to determine association of a node to the source nodes of the partitions. The reader is referred to Johnson et al. (*9*) for more information on this algorithm.

Spectral Partitioning. Spectral graph theory is used to study network properties using the graph Laplacian. The eigenvalues and eigenvectors of the Laplacian matrix can be used to identify low cost graph cuts. The cost of a cut is defined as a ratio of the weights on cut links to the size of the smaller subnetwork separated by the cut (*23, 25*).

The eigenvalues of an undirected graph Laplacian are real because the matrix is symmetric (*22*). Let φ represent the eigenvectors and λ the eigenvalues. The relation between the eigenvectors and eigenvalues for the graph Laplacian is shown in Equation 2. According to Spielman (*22*), the eigenvalues can be defined using Equation 3, where S is a vector space of dimension i, and i is the index of eigenvalue λ_i arranged in an ascending order. The eigenvector for the corresponding eigenvalue can be found using Equation 4:

$$L_G \varphi_i = \lambda_i \varphi_i \tag{2}$$

$$\lambda_i = \min_{S\, of\, \dim\, i} \max_{x \in S} \frac{x^T L_G x}{x^T x} \tag{3}$$

$$\varphi_i = \operatorname*{argmin}_{S\, of\, \dim\, i} \max_{x \in S} \frac{x^T L_G x}{x^T x} \tag{4}$$

The Laplacian matrix L_G is also positive definite and thus the eigenvalues are non-negative. The second smallest eigenvalue and associated eigenvector obtained from Equations 3 and 4 can be used to partition the graph. The resulting partition is an approximation of the cut that minimizes the ratio cut in Equation 5, where $\text{cut}(A, \bar{A})$ is the sum of the weights on the links separating the subnetworks A and $\bar{A}$ that are generated from the cut. The denominator of the ratio cut is the size of the smaller subnetwork A, where the size is determined by the number of nodes in A. Minimizing the ratio cut aims to find a cut with minimal weights on the links separating the subnetworks, and to maintain a balance in size of the generated subnetworks (*25*):

$$\text{ratio cut} = \frac{\text{cut}(A, \bar{A})}{|A|} \tag{5}$$

To improve the efficiency of DSTAP, a flow weighted version of the Laplacian is used such that the cut cost in Equation 5 represents the interflow between subnetworks. This will improve the convergence rate of DSTAP. The Laplacian matrix is also normalized using Equation 6 in a manner similar to (*22, 24–26*). This normalization will generate partitions that are balanced by the total flow within the partitions instead of the number of nodes in Equation 5. In the DSTAP framework, balancing the partitions by flow would reduce the per iteration computation time needed to solve the subproblems in parallel:

$$L_{\text{symm}} = D_G^{-1/2} L_G D_G^{-1/2} \tag{6}$$

After calculating the second smallest eigenvalue and associated eigenvector of the normalized Laplacian, the nodes of the network are sorted based on the magnitude of the corresponding element in the eigenvector. The sorted list of nodes is then divided into two parts based on the signs of the corresponding eigenvector elements. This will generate the required partitions (*26*, *27*). The full algorithm for the flow weighted spectral partitioning is shown in Algorithm 6.

Because the spectral partitioning method proposed is based on link flows, a few implementation issues need to be considered. The use of the second smallest eigenvalue as the basis for partitioning requires the graph to be connected. Specifically, the weighted adjacency matrix M_G should result in a connected graph. Otherwise, the second smallest eigenvalue will be zero. To ensure that the component being partitioned is connected, a preprocessing stage precedes the spectral analysis. In this stage, the links with zero flow are identified. If those links separate the network into components such that each component has positive intraflow, the spectral partitioning is performed for each component separately. However, in transportation networks, it is more likely to observe multiple components where only one component has flow. In the study's analysis, this occurred because of the existence of peripheral links that do not have any flow but are included in the network geometry. In this case, those links are ignored because they are not used, and should not influence the partitioning of the main component.

Another consideration is the availability of flow values for the links in the network. In the case in which the TAP should be solved multiple times, solving the full network once to obtain link flows is worthwhile because the flows could be used to partition the network in subsequent iterations. If the flow values on the links change each time TAP is solved, the partitions could be updated iteratively. Alternatively, an approximate link flow solution could be obtained by solving centralized traffic assignment to a high gap value.

Demonstrations

The performance of algorithms is compared on a hypothetical network consisting of two copies of the Sioux Falls network and on three standard test networks: Anaheim, Austin, and Chicago sketch (*31*). Considering the previous discussion on the required computation time in the DSTAP section, the analysis is divided into a section on the computation time per iteration and another section on the DSTAP convergence rate. It is noted that computation time needed for partitioning is insignificant for both SDDA and flow weighted spectral partitioning (less than 1 s on a 3.3 GHz machine with 8 GB RAM), and is thus not included in the analysis.

Computation Time per DSTAP Iteration

As mentioned in the section on the decomposition approach for static traffic assignment, the computation time per iteration of DSTAP is dominated by the number of artificial links created and the time required to solve the subproblems in parallel.

The number of regional artificial links created is determined by the number of boundary nodes in the subnetworks. Therefore, the number of boundary nodes generated by each algorithm is compared. Note that the primary objective of the SDDA algorithm is to reduce the number of boundary nodes between the subnetworks.

The computation time required to solve the subproblems in parallel could be reduced by balancing the size of the subproblems. The flow weighted spectral partitioning method aims to minimize the flow balanced cut cost. If the cut cost is always equal to 1, the flow weighted spectral partitioning method will divide the flow equally among the subnetworks. This reduces the computation time needed to solve the subproblems in parallel. The SDDA algorithm creates subnetworks that are balanced by number of nodes as a secondary objective.

Table 1 shows the results for the number of subnetwork boundary nodes generated by the algorithms and the computation time needed to solve the subproblems in parallel. Unless mentioned otherwise, the process is to generate two subnetworks from each network. For minimizing the number of boundary nodes SDDA performed better than the flow weighted spectral partitioning method for the Austin and Anaheim networks. This result is expected because the objective of the flow weighted spectral partitioning method is to minimize the

Algorithm 2 Flow weighted spectral partitioning

Step 1: Pre-process the network to remove links with zero flow
If removing zero flow links creates multiple components with positive flow, then partition each component separately.
Step 2: Calculate the flow weighted graph Laplacian
Step 3: Normalize the graph Laplacian using Equation 6
Step 4: Get the eigenvector to be used for partitioning
Step 5: Order the nodes of the graph based on the eigenvector
Step 6: Partition the network by dividing the ordered node list based on the sign of the corresponding eigenvector elements

Table 1. Comparison of Network Partitioning Algorithms

Network	Boundary nodes	Subnet computation time (s)	Interflow
Austin (SDDA)	174	632.83	186161
Austin (spectral)	329	746.81	137940
Austin (4 subnets, SDDA)	296	290.97	368718
Austin (4 subnets, spectral)	440	82.50	296870
Anaheim (SDDA)	46	0.10	81991
Anaheim (spectral)	48	0.13	56539
Chicago sketch (SDDA)	74	9.79	154791
Chicago sketch (spectral)	50	7.42	201603

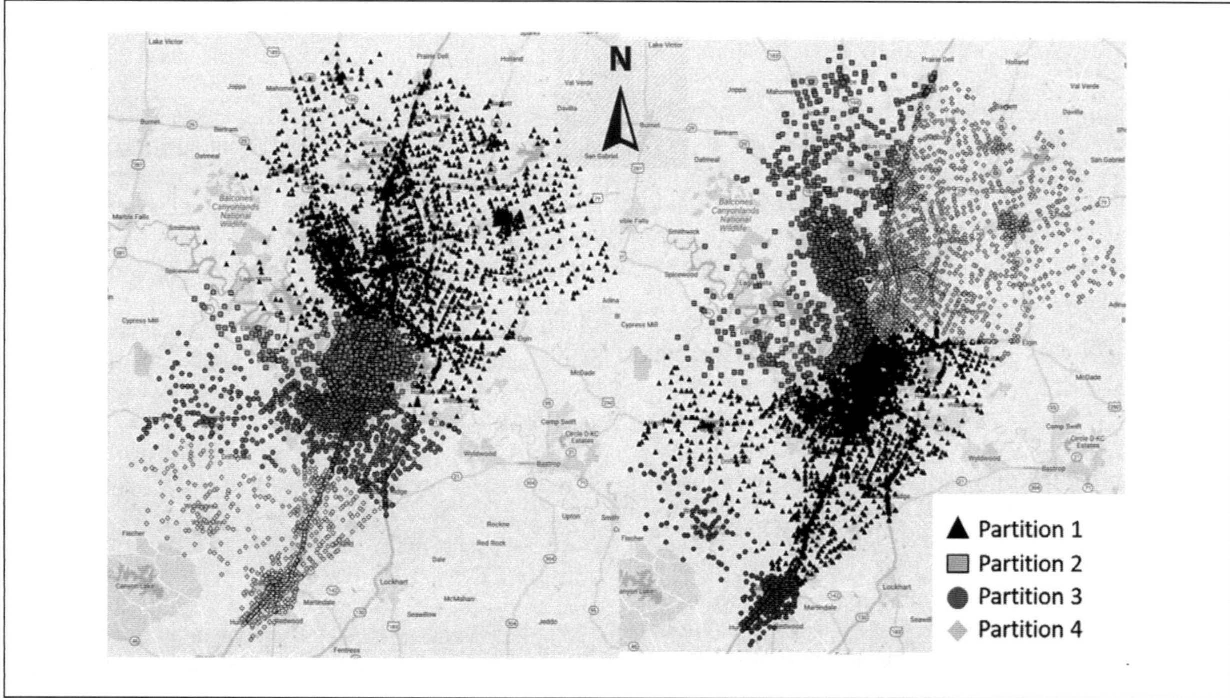

Figure 2. Partitioning of Austin regional network into four partitions: flow weighted spectral partitioning (left); SDDA (right).

balanced interflow whereas SDDA minimizes the boundary nodes. This implies that the number of regional artificial links generated by an SDDA partition will be lower.

In terms of creating balanced subproblems, the flow weighted spectral partitioning method performed better than SDDA. The importance of balancing subproblems by flow is demonstrated by the partitioning of the Austin network into four subnetworks. The computation time needed to solve the subproblems in parallel using the SDDA partitions was approximately 3.5 times the corresponding time resulting from the flow weighted spectral partitioning algorithm. Figure 2 shows the partitions generated for the Austin network. Subnetwork 1 in the SDDA partition contains 65% of the flow. The computation time associated with this subnetwork determines the computation time needed to solve the lower level subproblems at each iteration of DSTAP. The maximum share of network flow within a subnetwork resulting from the flow weighted spectral partitioning algorithm is 39%. For the Chicago sketch network, SDDA also creates heavily imbalanced subnetworks with one subnetwork containing 90% of the flow. If this network was larger, the difference in subnetwork computation time would be significant.

DSTAP Convergence Rate

The study also measures the rate at which the DSTAP algorithm converges towards a global equilibrium given

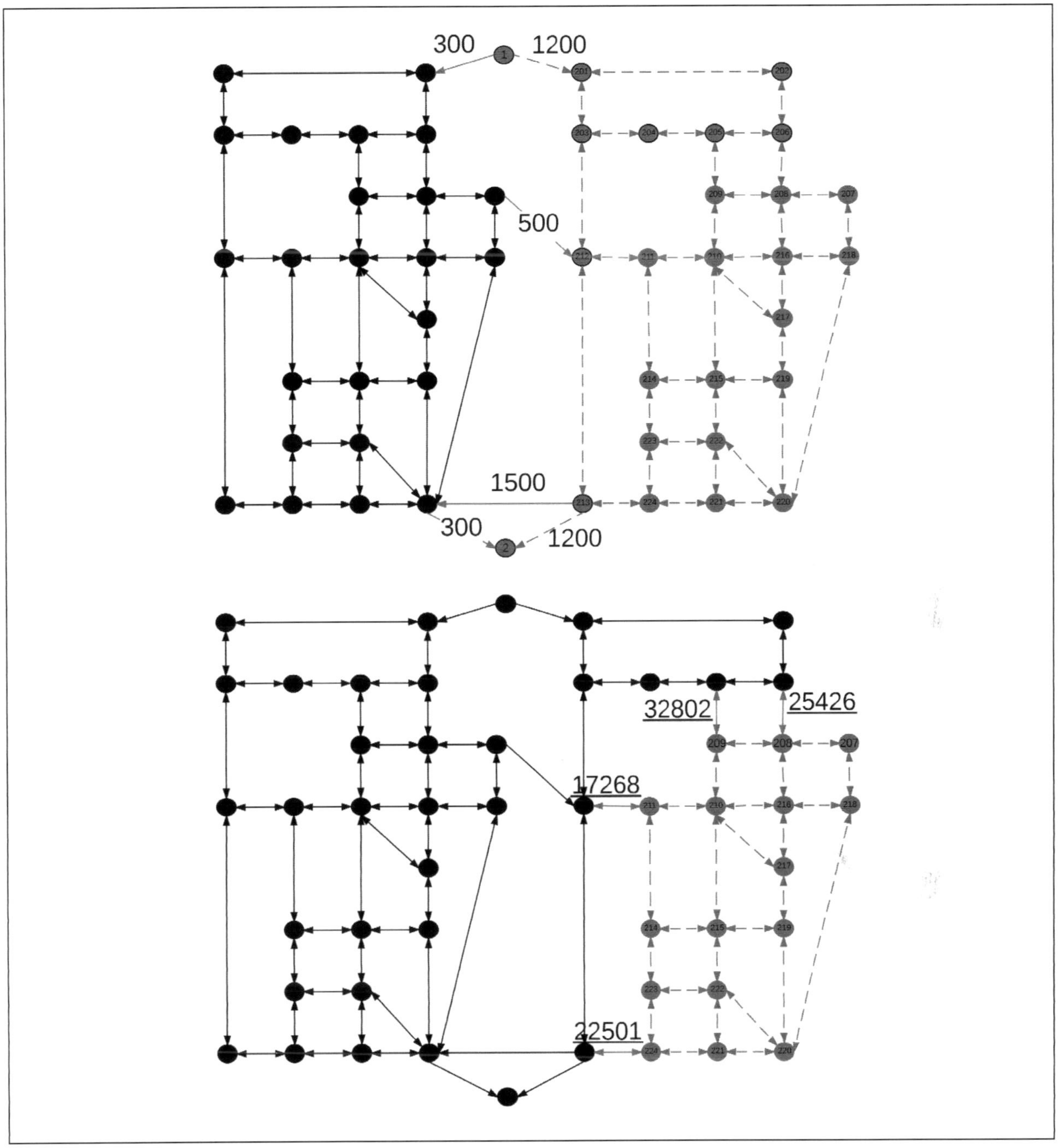

Figure 3. Partitioning of double Sioux Falls hypothetical network: flow weighted spectral partitioning (top); SDDA (bottom). The line type defines different partitions.

the subnetworks generated from a specific partitioning procedure. This convergence rate is tested using a hypothetical network with two copies of Sioux Falls. The network was created by replicating the Sioux Falls network and adding artificial demand between the two copies as shown in Figure 3. The artificial demand was kept low at 1.5% of the total demand within each network.

Figure 3 also shows the subnetworks generated by the flow weighted spectral algorithm and by SDDA. The generated partitions demonstrate the importance of flow weighted spectral partitioning for networks which have

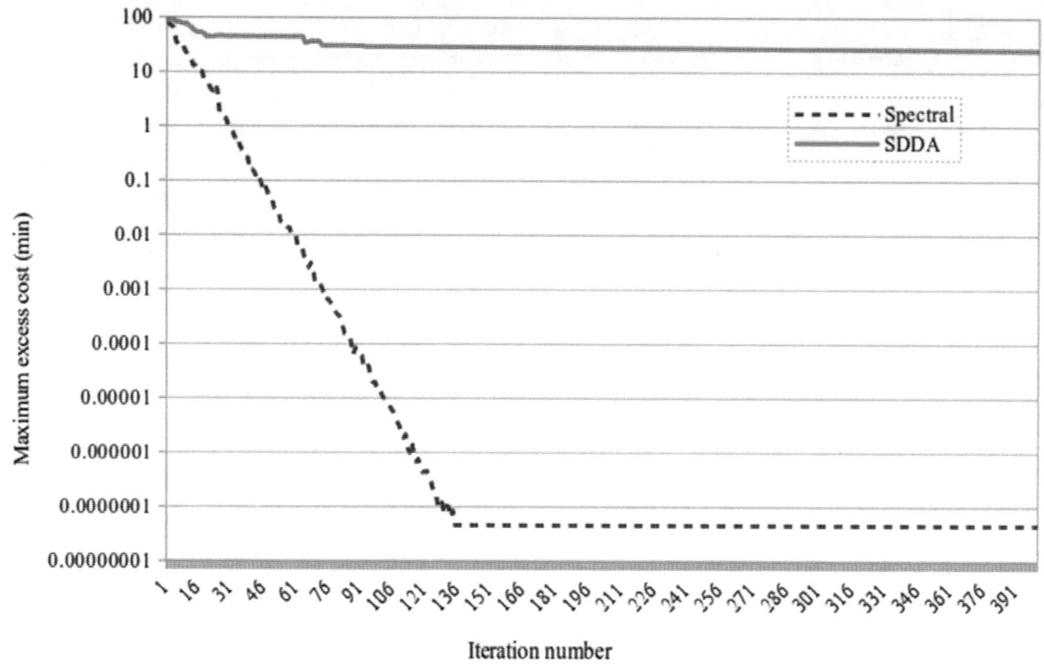

Figure 4. Iterative change of the maximum excess cost of the DSTAP algorithm when used with flow weighted spectral partitions and SDDA partitions of the hypothetical double Sioux Falls network.

intuitive geographic concentrations such as networks in statewide planning models with concentrated flow density in each city. The flow weighted spectral partitioning method was able to identify each Sioux Falls network as a separate component, as opposed to partitions generated by SDDA. In terms of subnetwork boundary nodes, both partitions are equivalent.

In the DSTAP section, it was shown that faster convergence could be achieved if the interflow between subnetworks was minimized. The results in Table 1 and in Figure 3 indicate that the flow weighted spectral partitioning method is superior to the SDDA algorithm for minimizing interflow. The only exception is for the Chicago sketch network. However, the partition generated by SDDA for the Chicago sketch network was heavily imbalanced with one partition containing 90% of the flow. It is expected that the spectral partitioning method will avoid such cuts because of the flow balancing requirement.

Figure 4 shows the convergence rate of DSTAP for the hypothetical double Sioux Falls network when partitioned using the flow weighted spectral partitioning method and SDDA. DSTAP converges to the global equilibrium solution after approximately 135 iterations using partitions generated from the flow weighted spectral partitioning algorithm. As for the SDDA partitions, the convergence rate of the DSTAP algorithm was low. This demonstrates the importance of minimizing the interflow between the subnetworks.

Conclusions

This paper evaluates the performance of different partitioning algorithms used for spatial parallelization of the static TAP. The partitioning objective is to minimize the computation time needed to solve the static traffic assignment using a decomposition approach. The computation time per DSTAP iteration could be reduced by minimizing the number of subnetwork boundary nodes and the time required to solve the TAP for the subnetworks in parallel. The convergence rate of DSTAP depends on the interflow between subnetworks.

Two different methods for partitioning are tested. The first approach is an agglomerative clustering algorithm developed by Johnson et al. (*9*) to minimize the number of boundary nodes between the subnetworks and to create partitions that are balanced in size. The second approach developed is based on flow weighted spectral partitioning. The results indicate that the agglomerative clustering algorithm generates subnetworks that have a low number of boundary nodes. However, the subnetworks generated from this method may be heavily imbalanced as shown for the Austin and Chicago sketch networks. This leads to higher computation time for solving the DSTAP subproblems in parallel. The flow weighted spectral partitioning method generates flow balanced subnetworks which reduce the per iteration computation time. In addition, the interflow between subnetworks is minimized by the spectral partitioning algorithm, which leads to a faster convergence rate of the DSTAP algorithm.

Future work will further assess the trade-offs between minimizing the per iteration computation time and maximizing the convergence rate of DSTAP. Partitioning methods that aim to simultaneously minimize boundary nodes and interflow will be explored. Alternative approximations will be sought that reduce the number of artificial links generated by DSTAP.

Acknowledgments

The authors would like to thank Dr. Klaus Nökel, PTV Group, for his suggestions and comments. This material is based on work supported by the National Science Foundation under Grant No. 1254921. Partial support was provided by the Data-Supported Transportation Operations and Planning center (D-STOP) and the Cooperative Mobility for Competitive Megaregions center (CM2). The authors are grateful for this support.

Author Contributions

The authors confirm contribution to the paper as follows: study conception and design: Cesar N. Yahia, Venktesh Pandey, Stephen D. Boyles; analysis and interpretation of results: Cesar N. Yahia, Venktesh Pandey, Stephen D. Boyles; draft manuscript preparation: Cesar N. Yahia, Venktesh Pandey, Stephen D. Boyles. All authors reviewed the results and approved the final version of the manuscript.

References

1. Beckmann, M., C. B. McGuire, and C. B. Winsten. *Studies in the Economics of Transportation*. Yale University Press, New Haven, CT, 1956.
2. Bar-Gera, H. Origin-Based Algorithm for the Traffic Assignment Problem. *Transportation Science*, Vol. 36, No. 4, 2002, pp. 398–417.
3. Dial, R. B. A Path-Based User-Equilibrium Traffic Assignment Algorithm that Obviates Path Storage and Enumeration. *Transportation Research Part B: Methodological*, Vol. 40, No. 10, 2006, pp. 917–936.
4. Jafari, E., V. Pandey, and S. D. Boyles. A Decomposition Approach to the Static Traffic Assignment Problem. *Transportation Research Part B: Methodological*, Vol. 105, 2017, pp. 270–296.
5. Colson, B., P. Marcotte, and G. Savard. An Overview of Bilevel Optimization. *Annals of Operations Research*, Vol. 153, No. 1, 2007, pp. 235–256.
6. Bar-Gera, H. Traffic Assignment by Paired Alternative Segments. *Transportation Research Part B: Methodological*, Vol. 44, No. 8–9, 2010, pp. 1022–1046.
7. Abdelghany, K., H. Hashemi, and A. Alnawaiseh. Parallel All-Pairs Shortest Path Algorithm: Network Decomposition Approach. *Transportation Research Record: Journal of the Transportation Research Board*, 2016. 2567: 95–104.
8. Boyles, S. D. Bush-Based Sensitivity Analysis for Approximating Subnetwork Diversion. *Transportation Research Part B: Methodological*, Vol. 46, No. 1, 2012, pp. 139–155.
9. Johnson, P., D. Nguyen, and M. Ng. Large-Scale Network Partitioning for Decentralized Traffic Management and Other Transportation Applications. *Journal of Intelligent Transportation Systems*, Vol. 20, No. 5, 2016, p. 461.
10. Frank, M., and P. Wolfe. An Algorithm for Quadratic Programming. *Naval Research Logistics*, Vol. 3, 1956, pp. 95–110.
11. Jayakrishnan, R., W. Tsai, J. Prasker, and S. Rajadhyaksha. A Faster Path-Based Algorithm for Traffic Assignment. *Transportation Research Record: Journal of the Transportation Research Board*, 1994. 1443: 75–83.
12. Nie, Y. M. A Class of Bush-Based Algorithms for the Traffic Assignment Problem. *Transportation Research Part B: Methodological*, Vol. 44, No. 1, 2010, pp. 73–89.
13. Gentile, G. Local User Cost Equilibrium: A Bush-Based Algorithm for Traffic Assignment. *Transportmetrica A: Transport Science*, Vol. 10, No. 1, 2014, pp. 1–40.
14. Zheng, H., and S. Peeta. Cost Scaling Based Successive Approximation Algorithm for the Traffic Assignment Problem. *Transportation Research Part B: Methodological*, Vol. 68, 2014, pp. 17–30.
15. Josefsson, M., and M. Patriksson. Sensitivity Analysis of Separable Traffic Equilibrium Equilibria with Application to Bilevel Optimization in Network Design. *Transportation Research Part B: Methodological*, Vol. 41, No. 1, 2007, pp. 4–31.
16. Chen, R., and R. R. Meyer. Parallel Optimization for Traffic Assignment. *Mathematical Programming*, Vol. 42, No. 1, 1988, pp. 327–345.
17. Lotito, P. A. Issues in the Implementation of the DSD Algorithm for the Traffic Assignment Problem. *European Journal of Operational Research*, Vol. 175, No. 3, 2006, pp. 1577–1587.
18. Etemadnia, H., K. Abdelghany, and A. Hassan. A Network Partitioning Methodology for Distributed Traffic Management Applications. *Transportmetrica A: Transport Science*, Vol. 10, No. 6, 2014, pp. 518–532.
19. Garg, N., V. V. Vazirani, and M. Yannakakis. Approximate Max-Flow Min-(Multi) Cut Theorems and Their Applications. *SIAM Journal on Computing*, Vol. 25, No. 2, 1996, pp. 235–251.
20. Saeedmanesh, M., and N. Geroliminis. Clustering of Heterogeneous Networks with Directional Flows Based on "Snake" Similarities. *Transportation Research Part B: Methodological*, Vol. 91, 2016, pp. 250–269.
21. Karypis, G., and V. Kumar. A Fast and High Quality Multilevel Scheme for Partitioning Irregular Graphs. *SIAM*

Journal on Scientific Computing, Vol. 20, No. 1, 1998, pp. 359–392.
22. Spielman, D. A. Spectral Graph Theory and Its Applications. *Proc., 48th Annual IEEE Symposium on Foundations of Computer Science*, IEEE, Providence, RI, 2007, pp. 29–38.
23. Spielman, D. A., and S. Teng. Spectral Partitioning Works: Planar Graphs and Finite Element Meshes. *Linear Algebra and its Applications*, Vol. 421, No. 2–3, 2007, pp. 96–105.
24. Newman, M. E. J. Spectral Methods for Community Detection and Graph Partitioning. *Physical Review E*, Vol. 88, No. 4, 2013, p. 042822.
25. Von Luxburg, U. A Tutorial on Spectral Clustering. *Statistics and Computing*, Vol. 17, No. 4, 2007, pp. 395–416.
26. Shi, J., and J. Malik. Normalized Cuts and Image Segmentation. *IEEE Transactions on Pattern Analysis and Machine Intelligence*, Vol. 22, No. 8, 2000, pp. 888–905.
27. Bell, M. G. H., F. Kurauchi, S. Perera, and W. Wong. Investigating Transport Network Vulnerability by Capacity Weighted Spectral Analysis. *Transportation Research Part B: Methodological*, Vol. 99, 2017, pp. 251–266.
28. Martinez, S., G. Chatterji, D. Sun, and A. M. Bayen. A Weighted-Graph Approach for Dynamic Airspace Configuration. *Proc., AIAA Conference on Guidance, Navigation, and Control (GNC)*. American Institute of Aeronautics and Astronautics, Reston, VA, 2007.
29. Ma, Y., Y. Chiu, and X. Yang. Urban Traffic Signal Control Network Automatic Partitioning using Laplacian Eigenvectors. *Proc., 12th International IEEE Conference on Intelligent Transportation Systems, ITSC'09*. IEEE, Maui, HI, 2009, pp. 1–6.
30. Jafari, E., and S. D. Boyles. Improved Bush-Based Methods for Network Contraction. *Transportation Research Part B: Methodological*, Vol. 83, 2016, pp. 298–313.
31. Stabler, B. *Transportation Networks for Research*, 2018. http://www.bgu.ac.il/bargera/tntp. Accessed August 1, 2017.

The Standing Committee on Transportation Network Modeling (ADB30) peer-reviewed this paper (18-06570).

Article

TRR
JOURNAL OF THE TRANSPORTATION RESEARCH BOARD

Transportation Research Record
2018, Vol. 2672(48) 127–138

DOI: 10.1177/0361198118799165
journals.sagepub.com/home/trr
SAGE

Characterization of Taxi Fleet Operational Networks and Vehicle Efficiency: Chicago Case Study

Ying Chen[1], Michael Hyland[1], Michael Patrick Wilbur[1], and Hani S. Mahmassani[2]

Abstract

Taxi fleets serve a significant and important subset of travel demand in major cities around the world. This paper characterizes the Chicago taxi fleet operational network using complex network metrics and analyzes the operational efficiency of individual taxis over the past four years using an extensive taxi-trip dataset. The dataset, recently released by the city of Chicago, includes the pickup and drop-off census tracts and time stamps for over 100 million taxi trips. The paper explores year-over-year changes in the spatial distribution of Chicago taxi travel demand. The taxi pickup and drop-off census tract locations are modeled as nodes, and links are generated between unique pickup and drop-off node pairs. The analysis shows that high-demand pickup and drop-off location pairs in 2013 generated similar trip volumes in 2016; however, the low-demand pairs in 2013 generated significantly fewer trips in 2016. Additionally, this paper presents temporal efficiency and spatial efficiency metrics. The temporal efficiency metric determines the percentage of in-service time taxis are productive (i.e., transporting travelers), rather than empty. The spatial efficiency metric measures the percentage of taxi miles that are productive (i.e., loaded), rather than empty. The efficiency analysis of the Chicago taxi fleet shows that, for most taxis, around 50% of their in-service time and travel distance are unproductive. This inefficiency negatively affects the profitability of individual drivers and the fleet, traffic congestion, vehicle emissions, the service quality provided to customers, and the ability of taxi services to compete with emerging mobility services.

The emergence of carsharing, ridesharing, and particularly ridesourcing services over the past half-decade, and the impending advent of fully autonomous vehicles (AVs) has led many transportation industry experts to suggest that we are moving toward a future where a much larger percentage of travel demand is served by transportation service providers. There are many potential benefits and pitfalls associated with this transportation future. While outcomes depend on numerous factors, positive outcomes require the efficient use of vehicles and roadways.

Recently, several researchers have begun analyzing the impacts of shared-use AV mobility services (SAMSs) on vehicle miles traveled, car ownership, parking, and vehicle emissions (*1*). Other researchers, including the authors of this paper, are working on developing efficient strategies to operate and manage AV fleets in real time (*2*). These research studies employ simulation methods to analyze the operational efficiency of SAMSs. To provide an operational efficiency benchmark using real-world data, rather than simulation results, this paper analyzes the operational efficiency of the Chicago taxi fleet.

Studying taxi supply and demand, and analyzing the operational efficiency of taxi fleets are important research aims, independent of their relation to SAMSs. Taxis serve a sizable amount of travel demand in many large urban areas. The operational efficiency of taxi fleets and individual taxis impacts not only the competitiveness and profitability of taxi drivers and taxi fleets, but also the efficiency of roadway networks and the service quality offered to travelers. Additionally, more efficient operations can help taxis compete with the emerging passenger transportation services that have caused considerable losses in the market share of taxi services (*3*). Motivated by the importance of operational efficiency, as well as the emergence and growth of

[1]Department of Civil and Environmental Engineering, Northwestern University, Evanston, IL
[2]Transportation Center, Northwestern University, Evanston, IL

Corresponding Author:
Address correspondence to Hani S. Mahmassani: masmah@northwestern.edu

ridesourcing services, this paper characterizes and quantifies the spatial and temporal efficiency of individual taxis in the Chicago taxi fleet.

This paper presents a spatial efficiency metric and a temporal efficiency metric. The spatial efficiency metric determines the percentage of a taxi's total miles that are productive. Productive miles are defined as miles where the taxi is transporting passengers. Unproductive miles include miles spent roaming for passengers and traveling to pick up travelers. The temporal efficiency metric determines the percentage of in-service time taxis are productive. In-service time includes all the time a taxi is looking for passengers or serving passengers, that is, all the time a driver is in the taxi. The temporal and spatial efficiency measures allow the analyst, or fleet operator, to assess the efficiency of individual taxi trips, individual taxis, and the entire taxi fleet.

In addition to characterizing and quantifying the efficiency of taxis, this paper also characterizes the spatial distribution of the Chicago taxi operational network (i.e., the spatial distribution of taxi requests in Chicago) using complex network metrics. Additionally, the paper clusters individual taxis based on their daily usage rates over the entire year.

This study utilizes the recently released taxi data from the city of Chicago (*4*). Chicago followed in the footsteps of New York City who released 2009 through 2015 taxi data that covers 1.1 billion taxi trips. The New York dataset has spurred a significant volume of research. Schneider presents a comprehensive exploratory analysis and visualization of the New York City taxi data (*5*). Qian and Ukkusuri (*6*) examine the spatial variation of taxi demand in New York via a geographically weighted regression model to assess factors that impact taxi demand. Haggag et al. examine 2009 New York City taxi data in an effort to analyze learning by doing (*7*). The research finds that more experienced drivers are significantly more efficient than new drivers. The most interesting finding is that a driver's time between dropping off one customer and picking up the next customer (i.e., the driver's spatial efficiency) depends on the driver's history of drop-offs in the current region but is unaffected by the driver's total city-wide experience. King and Saldarriaga (*8*) find that spatial regulation of taxi services in New York City increases the number of unproductive trips by 20%. The paper estimates that the cumulative effect of this regulation is 300,000 extra fleet miles per week. Yang and Gonzalez (*9*) use the NYC taxi data to identify spatio-temporal mismatches in the supply of taxis and demand for taxi services. Zhan et al. (*10, 11*) extensively study the efficiency of urban taxi fleets and present a graph-based approach to evaluate the efficiency of a taxi fleet. Liu et al. use taxi-trip data and network science methods to examine travel patterns and city structure (*12*).

Chicago Taxi-Trip Data

The taxi-trip data used in this study is a subset of the taxi dataset available through the Chicago Data Portal (*4*). The dataset includes information on over 105 million taxi trips made in the city of Chicago between January 2013 and 2017. The full dataset includes a total of 23 columns with information about each taxi trip. This study makes use of the following columns:

- Taxi ID
- Trip Start Timestamp (15-minute interval)
- Trip End Timestamp (15-minute interval)
- Trip Duration (seconds)
- Trip Distance (miles)
- Pickup Location (census tract centroid or community area centroid)
- Drop-off Location (census tract centroid or community area centroid)

Due to privacy concerns, the dataset does not provide exact geographical coordinates for pickup or drop-off locations; rather the dataset provides census tract or community area centroids.

Table 1 displays the number of taxi trips recorded, and the number of unique taxis that made at least one trip in each year between 2013 and 2016. Table 2 displays the distribution of taxi-trip distances, durations, and trip counts.

This paper presents several interrelated analyses using the Chicago taxi data. Taxi trips with distances (durations) longer than 40 miles (90 minutes) were either removed from each analysis, replaced with the shortest path distance (time) between the trip's pickup and drop-off locations using the Google Maps API, or simply replaced with a distance (time) of 40 miles (90 minutes). The 40-mile cut-off was selected because this is

Table 1. Chicago Taxi-Trip Dataset General Statistics

	2013	2014	2015	2016
Number of taxis	5,557	7,582	7,552	7,667
Number of trips	26,870,287	31,021,726	27,400744	19,878,249

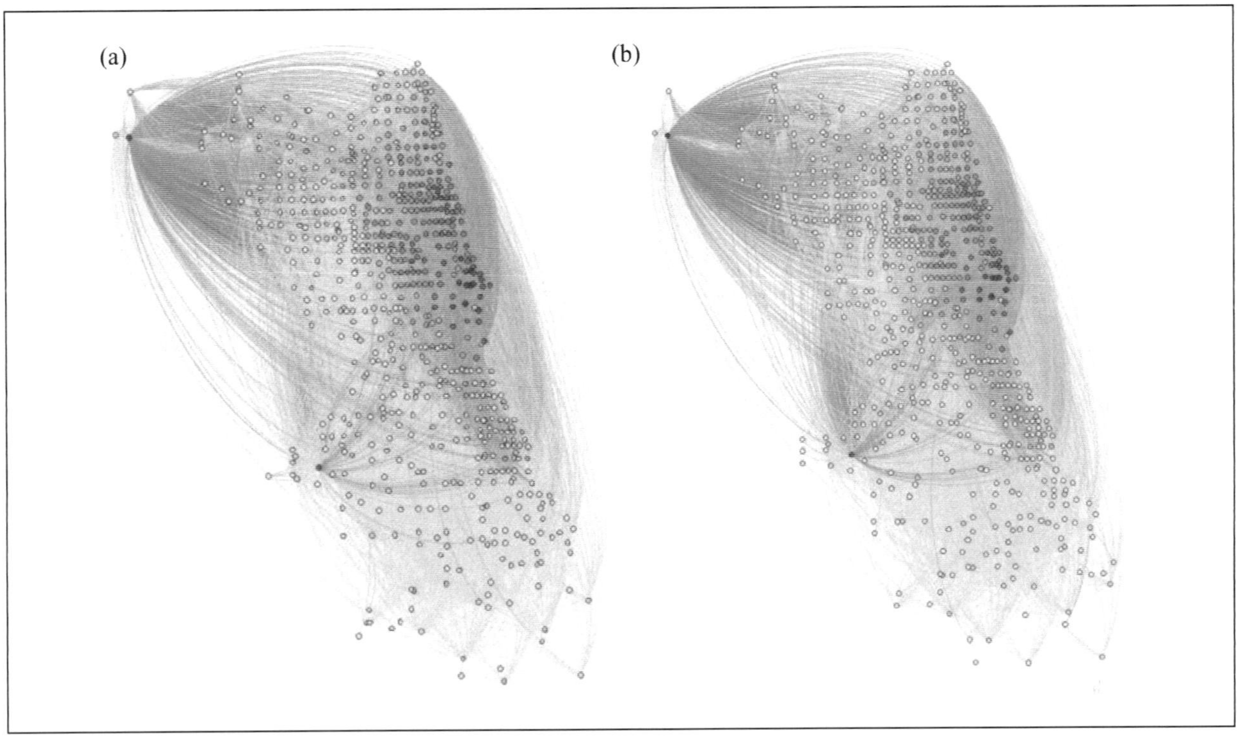

Figure 1. Complex network structure of taxi trips in the Chicago metropolitan area in (*a*) 2014 and (*b*) 2015.

Table 2. Individual Taxi-Trip Statistics

	Distance (mile)	Duration (minute)	Trip count per taxi (2013~2016)
Min.	0.11	1	1
1st	0.9	6	429
Median	1.7	10	14,210
Mean	3.8	14	13,150
3rd	3.9	17	22,160
Max.	40	90	48,020

approximately the longest possible travel distance between any two locations in the Chicago metropolitan region. Similarly, the 90-minute cut-off was selected because it is approximately the travel time from the northern-most edge to the southern-most edge of the Chicago metropolitan region, during the off-peak period.

Complex Network Analysis

Taxi Trips Network Analysis

This section analyzes the spatial properties of the Chicago taxi fleet operational network and taxi demand using complex network metrics. Figure 1*a* shows the regional coverage and qualitative structure of the taxi-trip network in 2014. The 2014 network consists of 679 distinct nodes (i.e., census tracts) and 56,671 directed edges connecting nodes. Edges (i.e., links) are generated for each unique taxi-trip origin-destination node pair in each year. Figure 1*b* shows the taxi-trip network in 2015. Dark-blue (light-green) nodes and links represent high- (low-)volume nodes and links. The high-density nodes include O'Hare international airport in the northwest corner of Figure 1, *a* and *b*, Midway airport in the southwest corner, and the nodes in the central business district on the eastern side.

Table 3 presents several complex network metrics to characterize the Chicago taxi operational network during the past four years. The number of links (L) and total number of trips (T) decreased significantly (36% each) between 2014 and 2016. However, the number of nodes (N) remained relatively constant between 2013 and 2016 due to the spatial aggregation of trip origin and destination locations.

Two common complex network metrics used to measure network connectivity are $\delta = 2L/(N\hat{}2)$ and L/N. Higher values for both metrics indicate higher network connectivity. The values in Table 3 indicate that taxi network connectivity increased between 2013 and 2014 but decreased significantly between 2014 and 2016.

The mean trip length ($<l>$) and mean trip duration ($<d>$) of Chicago taxi trips were consistent between

Table 3. Taxi Demand Network Characteristics in Chicago

Data	Description	Chicago taxi trips network			
		2013	2014	2015	2016
N	# of nodes	650	679	685	653
L	# of links	51,743	56,671	49,572	36,528
L/N		79.6	83.5	72.4	55.9
δ	2L/(N^2)	0.24	0.25	0.21	0.17
T	Total # of trips	21.3M	25.8M	22.5M	16.6M
$<l>$	Mean trip length (km)	6.75	6.91	7.17	7.06
$CV<l>$	Coeff. of variation $<l>$	1.1	1.1	1.1	1.2
$<d>$	Mean trip duration (sec)	743	746	754	745
$CV<d>$	Coeff. of variation $<d>$	0.55	0.56	0.58	0.58
$<k>$	Mean node degree	79.6	83.5	72.4	55.9
$CV<k>$	Coeff. of variation $<k>$	1.1	1.1	1.2	1.3
$<F>$	Mean node flux	32,776	37,997	32,847	25,421
$CV<F>$	Coeff. of variation $<F>$	4.1	4.1	4.3	4.6
$<w>$	Mean link weight	411	455	453	454
$CV<w>$	Coeff. of variation $<w>$	9.5	9.8	9.7	9.7
C	Clustering coefficient	0.71	0.71	0.69	0.66

2013 and 2016, as were the coefficients of variation for both metrics. Mean trip length ($<l>$) remained around 7 kilometers with a standard deviation greater than 7 kilometers. Mean taxi-trip duration hovered around 12.5 minutes, with a standard deviation of 7 minutes.

The mean node degree ($<k>$), which represents the average number of links connected to a node, increased from 2013 to 2014 for taxi trips; then declined after 2014. The same is true of the mean node flux ($<F>$), which measures the number of trips starting or ending at a node. These findings are unsurprising given the reduction in both taxi trips (T) and links (L) along with the consistency in the number of nodes (N) between 2014 and 2016.

The node degree coefficient of variation ($CV<k>$) and node flux coefficient of variation ($CV<F$) both remained relatively constant between 2013 and 2016 with slight increases over time. The node flux coefficient of variation ($CV<F>$) is very high indicating significant variation in the number of trips originating and terminating at nodes. This is likely a result of the disproportionate share of trips at the O'Hare and Midway airport nodes.

The mean link weight ($<w>$) measures the average number of trips across each link. The mean link weight ($<w>$) remains relatively constant between 2013 and 2016 indicating that the number of taxi trips (T) decreased at nearly the same rate as the number of links (L) in the Chicago taxi network. The link weight's coefficient of variation ($CV<w>$) is relatively small.

The network clustering coefficient (C) measures the extent all nodes in a network are clustered. C is defined in Equation 1, where a triplet consists of three connected nodes, and a connected triplet is a connected subgraph that consists of three vertices and two edges (*13*). Table 3 shows that C did not vary significantly between 2013 and 2016:

$$C = \frac{\textit{number of closed triplets}}{\textit{number of connected triplets of vertices}} \tag{1}$$

Table 3 displays a significant decrease in annual taxi trips over time. Further analysis of the data shows that most of lost demand came from low-demand origin-destination pairs in 2014. This finding might be due to the emergence of ridesourcing companies, Uber and Lyft, and particularly their ability to better serve customers in remote areas.

Taxi Fleet Usage Analysis

This section analyzes the usage rate of individual taxis.

Histogram of Taxi Fleet by Average Daily Trips

This section presents a histogram of the average number of trips taxis made per in-service day. Let I_y denote the set of all taxis, in year y, indexed by taxi $i \in I_y$. Let $N_{i,y}$ denote the set of trips completed by taxi $i \in I_y$ in year y, indexed by trip $n \in N_{i,y}$. For notational simplicity, and the fact that we perform a separate analysis for each year y, the subscript y is removed from these sets throughout the remainder of the paper. Equation 2 determines the average number of daily trips for taxi i (τ_i) in year y, where σ_i is the number of days taxi i is in service during year y:

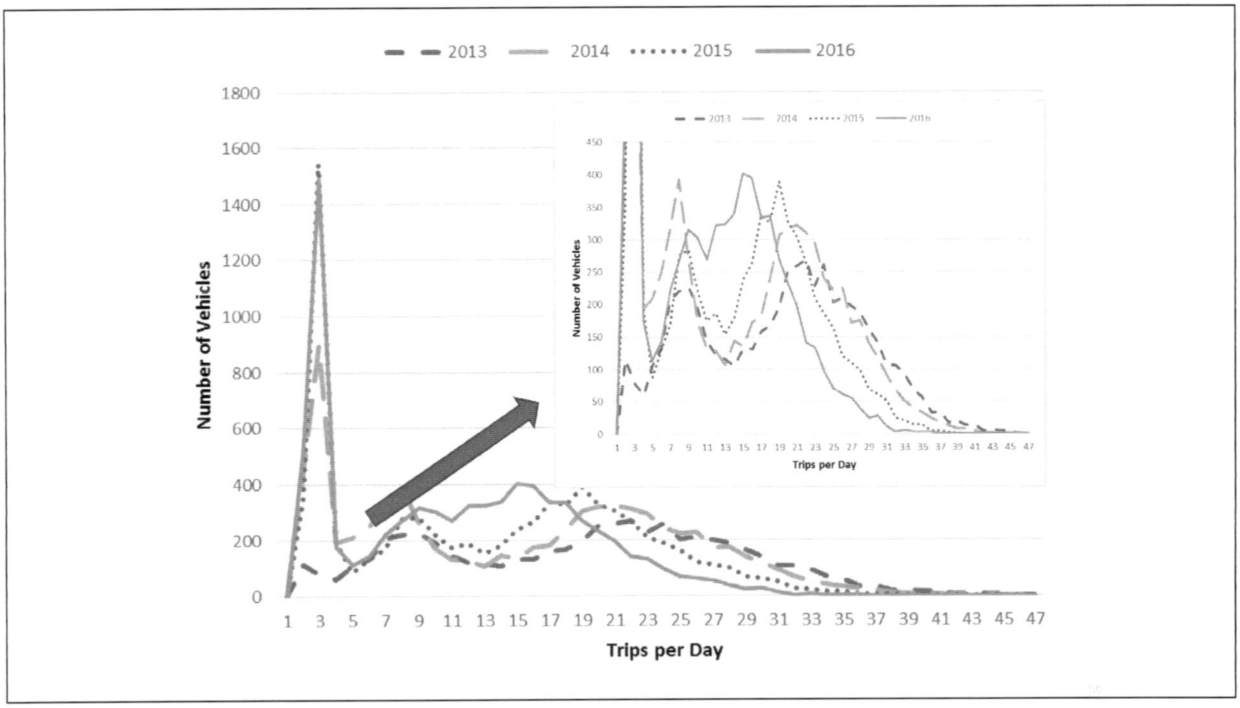

Figure 2. Distribution of the average number of daily trips across taxis.

$$\tau_i = \frac{|N_i|}{\sigma_i} \quad (2)$$

Figure 2 displays a histogram of τ_i for each year y. In 2015 and 2016, around 1,500 taxis only made three trips per *in-service* day on average. However, most taxis completed between seven and 25 trips per in-service day, with a few taxis completing more than 40 trips per in-service day.

The inset graph in Figure 2 indicates a clear leftward shift in the average trips per in-service day distribution between 2013 and 2016. This indicates that high-usage taxis made fewer and fewer trips per in-service day over the past four years. This shift likely represents declining revenue and profits for individual taxi drivers.

Clustering Taxis by Daily Trip Count

K-means Clustering. This section describes the K-means clustering algorithm employed to cluster taxis based on their daily trip counts. Let D be the set of days in a year, indexed by $d \in D$, where $|D|$ is typically 365 (or 366 in a leap year). Let γ_{id} denote the number of trips made by taxi $i \in I$ on day $d \in D$, and let Γ_i denote the taxi daily trip count vector of length $|D|$ for taxi i ($\gamma_{id} \in \Gamma_i$).

The K-means clustering problem involves assigning each taxi $i \in I$ into one and only one cluster. Let S denote the set of clusters, where $|S| = K$. The elements of S are denoted $S_k, k = 1, 2, \ldots, K$, where S_k is a set of taxis ($S_k \subset I$). Let $\pi_{S_k,d}$ denote the mean number of trips for the taxis in cluster S_k on day d, and Π_{S_k} denote the mean daily trip count vector for the taxis assigned to cluster S_k ($\pi_{S_k,d} \in \Pi_{S_k}$), where $|\Pi_{S_k}| = |\Gamma_i| = |D|$. Equation 3 displays the squared error between the mean daily trip values for cluster S_k ($\pi_{S_k,d}$) and the daily trip values for each taxi i assigned to cluster S_k ($\gamma_{id}|i \in S_k$):

$$SE(S_k) = \sum_{i \in S_k} \| \Gamma_i - \Pi_{S_k}{}^2 \| = \sum_{i \in S_k} \sum_{d=1}^{|D|} (\gamma_{id} - \pi_{S_k,d})^2 \quad (3)$$

K-means clustering algorithm identifies the set of clusters S that minimize the within-cluster sum of squares (WCSS) as defined in Equation 4:

$$J(K) = \min_S \sum_{k=1}^{K} SE(S_k) = \min_S \sum_{k=1}^{K} \sum_{i \in S_k} \| \Gamma_i - \Pi_{S_k} \|^2 \quad (4)$$

where, K is the number of clusters and $J(K)$ is the minimal WCSS. $J(K)$ is a monotonically decreasing function of K. Equation 5 displays the metric used in this study to help select the number of clusters (K), where R is the rate of change (decrease) in $J(K)$ as K increases:

$$R = \frac{J(K) - J(K+1)}{J(K)} \times 100\% \quad (5)$$

Table 4. Percentage of Taxis in Each Cluster

Cluster	2013	2014	2015	2016
Most trips	17.06%	15.18%	8.47%	7.54%
More trips	11.77%	24.48%	20.19%	17.54%
Normal	25.52%	7.24%	20.52%	18.49%
Less trips	22.10%	20.76%	18.78%	17.67%
Least trips	23.56%	32.34%	32.03%	38.75%

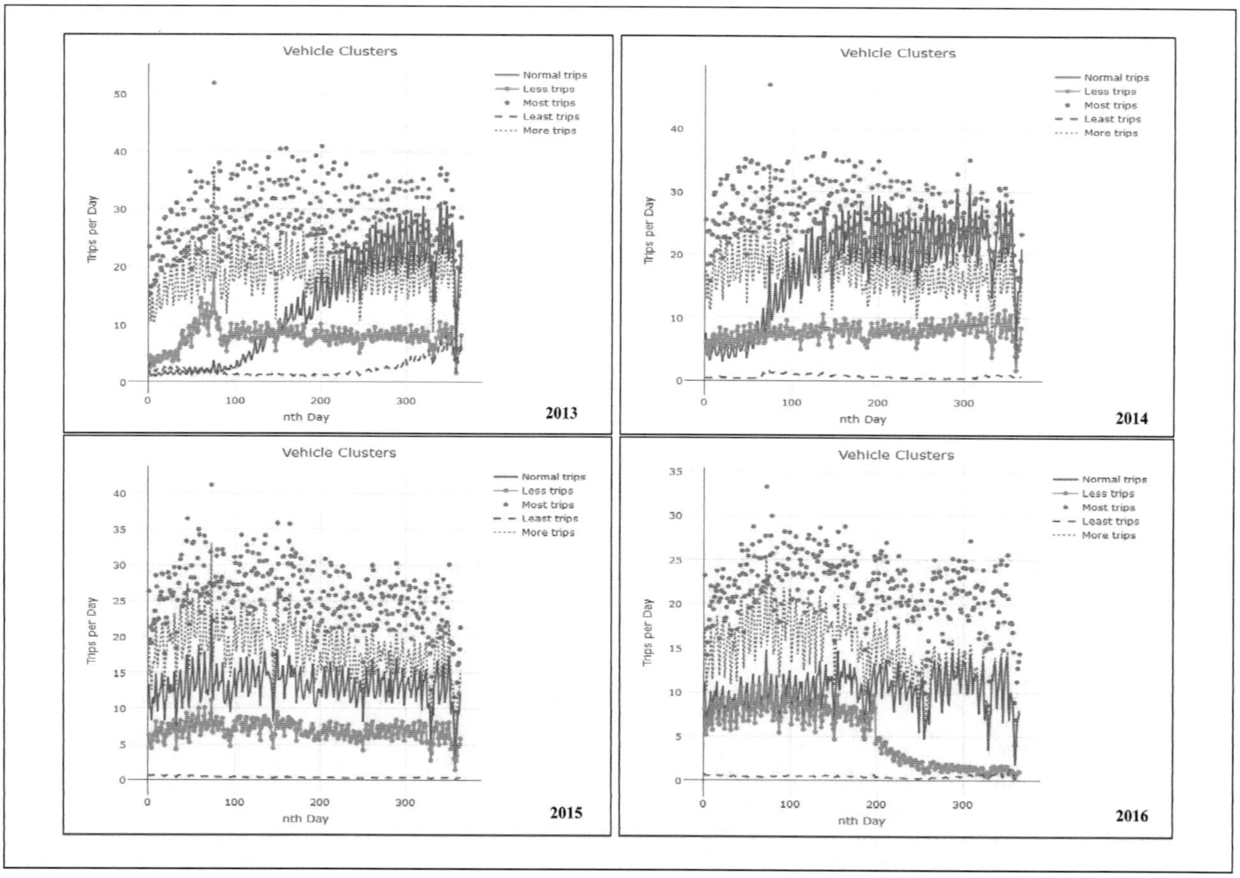

Figure 3. Taxi clusters results.

Cluster Results. For the Chicago taxi data, Equation 5 yields five taxi clusters. Table 4 shows the percentage of taxis in each of the five taxi clusters by year.

Figure 3 presents the daily taxi-cluster centroids for each year, where the x-axis displays d and the y-axis displays $\pi_{S_k,d}$. In 2013, the range of average daily trip counts for taxis making the *most trips* was between 20 and 35; the range for taxis making *more trips* was between 15 and 25; the range for taxis with *normal trips* was 1–2 for the first 90 days, before increasing steadily over the next 150 days, and then plateauing between 15 and 25 for the last part of the year; the range for taxis making *less trips* was between 5 and 10; and the range for taxis making the *least trips* was one or two for the first 250 days, before it increased to between 5 and 10. The *least trips* and *less trips* clusters likely correspond to the three trips per in-service day peak in Figure 2. The 2014 daily trips ranges are like the 2013 daily trip ranges for all five clusters.

In 2015, the daily trip counts in each cluster were stable; that is, no systematic increases or decreases in daily trip counts after a certain day of the year. The same was largely true for 2016, except the *less trips* cluster range decreased from five to 10 daily trips the first 200 days of the year to one to three daily trips during the last 165 days.

The results in Figure 3 clearly indicate that there are significant differences in taxi trips per day in the Chicago taxi fleet. Some taxis consistently made 20 to 35 trips per day and others only made one to two trips per day.

Histogram of Days in Service by Taxi Cluster. This subsection presents histograms of the percentage of in-service days for the taxis in each cluster, independent of the average number of fares/trips per day (see Figure 4). This analysis was conducted to validate the cluster analysis results in Figure 3.

In Figure 4, taxis in the *most trips* cluster were in-service 70% to 100% of the days in each year. Similarly, as expected, most taxis in the *least trips* cluster were only in-service 0% to 40% of the days each year. The other graphs in Figure 4 also largely follow what one would expect based on the results in Figure 3. One exception is that the taxis in the *less trips* cluster were in-service a large percentage of the days. Based on Figure 3, they must have served only a few trips per day.

Taxi Efficiency Analysis

The last section focused on the usage rate of individual taxis, whereas this section focuses on the efficiency of individual taxis.

Efficiency Metrics

Temporal Efficiency Metric. Let $p_{i,n}$ denote the productive time associated with taxi-trip i, n, and let and $u_{i,n}$ denote the unproductive time between taxi-trip i, n and taxi-trip $i, n+1$. Let $durai_{i,n}$ denote the trip duration value in the taxi dataset and let $b_{i,n}$ denote a binary variable equal to 1 if a trip duration value exists for taxi-trip i, n. Let $pos_{i,n,p}$ and $pos_{i,n,d}$ denote the pickup and drop-off locations (longitude and latitude) of taxi-trip i, n, respectively. Finally, let $A_{i,n,p}$ and $A_{i,n,d}$ denote binary variables equal to 1 if the pickup and drop-off locations of taxi-trip i, n exist in the dataset, respectively. Algorithm 1*a* assigns values to $p_{i,n}$ (the Google API function is described below).

Similarly, algorithm 1*b* assigns values to $u_{i,n}$. Let $t_{i,n,p}$ and $t_{i,n,d}$ denote the trip pickup and drop-off time for taxi-trip combination i, n, respectively. Moreover, let $B_{i,n,p}$ and $B_{i,n,d}$ denote binary variables equal to 1 if these values exist in the dataset for pickups and drop-offs, respectively.

Equation 6 displays the temporal efficiency metric for taxi-trip i, n ($\varphi_{i,n}$) as a function of $p_{i,n}$ and $u_{i,n}$ obtained via algorithm 1*a* and *b*, respectively. Values close to 100% represent temporally efficient taxi trips; whereas, values close to 0% represent temporally inefficient taxi trips:

$$\varphi_{i,n} = \left(\frac{p_{i,n}}{u_{i,n} + p_{i,n}}\right) \times 100\% \tag{6}$$

Spatial Efficiency Metric. Let $l_{i,n}$ denote the loaded distance associated with taxi-trip i, n. Similarly, let $e_{i,n}$ denote the empty distance between the drop-off location of taxi-trip i, n and the pickup location of taxi-trip $i, n+1$. Let $dist_{i,n}$ denote the trip distance value in the taxi dataset and let $a_{i,n}$ denote a binary variable equal to 1 if a trip distance value exists for taxi-trip i, n. Algorithm 2*a* and *b* assign values to $l_{i,n}$ and $e_{i,n}$, respectively.

Equation 7 displays the spatial efficiency metric for taxi-trip i, n ($\xi_{i,n}$) as a function of $l_{i,n}$ and $e_{i,n}$ obtained via algorithm 2*a* and *b*, respectively. Values close to 100% represent spatially efficient taxi trips; whereas, values close to 0% represent spatially inefficient taxi trips:

$$\xi_{i,n} = \left(\frac{l_{i,n}}{e_{i,n} + l_{i,n}}\right) \times 100\% \tag{7}$$

Google Maps Distance Matrix API. The Google Maps Distance Matrix API (https://developers.google.com/maps/) is a service provided by Google Inc. that estimates travel distance and time for a recommended route between origin and destination points (*14*). API requests must include origin and destination locations, a unique Google API key, and a transport mode. Optional parameters include arrival or departure time, and traffic model (best guess, pessimistic, optimistic). In this paper, we use the default driving mode and we do not provide specific arrival or departure times. The API returns trip distance and duration for any feasible origin-destination coordinate pair input. For the current study, we calculate the shortest travel time and distance between each unique pickup and drop-off location pair. Therefore, the algorithms presented in the last two subsections use lookup table values for API-generated trip distances and trip durations.

Efficiency Results

Temporal Efficiency Results. This section presents temporal efficiency measures for the Chicago taxi fleet. Figure 7*a* presents a histogram of temporal efficiency for the taxis in the *most trips* cluster. Similarly, Figure 7*b* presents a histogram of temporal efficiency for the taxis in the *normal trip* count cluster. The x-axis is the percentage of loaded time in bin intervals of 10%, and the y-axis is the number of taxis in each bin.

Figure 7, *a* and *b* show that around 50% of taxi in-service time is unproductive; that is time not spent

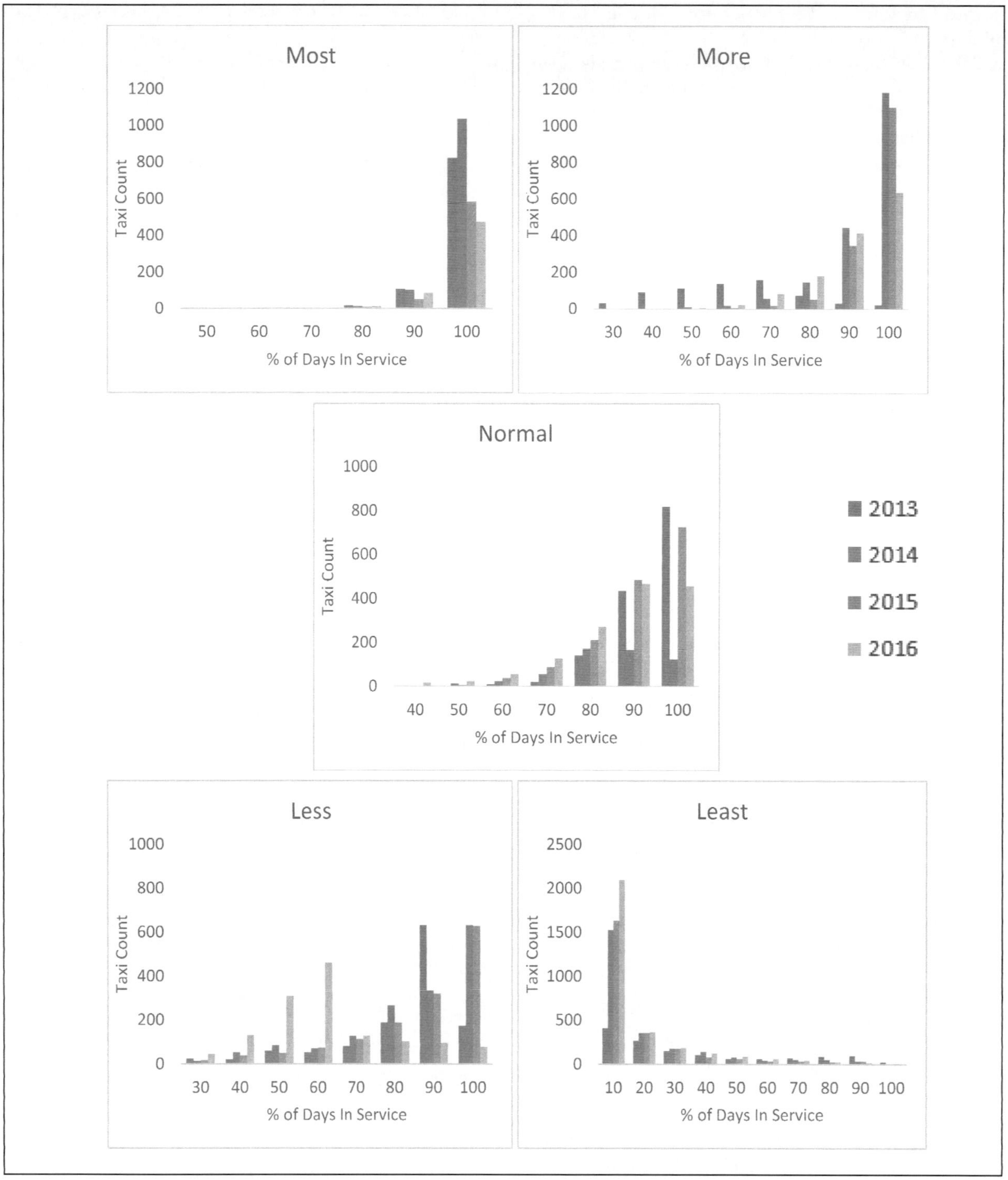

Figure 4. Histograms for the percentage of days in service for each taxi cluster.

transporting travelers and collecting fares. For the taxis in the *most trips* cluster in Figure 7*a*, the most-productive taxis are productive 70%–80% of their in-service time. The least-productive taxis are productive 20%–40% of their in-service time. However, a large majority of the taxis are productive 40%–60% of their in-service time.

Like the taxis in the *most trips* cluster, only a few of the taxis in the *normal trips* cluster are productive 70%–

(a)

Algorithm 1a

for all $i \in I$: //All taxis

  for all $n \in N_i$: //All taxi-trips

    if $b_{i,n} == 1$ **and** $dura_{i,n} < 90min$: //If trip duration exists and is less than 90 minutes

      $p_{i,n} = dura_{i,n}$ // Assign trip duration value from taxi dataset

    else if $b_{i,n} == 0$ **and** $(A_{i,n,p} == 0\ or\ A_{i,n,d} == 0)$: // If trip duration and pickup or drop-off //location do not exist in dataset

      Remove i, n **from temporal efficiency analysis** //Necessary data not available

    else:

      if $pos_{i,n,p} == pos_{i,n,d}$: //If n's pickup and drop-off census tract are the same

        Remove i, n **from temporal efficiency analysis** //Same census tract, duration equals zero

      else:

        $temp1 = GoogleAPIdura(pos_{i,n,p}, pos_{i,n,d})$ //Call Google API

        $p_{i,n} = \min(90min,\ temp1)$ //Assign minimum of API call or 90 minutes

(b)

Algorithm 1b

for all $i \in I$: //All taxis

  for all $n \in N_i$: //All taxi-trips

    if $B_{i,n,d} == 1\ and\ B_{i,n+1,p} == 1$: //If drop-off time of trip n and pickup time of trip $n+1$ exist

      $\Delta t = (t_{i,n+1,p} - t_{i,n,d})$ //Time gap between drop-off time of trip n and pickup time of trip $n+1$

      if $\Delta t < 90min$: //If time gap less than 90 minutes

        $u_{i,n} = \Delta t$ //Assign time gap as unproductive+ time

      else if $\Delta t > 120min$: //If time gap greater than 120 minutes

        Remove i, n **from temporal efficiency analysis** //Driver is on a break or went home

      else:

        if $pos_{i,n,d} == pos_{i,n+1,p}$: //If n's drop-off census tract the same as $n+1$'s pickup tract

          Remove i, n **from temporal efficiency analysis** //Same census tract, duration is zero

        else:

          $temp2 = GoogleAPIdura(pos_{i,n,d}, pos_{i,n+1,p})$ //Call Google API

          $u_{i,n} = \min(90min,\ temp2)$ //Assign minimum of API call or 90 minutes

    else if $A_{i,n,d} == 1\ and\ A_{i,n+1,p} == 1$: // If drop-off location of n and pickup location of $n+1$ exist

      if $pos_{i,n,d} == pos_{i,n+1,p}$: //If n's drop-off census tract the same as $n+1$'s pickup census tract

        Remove i, n **from temporal efficiency analysis** //Same census tract, duration equals zero

      else:

        $temp3 = GoogleAPIdura(pos_{i,n,d}, pos_{i,n+1,p})$ //Call Google API

        $u_{i,n} = \min(90min,\ temp3)$ //Assign minimum of API call or 90 minutes

    else:

      Remove i, n **from temporal efficiency analysis** //Necessary data not available

Figure 5. Algorithm to obtain (*a*) the productive time of taxi-trip i, n, and (*b*) the unproductive time between taxi-trip i, n and taxi-trip $i, n+1$.

80% of their in-service time. A large majority of the *normal trips* cluster taxis were productive 40%–60% of their in-service time. Figure 7*b* also shows that in 2013 and 2015 many *normal trip* taxis were productive only 10% of their in-service time.

The results in Figure 7, *a* and *b* indicate that most taxis in the Chicago fleet are highly inefficient. Most drivers spend nearly half of their in-service time not generating revenue. This suggests that another mobility service provider (e.g., Uber or Lyft) could dominate the passenger transportation market by operating a more efficient vehicle fleet and passing the cost savings on to customers. Therefore, taxi services should aim to increase their operational efficiency to compete with emerging mobility services.

Spatial Efficiency Results. As an illustrative example, this section presents spatial efficiency measures for two taxis.

(a)

Algorithm 2a

for all $i \in I$**:** //All taxis
  for all $n \in N_i$**:** //All taxi-trips
    if $a_{i,n} == 1$ **and** $dist_{i,n} < 40\ miles$ **:** //If trip distance exists and is less than 40 miles
      $l_{i,n} = dist_{i,n}$ // Assign trip distance value from dataset
    else if $a_{i,n} == 0$ **and** $(A_{i,n,p} == 0\ or\ A_{i,n,d} == 0)$: // If trip distance and pickup or drop-off //location do not exist in dataset
      Remove i, n **from spatial efficiency analysis** //Necessary data not available
    else:
      if $pos_{i,n,p} == pos_{i,n,d}$: //If n's pickup and drop-off census tract are the same
        Remove i, n **from spatial efficiency analysis** //Same census tract, distance equals zero
      else:
        $temp4 = GoogleAPIdist(pos_{i,n,p}, pos_{i,n,d})$ //Call Google API
        $l_{i,n} = \min(40\ miles,\ temp4)$ //Assign minimum of API call or 40 miles

(b)

Algorithm 2b

for all $i \in I$**:** //All taxis
  for all $n \in N_i$**:** //All taxi-trips
    $\Delta t = (t_{i,n+1,p} - t_{i,n,d})$ //Time gap between drop-off time of trip n and pickup time of trip $n+1$
    if $A_{i,n,d} == 0\ and\ A_{i,n+1,p} == 0$ //If n's drop-off location and $n+1$'s pickup location do not exist
      Remove i, n **from spatial efficiency analysis** //Necessary data not available
    else if $\Delta t > 120min$**:** //If time gap greater than 120 minutes
      Remove i, n **from spatial efficiency analysis** //Driver is on a break or went home
    else:
      if $pos_{i,n,d} == pos_{i,n+1,p}$: //If n's drop-off census tract the same as $n+1$'s pickup census tract
        Remove i, n **from spatial efficiency analysis** //Same census tract, distance equals zero
      else:
        $temp5 = GoogleAPIdist(pos_{i,n,d}, pos_{i,n+1,p})$ //Call Google API
        $e_{i,n} = \min(40\ miles,\ temp5)$ //Assign minimum of API call or 40 miles

Figure 6. Algorithm to obtain (*a*) the loaded distance of taxi-trip i, n, and (*b*) the empty distance between taxi-trip i, n and taxi-trip $i, n+1$.

Figure 8*a* displays a histogram of the spatial efficiency of every trip made by Taxi #44 between 2013 and 2016. Taxi #44 is a member of the *most trips* taxi cluster. The figure shows a clear gaussian distribution centered around 50% for each year. This indicates that around 50% of Taxi #44's miles are unproductive (i.e., empty).

The histogram in Figure 8*b* is quite similar to the histogram in Figure 8*a*. Taxi #981, a member of the *less trips* taxi cluster, had a trip distribution slightly more uniform than Taxi #44; nevertheless, Taxi #981, on average, also seems to drive as many unproductive miles as productive miles. According to Figure 8*b*, many of Taxi #981's trips have a spatial efficiency value between 10% and 30%; this is quite inefficient.

Like the temporal efficiency measures for the Chicago taxi fleet, the spatial efficiency measures for these two representative taxis indicate that there is significant room for improvement in terms of the operational efficiency of individual taxis. It is important to note that the histograms in Figure 8, *a* and *b* exclude trips where $pos_{i,n,d} == pos_{i,n+1,p}$. This biases the overall results, because these trips would be highly efficient. However, the results are biased in the opposite direction because the empty distance ($e_{i,n}$) for each trip does not factor in any of the miles spent roaming for travelers.

Despite issues with the data, Figure 8, *a* and *b* show that many taxi trips are highly inefficient. As mentioned in the Introduction, this inefficiency negatively impacts the profit of taxi drivers and fleet operators, customer service quality, roadway congestion, and vehicle emissions.

Conclusion

The city of Chicago released taxi-trip data in December 2016. The current study utilizes the taxi data to spatially

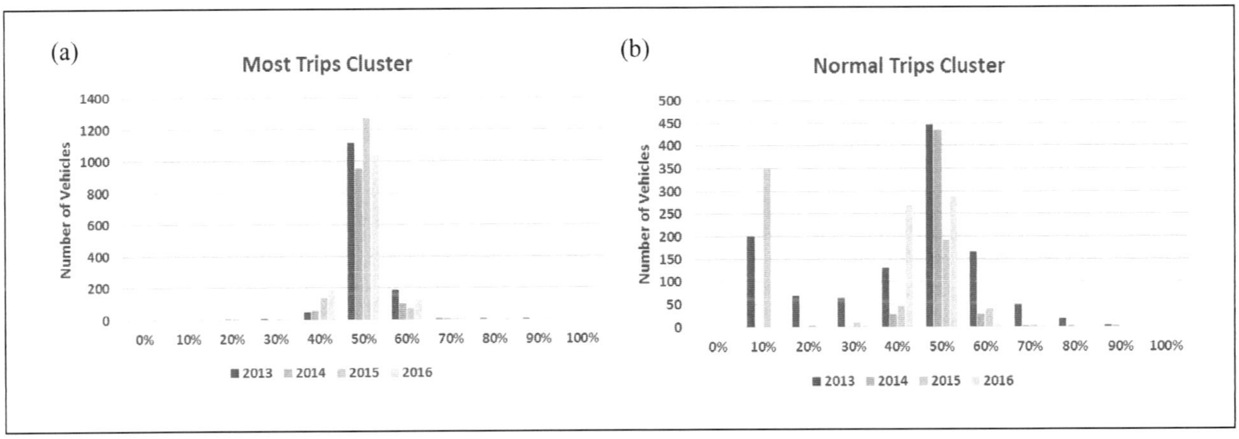

Figure 7. Histogram of the average temporal efficiency of taxis in (*a*) the *most trips* cluster by year and (*b*) the *normal trips* cluster by year.

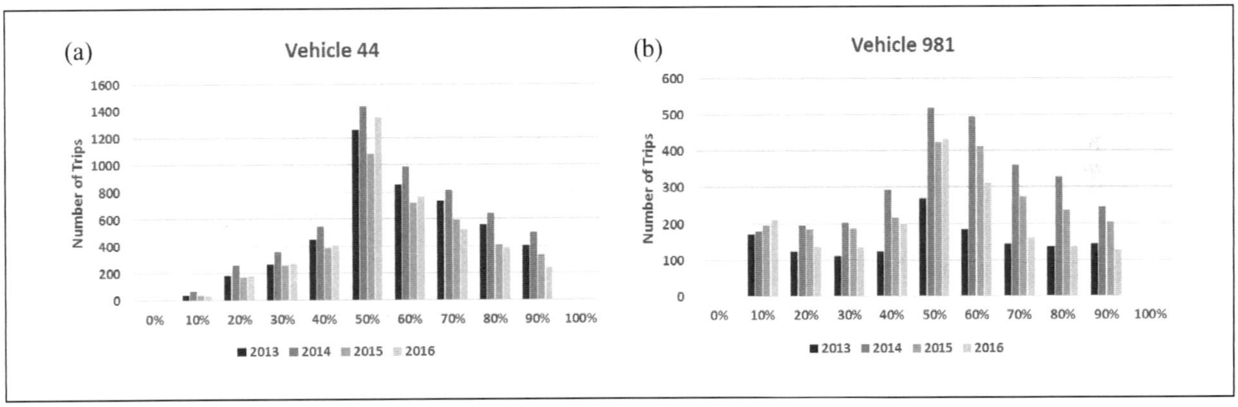

Figure 8. Histogram of the spatial efficiency of (*a*) taxi #44's trips from 2013 to 2016 and (*b*) taxi #981's trips from 2013 to 2016.

characterize the Chicago taxi market over the past four years, and to quantify the operational efficiency of taxis in the fleet.

To explore the taxi market in Chicago, we created a directed-graph connecting all unique taxi-trip origin-destination pairs. Using complex network metrics, we characterize the taxi-trip network. The characteristics provide insight into the spatial distribution of taxi demand in Chicago, as well as changes in the spatial distribution of taxi demand in Chicago over the past four years. The most interesting finding is that much of the decrease in overall taxi trips between 2014 and 2016 came from origin-destination pairs with low demand in 2013 and 2014. We posit that this is the result of emerging ride-sourcing services providing significantly better service than taxis in remote areas.

In addition, the paper presents two metrics to characterize and quantify the operational efficiency of individual taxis. The temporal efficiency metric determines the percentage of in-service time taxis spend transporting travelers. The spatial efficiency metric quantifies the percentage of a taxi's miles that are loaded. These metrics provide a means to measure the efficiency of individual taxis and taxi trips. Results indicate that most Chicago taxis and taxi trips generate as many empty miles as productive miles; that is, only 40–60% of taxi fleet miles are productive. Similarly, the temporal efficiency results indicate that taxis are unproductive 40–60% of their in-service time.

The efficiency results indicate that there is significant room for improvement in the efficiency of the Chicago taxi fleet. If mobility service providers can operate their fleets more efficiently than Chicago taxis, they can pass these cost savings onto travelers and provide lower cost transportation to customers. From a societal perspective, the spatial inefficiency of taxis is likely increasing traffic congestion and generating extra vehicle emissions.

Future research directions include further development of algorithms to filter taxi trips, estimating missing data values, and better handling of the spatial and temporal trip aggregation, to better evaluate the spatial and temporal efficiency of taxis. In addition, we plan to further explore the interesting finding that most of the taxi demand lost between 2013 and 2016 came from origin-destination pairs with low demand in 2013. Enriching the taxi data with census tract land use and socio-demographic data and developing taxi-trip count models for census tract origins, destinations, and origin-destination pairs is another interesting area of research. Lastly, authors of this paper are currently developing dispatching strategies to efficiently operate shared-use AV mobility services, with the goal of reducing temporal, and particularly spatial inefficiency.

Author Contributions

All authors contributed to all aspects of the study from conception and design, to data collection, analysis and interpretation of results, and manuscript preparation. All authors reviewed the results and approved the final version of the manuscript.

References

1. Fagnant, D. J., and K. M. Kockelman. The Travel and Environmental Implications of Shared Autonomous Vehicles, Using Agent-Based Model Scenarios. *Transportation Research Part C: Emerging Technologies*, Vol. 40, 2014, pp. 1–13.
2. Hyland, M. F., and H. S. Mahmassani. Taxonomy of Shared Autonomous Vehicle Fleet Management Problems to Inform Future Transportation Mobility. *Transportation Research Record: Journal of the Transportation Research Board*, 2017. 2653: 26–34.
3. Nie, Y. M. How Can the Taxi Industry Survive the Tide of Ridesourcing? Evidence from Shenzhen, China. *Transportation Research Part C: Emerging Technologies*, Vol. 79, 2017, pp. 242–256.
4. Chicago Data Portal. *Taxi Trips*. https://data.cityofchicago.org/Transportation/Taxi-Trips/wrvz-psew. Accessed July 29, 2017.
5. Schneider, T. W. *Analyzing 1.1 Billion NYC Taxi and Uber Trips, with a Vengeance*. http://toddwschneider.com/posts/analyzing-1-1-billion-nyc-taxi-and-uber-trips-with-a-vengeance/. Accessed July 27, 2017.
6. Qian, X., and S. V. Ukkusuri. Spatial Variation of the Urban Taxi Ridership using GPS Data. *Applied Geography*, Vol. 59, 2015, pp. 31–42.
7. Haggag, K., B. McManus, and G. Paci. Learning by Driving: Productivity Improvements by New York City Taxi Drivers. *American Economic Journal: Applied Economics*, Vol. 9, No. 1, 2017, pp. 70–95.
8. King, D. A., and J. F. Saldarriaga. Spatial Regulation of Taxicab Services: Measuring Empty Travel in New York City. *Journal of Transport and Land Use*, Vol. 11, No. 1, 2018. https://doi.org/10.5198/jtlu.2018.1063.
9. Yang, C., and E. J. Gonzales. Modeling Taxi Demand and Supply in New York City using Large-Scale Taxi GPS Data. In *Seeing Cities through Big Data: Research, Methods and Applications in Urban Informatics* (Thakuriah, P., N. Tilahun, and M. Zellner, eds.), Springer, pp. 405–425.
10. Zhan, X., X. Qian, and S. V. Ukkusuri. A Graph-Based Approach to Measuring the Efficiency of an Urban Taxi Service System. *IEEE Transactions on Intelligent Transportation Systems*, Vol. 17, No. 9, 2016, pp. 2479–2489.
11. Zhan, X., X. Qian, and S. V. Ukkusuri. Measuring the Efficiency of Urban Taxi Service System. *Proc., 3rd International Workshop on Urban Computing (UrbComp '14)*, pp. 1–9. New York, NY, 2014.
12. Liu, X., L. Gong, Y. Gong, and Y. Liu. Revealing Travel Patterns and City Structure with Taxi Trip Data. *Journal of Transport Geography*, Vol. 43, 2015, pp. 78–90.
13. Saberi, M., H. S. Mahmassani, D. Brockmann, and A. Hosseini. A Complex Network Perspective for Characterizing Urban Travel Demand Patterns: Graph Theoretical Analysis of Large-Scale Origin–Destination Demand Networks. *Transportation*, Vol. 44, No. 6, 2016, pp. 1–20.
14. Google. Google Maps API. https://developers.google.com/maps/documentation/distance-matrix/start. Accessed October 15, 2017.

The Standing Committee on Transportation Network Modeling (ADB30) peer-reviewed this paper (18-06576).

Article

TRR
JOURNAL OF THE TRANSPORTATION RESEARCH BOARD

Transportation Research Record
2018, Vol. 2672(48) 139–147

DOI: 10.1177/0361198118774753
journals.sagepub.com/home/trr
SAGE

Link Transmission Model-Based Linear Programming Formulation for Network Design

Shantanu Chakraborty[1], David Rey[1], Emily Moylan[1], and S. Travis Waller[1]

Abstract

This article presents a linear programming formulation to solve the network design problem using the link transmission model (LTM) as the underlying traffic flow model. The original LTM was adapted by incorporating link-sending and receiving flows using linear inequalities. Furthermore, route choice was relaxed, and transfer flow variables were used to model vehicles' routing decisions within the network. The objective function of the linear program aimed to minimize the total difference between the cumulative vehicle numbers (CVN) at the upstream and at the downstream boundaries of each link subject to flow-conservation and budget constraints. CVN were represented using transfer flows from connected links. The resulting formulation is a linear program that represents a dynamic system optimum traffic flow pattern, embedding the LTM's network loading procedure. In contrast to the single-destination system optimum dynamic traffic assignment, based on the cell transmission model, the proposed formulation requires considerably fewer decision variables, thus potentially providing a more scalable approach. The proposed formulation was implemented on an example network to illustrate the behavior of the model, and was compared with the cell-based formulation. We show that the model describes free-flow and congested traffic flow states accurately in terms of shock-wave propagation, queuing of vehicles, and optimal total system travel time.

Dynamic traffic assignment (DTA) models aim to develop a time-dependent traffic flow pattern from a time-varying travel demand on a transport network. Two types of traffic flow patterns are commonly examined: selfish routing (user equilibrium, UE) and social welfare (system optimum, SO). The former aims to represent an equilibrium in users' utilities, whereas the latter is beneficial for network-wide system performance—often measured by the total system travel time (TSTT). When the agents in a traffic network choose their routes to minimize their own expected travel time, the system progresses toward a state of UE, in which no further reduction in travel time is possible between each origin–destination pair (O-D) (*1*). The link flow rates corresponding to this state can be obtained by solving a nonlinear convex program minimizing the Beckmann function (*2*). The constraints of this optimization problem include the flow-conservation and nonnegativity constraints. As convex programming has appealing mathematical properties, this method has been widely studied for traffic assignment in the literature. However, UE flow patterns may lead to an underutilization of the network infrastructure compared with SO flow patterns.

System optimum dynamic traffic assignment (SODTA) allocates traffic on a network in such a way that the TSTT is minimized. In this case, not all agents in the traffic network experience equal and minimal travel time, which might incentivize drivers to switch routes in pursuit of a path with a shorter travel time. Hence, the SODTA flow pattern may not be sustainable by itself unless the commuters are cooperative in maximum social welfare. However, SODTA provides a measure of the optimal usage of network infrastructure and has attracted substantial attention over the years due to its application in various traffic management strategies such as marginal cost pricing, traffic control measures, emergency evacuations, and so forth. It also allows a transport modeler to internalize the externalities of congestion, accidents, and environmental impacts for a better assessment of network-level traffic conditions.

[1]Research Center for Integrated Transport Innovation (rCITI), School of Civil and Environmental Engineering, University of New South Wales, Sydney, NSW, Australia

Corresponding Author:
Address correspondence to Shantanu Chakraborty: shantanu.chakraborty@unsw.edu.au

Despite extensive exploration, SODTA remains an active research area as modeling and optimizing dynamic traffic networks is highly complex (*3*). Furthermore, the time-dependent traffic flow pattern solution of SODTA involves complex spatial and temporal interactions of traffic with nonconvex constraints. To account for this nonconvexity issue, various theoretical studies introduced a number of simplifications, such as networks with fixed route choice (*4*, *5*), networks with a single bottleneck (*6*, *7*), networks with only parallel routes (*8*) or the adoption of a path-based approach to eliminate the nonconvex constraints (*9–11*). Another simplified version of the SODTA problem involves analyzing a network with multiple origins and a single destination (*12–15*). This restrictive formulation attempts to avoid the basic nonconvexity that occurs due to the first-in-first-out requirement of traffic agents in a dynamic flow problem (*16*). Although single-destination SODTA deals with a single commodity and may not overcome the drawbacks of existing analytical models, it could provide valuable insights to eventually solve the multi-destination network problem. Even without this further development to multiple destinations, single-destination SODTA could be applicable to scenarios such as emergency evacuation toward a common assembly point, or modeling the morning peak traffic commute toward a high-demand centroid (e.g., central business districts) (*17*).

The computational complexity of single-destination SODTA for practical application purposes strongly depends on the underlying traffic flow model and its specific formulation. Various traffic flow models have been adopted in the formulation of the SODTA optimization problem such as link exit flow function (*12–14*), deterministic queuing models with no link interactions (*18*, *19*), and cell transmission models (CTM) (*15*). Ziliaskopoulos (*15*) proposed a linear programming formulation of the DTA problem based on CTM, introduced by Daganzo (*20*, *21*). This model uses cells to represent link flow, which captures traffic flow variability inside the link in contrast to models relying on link exit functions. The motivation behind this linear programming (LP) approach was to make use of the vast existing literature on LP for a better understanding and resolution of SODTA. Furthermore, the dual variables associated with the flow-conservation constraint of this LP represent the marginal contribution of an additional unit of demand at a given cell and in a given time interval, thereby providing means for developing marginal cost pricing strategies. This CTM-based LP approach has subsequently been adopted to solve single-level continuous network design problems (NDP) (*22*, *23*).

The objective function of an NDP minimizes the TSTT subject to flow-conservation as well as budget constraints. However, the CTM-based NDP formulation involves a large number of variables and constraints due to the presence of a large number of cells, even for small-size test networks. As single-destination NDP formulations represent traffic flow propagations explicitly by link-based variables, the number of variables and constraints of the optimization problem is directly proportional to the scale of the network and length of the study period (*17*). Hence, CTM-based NDP approaches may become computationally intractable for real-size traffic networks.

The link transmission model (LTM), proposed by Yperman (*24*), improves on CTM space discretization to a large extent by analyzing traffic flow at link boundaries only. This reduction in space discretization in turn reduces the computational complexity by n times for the same level of accuracy, where n is the average number of cells in a homogeneous network link (*24*). This reduction in computational complexity is expected to enable the proposed LP to scale to real-size traffic networks. In the literature, an LTM-based SODTA model has been developed with environmental objectives (*25*) and with dial-a-ride service constraints for shared autonomous vehicle routing problems (*26*). Considering the importance of developing a practical solution approach for single-destination NDP on realistic traffic networks, we proposed an LTM-based LP formulation for NDP.

The rest of this paper is structured as follows. The next section provides an overview of LTM and introduces the LP formulation for single-destination NDP. The following section presents the implementation of the proposed LP on a test network. The paper concludes by summarizing the findings of this study and discussing future research directions.

Link Transmission Model and LP Formulation for NDP

DTA models have two fundamental components: the route choice model and the dynamic network loading (DNL) model. DNL models represent the propagation of traffic flow over time in road networks based on kinematic wave theory, introduced by Lighthill and Whitham, and Richards (*27*, *28*). DNL models likewise consist of two components: a link model and a node model. Link models determine sending flows from upstream links and receiving capacities of downstream links at each time step based on traffic flow interactions within the links. Node models determine the actual transfer flow from the upstream link to the downstream link, which is the minimum of sending and receiving flows and the capacities of the corresponding links. Several solution approaches have been reported in the literature for DNL models such as link exit functions (*12*, *13*), CTM (*20*, *21*), and LTM (*24*). In this study, LTM was selected for network loading due to its simpler traffic flow modeling approach and reduced computational complexity compared with other solution methods.

LTM is a solution method for DNL models based on the simplified triangular fundamental diagram proposed by Newell (*29*). The model keeps track of the traffic flow rates at each time step in terms of cumulative vehicle numbers (CVN) at the link boundaries only, eliminating the minute space discretization used in CTM. Given a time-varying traffic demand and turning fractions at each node, LTM determines the dynamic link travel times on a traffic network based on the upstream and downstream CVN updated at each time step. The sending and receiving flows for each link pair are calculated based on the time-varying demand and the network and traffic flow characteristics. Consequently, traffic flow is propagated based on transfer flows at each node of the network.

LP Formulation for NDP

In this study, we propose a novel LP formulation for the single-destination NDP problem, using LTM as the underlying traffic flow model—we hereby refer to this formulation as LTM-NDP. The notations used in the LTM-NDP formulation are as follows.

Sets

A set of all physical links and centroid connectors:
A_R set of source centroid connectors,
A_S set of sink centroid connectors,
A_D set of physical links,
A_C set of physical links and A_R,
$\Gamma^-(i)$ set of predecessor links of link i,
$\Gamma^+(i)$ set of successor links of link i,
T set of discretized time steps.

Parameters

q_i capacity of link i
L_i length of link i,
k_i jam density of link i,
$v_{f,i}$ free-flow speed in link i,
w_i wave speed of in link i,
δ discretized time step,
$d_{i,t}$ demand inflow through source connector i at time t,
B Total available budget:
e_i increase in jam density per one unit of b_i,
f_i increase in capacity per one unit of b_i.

Variables

$x_{i,t}$ number of vehicles in link i at time t,
$y_{ij,t}$ transfer flow from link i to link j at time t,
$z^+_{i,t}$ CVN at upstream end of link i at time t,
$z^-_{i,t}$ CVN at downstream end of link i at time t,
b_i amount of budget allocated to link i.

The proposed LTM-NDP formulation is entirely based on links, and the nodes in the network are implicitly represented by single or multiple pairs of links. The set of links is partitioned into two categories: physical links and centroid connectors. The set of centroid connectors consists of source (A_R) and sink connectors (A_S), representative of source and sink nodes respectively, whereas the set of physical links (A_D) consists of existing network links. The set of predecessor and successor links are denoted by P and S respectively, whereas T denotes the set of discretized time steps.

The LTM works by updating CVN at both upstream and downstream locations of the network links. To develop the LTM-NDP formulation we introduced decision variables to represent CVN at the boundaries of each link. We denoted $z^+_{i,t}$ the upstream CVN and $z^-_{i,t}$ the downstream CVN of link i at time t.

LTM computes the maximum sending and receiving flows for each link at each time step (*24*). Let $t_s = t + \delta - \frac{L_i}{v_{f,i}}$, the maximum sending flow from link i at time t, be defined as

$$S_i(t) = \min\left\{(z^+_{i,t_s} - z^-_{i,t}), q_i\delta\right\} \tag{1}$$

Similarly, let $t_r = t + \delta - \frac{L_i}{w_i}$ the maximum receiving flow on link i at time t be defined as

$$R_i(t) = \min\left\{(k_j L_j - (z^+_{i,t} - z^-_{j,t_r})), q_i\delta\right\} \tag{2}$$

The sending and receiving flows are used in LTM to determine the transfer flow between a pair of connected links. In the original LTM formulation, route choice is governed by the turning ratios. In contrast, in SODTA route choice is optimized by the traffic assignment model, thus we relaxed the route choice component of the original LTM by introducing transfer flow variables $y_{ij,t} \geq 0$ that represent the flow traveling from link i to link j at time t. Let $\Gamma^-(i)$ ($\Gamma^+(i)$) denote the set of predecessors (successor) links of link i.

The sending flow, Equation 1, can then be modeled using the linear constraints:

$$\sum_{j\in\Gamma^+(i)} y_{ij,t} \leq (z^+_{i,t_s} - z^-_{i,t}) \qquad \forall i \in A_C, \forall t \in T \quad \text{where,}$$
$$t_s = t + \delta - \frac{L_i}{v_{f,i}} \tag{3}$$

$$\sum_{j\in\Gamma^+(i)} y_{ij,t} \leq q_i\delta \qquad \forall i \in A_C, \forall t \in T \tag{4}$$

To incorporate the scope for potential capacity improvement of f per unit of allocated budget b, Equation 4 is modified as follows:

$$\sum_{j \in \Gamma^+(i)} y_{ij,t} \leq \delta(q_i + f_i b_i) \qquad \forall i \in A_C, \forall t \in T \quad (5)$$

Similarly, the receiving flow, Equation 2, can be modeled using the linear constraints:

$$\sum_{i \in \Gamma^-(j)} y_{ij,t} \leq (k_j L_j - (z^+_{i,t} - z^-_{j,t_r})) \qquad \forall j \in A_D, \forall t \in T \quad (6)$$

$$\sum_{i \in \Gamma^-(j)} y_{ij,t} \leq q_j \delta \qquad \forall j \in A_D, \forall t \in T \quad (7)$$

Equations 6 and 7 are modified as follows to allow a possible improvement in capacity (f) and jam density (e) per unit of allocated budget b.

$$\sum_{i \in \Gamma^-(j)} y_{ij,t} \leq (k_j L_j + e_j b_j L_j - (z^+_{j,t} - z^-_{j,t_r})) \quad (8)$$
$$\forall j \in A_D, \forall t \in T \quad \text{where, } t_r = t + \delta - \frac{L_i}{w_i}$$

$$\sum_{i \in \Gamma^-(j)} y_{ij,t} \leq \delta(q_j + f_j b_j) \qquad \forall j \in A_D, \forall t \in T \quad (9)$$

Accordingly, CVN variables $z^+_{i,t}$ and $z^-_{i,t}$ can be defined using the constraints:

$$z^+_{i,t} = \sum_{t' < t} \sum_{h \in \Gamma^-(i)} y_{hi,t'} \qquad \forall i \in A_C, \forall t \in T \quad (10)$$

$$z^-_{i,t} = \sum_{t' < t} \sum_{j \in \Gamma^+(i)} y_{ij,t'} \qquad \forall i \in A_C, \forall t \in T \quad (11)$$

Note that the left-hand side of constraints 3–7 represents the total flow that could potentially exit link i (sending) or enter link i (receiving) while accounting for the topology of the network. In addition, it should be noted that the min functions in 1 and 2 are relaxed within constraints 3–7. While this linearization technique is commonly used (*15*), it does not impose a strict minimum, thus relaxing the original LTM formulation. This relaxation will be tightened by using an objective function that penalizes the travel time of vehicles in the network, thus incentivizing vehicles to "move along" instead of waiting. Furthermore, the numerical experiments conducted in this study show that a strict minimum is always observed when implementing the proposed LTM-NDP formulation.

The time-dependent travel demand can be incorporated in the formulation using the constraint,

$$z^+_{i,t} = \sum_{t' < t} d_{i,t'} \qquad \forall i \in A_R, \forall t \in T \quad (12)$$

Finally, the budget constraint is presented in Equation 13, which specifies the upper limit of the sum of allocated budget as the total available budget B for the network design.

$$\sum_{i \in A_C} b_i \leq B \quad (13)$$

The traditional objective function in SODTA formulations is to minimize TSTT, which is the sum of the experienced travel times of all the users in the network summed over the entire analysis period. The density of link i at t is $x_{i,t} = z^+_{i,t} - z^-_{i,t}$. At each time interval, the experienced travel time of vehicles in link i is $\delta x_{i,t}$, as per the Courant–Friedrichs–Lewy condition (*30*) of LTM, which requires that each time step should be shorter than the free-flow travel time of the shortest link in the network to prevent vehicles from moving out of the link during the time interval ($\delta \leq \frac{L_i}{v_{f,i}}$). Hence, the objective function of LP can be presented as follows (14).

$$\min \delta \sum_{t \in T} \sum_{i \in A_C} x_{i,t} \quad (14)$$

Note that the objective function (Equation 14) also minimizes the waiting time of the vehicles within the source centroid connector that are queued up due to congestion in the network. Hence, the combination of Constraint 12, the objective function, ensures that the travel demand departing from centroid $i \in A_R$ at time t is accounted for in the computation of TSTT.

The single-destination LTM-NDP formulation is presented in 15–30.

$$\min \delta \sum_{t \in T} \sum_{i \in A_C} x_{i,t} \quad (15)$$

subject to,

$$z^+_{i,t} = \sum_{t' < t} \sum_{h \in \Gamma^-(i)} y_{hi,t'} \qquad \forall i \in A_D, \forall t \in T \quad (16)$$

$$z^-_{i,t} = \sum_{t' < t} \sum_{j \in \Gamma^+(i)} y_{ij,t'} \qquad \forall i \in A_C, \forall t \in T \quad (17)$$

$$x_{i,t} = z^+_{i,t} - z^-_{i,t} \qquad \forall i \in A_C, \forall t \in T \quad (18)$$

$$\sum_{j \in \Gamma^+(i)} y_{ij,t} \leq (z^+_{i,t_s} - z^-_{i,t}) \qquad \forall i \in A_C, \forall t \in T \quad \text{where,}$$
$$t_s = t + \delta - \frac{L_i}{v_{f,i}} \quad (19)$$

$$\sum_{j \in \Gamma^+(i)} y_{ij,t} \leq \delta(q_i + f_i b_i) \qquad \forall i \in A_C, \forall t \in T \quad (20)$$

$$\sum_{i \in \Gamma^-(j)} y_{ij,t} \leq (k_j L_j + e_j b_j L_j - (z^+_{j,t} - z^-_{j,t_r})) \quad (21)$$
$$\forall j \in A_D, \forall t \in T \quad \text{where, } t_r = t + \delta - \frac{L_i}{w_i}$$

$$\sum_{i \in \Gamma^-(j)} y_{ij,t} \le \delta(q_j + f_j b_j) \qquad \forall j \in A_D, \forall t \in T \tag{22}$$

$$z_{i,t}^+ = \sum_{t'<t} d_{i,t'} \qquad \forall i \in A_R, \forall t \in T \tag{23}$$

$$\sum_{i \in A_C} b_i \le B \tag{24}$$

$$y_{ij,0} = 0 \qquad \forall i \in A, \forall j \in A \tag{25}$$

$$y_{ij,t} \ge 0 \qquad \forall i \in A_C, \forall t \in T \tag{26}$$

$$x_{i,t} \ge 0 \qquad \forall i \in A_C, \forall t \in T \tag{27}$$

$$z_{i,t}^+ \ge 0 \qquad \forall i \in A_C, \forall t \in T \tag{28}$$

$$z_{i,t}^- \ge 0 \qquad \forall i \in A_C, \forall t \in T \tag{29}$$

$$b_i \ge 0 \qquad \forall i \in A_C \tag{30}$$

Constraints 16–18 are definitional constraints. Constraints 16 and 17 define CVN at the upstream and downstream ends of a link at each time step as the total transfer flow from the predecessor links and the total transfer flow to the successor links respectively until the previous time step. Constraint 18 defines the number of vehicles in a link as the difference of these CVN at the upstream and downstream ends of the link at each time step.

Constraints 19 and 20 refer to the sending flow constraints of the original LTM model. According to LTM theory, if a free-flow state of traffic is observed at the downstream end of a link at time t, this state must have emerged from the upstream end and traveled at the free-flow speed of the link, $v_{f,i}$ (*24*). Constraint 19 is formulated based on this concept, which describes that the transfer flow from each incoming link summed over all outgoing links during each time interval $[t, t+\delta]$ must be less than or equal to the difference in CVN upstream at $t_s = t + \delta - \frac{L_i}{v_{f,i}}$ and downstream at t. As the value of t_s is likely to fall between two time steps, CVN is obtained using a linear interpolation method between these two corresponding time steps. Constraint 20 presents the upper limit of this sending flow as the capacity of the incoming link with a potential capacity improvement function.

Constraints 21 and 22 refer to the receiving flow capacity of each link accounting for congestion effects. As per LTM theory, if a congested traffic state is observed at the upstream end of a link, this state must have emerged from the downstream end, traveling backward at a speed equal to the congestion wave speed w (*24*). Based on this concept, Constraint 21 describes the amount of empty spaces left in a receiving link at each time step accounting for potential improvement in the jam density. In this constraint, the number of empty spaces is presented as the difference between the total spaces in the link (product of jam density and length of the link) and the occupied spaces (difference in CVN accounting for congestion). The number of occupied spaces is determined by subtracting CVN downstream at $t_r = t + \delta - \frac{L_i}{w_i}$ from CVN upstream at t. CVN at t_r is obtained following a similar interpolation method as that of t_s. According to Constraint 21, the transfer flow to each outgoing link summed over all incoming links must be less than or equal to the remaining empty spaces in the outgoing link. Constraint 22 presents the upper limit of the receiving flow as the capacity of the receiving link with a potential capacity improvement function. Constraint 23 expresses CVN at the upstream of the source centroid connector as the sum of demand expected to enter into the network over time. Equation 24 is related to the budget constraint. Constraint 25 specifies the initial flow conditions. Constraints 26–30 state the nonnegativity conditions.

It should be noted that the proposed LP does not define a separate set of constraints for merge and diverge situations. Constraints 19–22 are formulated in such a way that they apply to both merge and diverge nodes. This combined set of constraints allows the proposed LP to have a simpler form compared with other existing solution methods.

In the following section, the proposed LP is implemented on a test network.

Numerical Experiments

In this section we illustrate the proposed LP formulation by solving an example network with six nodes and eight links. The network characteristics are presented in Table 1. Figure 1 presents the schematic diagram of the network.

The network consists of a single origin R, and a single destination S, which are connected to the physical network by centroid connectors 1 and 10. There are four paths between this O-D pair: R-1-2-4-8-10-S, R-1-2-5-9-10-S, R-1-3-7-9-10-S and R-1-3-6-8-10-S. The length of the centroid connectors 1 and 10 are specified as zero, whereas the capacity, jam density, and free-flow speed

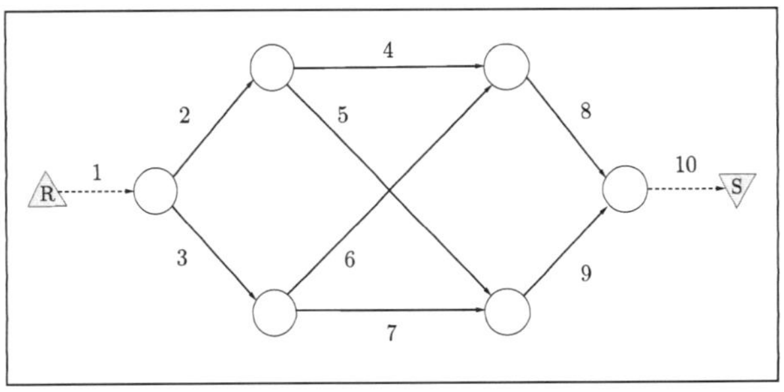

Figure 1. Example network for LTM-NDP.

Table 1. Characteristics of the Example Network

Links	Length (meters)	Capacity (vehicles/hour)	Jam density (vehicles/km)	Free-flow speed (meters/second)
1	0	100,000	100,000	100,000
2	900	3,600	300	30
3	900	3,600	300	30
4	900	1,800	150	30
5	1200	3,600	300	30
6	1200	3,600	300	30
7	1200	1,800	150	30
8	1200	3,600	300	30
9	900	3,600	300	30
10	0	100,000	100,000	100,000

Table 2. Optimal Occupancy $\mathbf{x}_{i,t}$ (in Number of Vehicles)

	Time steps														
Links	0	1	2	3	4	5	6	7	8	9	10	11	12	13	14
Demand (veh)	40	20	20	0	0	0	0	0	0	0	0	0	0	0	0
1	0	40	30.4	20.8	0	0	0	0	0	0	0	0	0	0	0
2	0	0	19.6	39.2	58.8	39.2	19.6	0	0	0	0	0	0	0	0
3	0	0	10	20	21.2	11.2	1.5	0	0	0	0	0	0	0	0
4	0	0	0	0	0	9.6	19.2	28.8	19.2	9.6	0	0	0	0	0
5	0	0	0	0	0	10	20	30	30	20	10	0	0	0	0
6	0	0	0	0	0	0.4	0.4	0.4	0.4	0	0	0	0	0	0
7	0	0	0	0	0	9.6	19.2	20.8	20.8	11.2	1.5	0	0	0	0
8	0	0	0	0	0	0	0	0	9.6	19.6	29.2	29.2	19.6	9.6	0
9	0	0	0	0	0	0	0	0	0	19.6	39.2	50.8	31.2	11.5	0

are set at very high values. The physical links 2, 3, 4, and 9 have a link length of 900 m, and links 5, 6, 7, and 8 have a length of 1,200 m. The capacity and jam density of links 4 and 7 are 1,800 vph (5 veh/time step) and 150 veh/km respectively. The remaining links have a capacity of 3,600 vph (10 veh/time step) and a jam density of 300 veh/km. We assumed the free-flow speed to be 30 m/s throughout the network and a total budget of nine units was allocated for potential improvement of capacity as well as jam density. The increase in jam density (e_i) and capacity (f_i) per one unit of allocated budget (b_i) were link-specific parameters with a value of 0.33 for links 2,3,4, and 9 and 0.25 for links 5,6,7, and 8.

The total period of the analysis was discretized into time steps. In this network, we assumed a time step of 10 s and an assignment period of 140 s. The time-varying demand was loaded into the network at each time step through Connector 1. It should be noted that the demand at time step t specifies the number of vehicles loaded into the network during the time interval $[t, t + \delta]$.

The proposed LP was implemented on a RedHat CentOS 5.9 Linux server with eight 3.60-GHz Intel Xeon central processing units with a 188-GB physical memory using AMPL and CPLEX 12.4.0.1 as the optimization solver. The optimal occupancies in each link at each time step are presented in Table 2. The LP formulation for this case resulted in a program with 595 variables, 899 constraints, and 99 dual simplex iterations, and 0.006 s to converge to a TSTT value of 8919.23 vehicle-seconds. Tables 2 and 3 present the output of the LP as follows.

Table 2 presents the optimal occupancy at each link of the network at each time step in terms of number of vehicles. The first row presents the time-varying demand loaded into the network. It should be noted that the demand at the first time step (40) is more than the total capacity of the two downstream links (20). As a result, the capacity of Link 2 is increased from 10 to 19.6 veh/time step. This is detailed in Table 3, which gives the optimal budget allocation. At the optimum, 2.88 units of budget are allocated to Link 2, and 1.38, 1.85, and 2.88 units are allocated to links 4, 7, and 9 respectively. Vehicles are found to queue up in Link 2, reaching a value of 58.8 at Time Step 4 and then clearing up completely by the end of the 6th time step. This queue spillover phenomenon is observed in the other links in the network as well, except in Link 6, which experiences an insignificant amount of flow throughout the simulation period.

Table 3. Budget Allocation in Each Link and Corresponding Cells

Links	Allocated budget (unit)	Cells	Allocated budget (unit)
1	0.00	1	0.00
2	2.88	2	0.96
		3	0.96
		4	0.96
3	0.00	5	0.00
		6	0.00
		7	0.00
4	1.38	8	0.46
		9	0.46
		10	0.46
5	0.00	11	0.00
		12	0.00
		13	0.00
		14	0.00
6	0.00	15	0.00
		16	0.00
		17	0.00
		18	0.00
7	1.85	19	0.46
		20	0.46
		21	0.46
		22	0.46
8	0.00	23	0.00
		24	0.00
		25	0.00
		26	0.00
9	2.88	27	0.96
		28	0.96
		29	0.96
10	0.00	30	0.00

Comparison with CTM-Based NDP

In this section, we compare the efficiency and the accuracy of the proposed LTM-NDP formulation with a CTM-based NDP formulation. We assumed that the cells inside the same link had uniform traffic characteristics. In this scenario, we postulated that the budget would be allocated uniformly among the cells in the same link in the network. This claim is based on the properties of the transfer flow between two adjacent cells in the network. According to CTM theory, the number of vehicles to be transferred (y_{ij}) from cell i to cell j should be less than or equal to the capacities of both the links Q_i and Q_j respectively at each time step. Hence, if the budget is allocated non-uniformly among the cells representative of a link, then there must exist at least one pair of adjacent cells with different capacities. For this pair of cells, the flow transferred is, at most, the minimum of these capacities, hence the throughput of their corresponding link will remain limited by the smallest cell capacity of this link. This implies that a non-uniform allocation of the budget among the cells representative of a link may result in sub-optimal network performance. We analyzed this claim by implementing CTM-based NDP on the same network as follows.

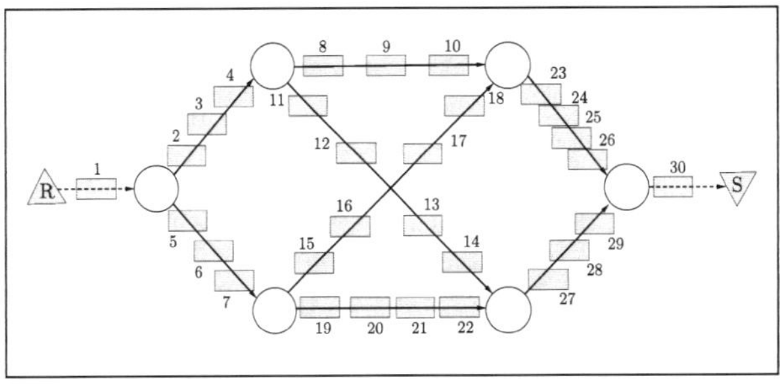

Figure 2. Example network for CTM-NDP.

The cell representation of the example network is presented in Figure 2. The CTM network consists of 30 cells, each with a length of 300 m, which is equal to the distance traveled by a vehicle in one time step. The network characteristics remained unchanged. In order to obtain equivalent e_i and f_i values for the CTM-NDP formulation, these values were multiplied by the number of cells in each link. However, the program consisted of a significantly higher number of variables (897), constraints (2,031), and took 244 dual simplex iterations and 0.012 s to converge to the same solution (TSTT = 8,919.23 vehicle-seconds) as the LTM-NDP formulation. Furthermore, the sum of the optimal cell occupancies in a link were found to be exactly the same as the optimal link occupancies obtained from the LTM-NDP formulation, thus verifying the consistency of both the models. The allocated budget for each cell is presented in Table 3.

It is evident from Table 3 that the total budget allocated to a link based on the LTM-NDP formulation is the sum of the allocated budget in its consisting cells. From these results, it is inferred that the cell representation of a network may not necessarily provide additional valuable degrees of freedom for NDP, as improving only a segment of a link may not increase the throughput of vehicles in the network.

Conclusions

In this paper, an LTM-based LP formulation was proposed and compared with its CTM counterpart on an example network to solve a dynamic NDP under SO traffic conditions. Compared with the original LTM, the proposed LTM-NDP approach optimized the turning ratios at each node of the network at each time step for an optimal system performance. The optimal solution flows from the two formulations were found to be consistent for accurately describing free-flow as well as congested states of traffic in terms of propagation of backward shock-waves, queuing of vehicles, and optimal TSTT values. The objective function of the model also accounted for the waiting time of vehicles queued up at the entrance of a network. The proposed LP relaxed the strict minimum constraints on sending and receiving flows described in the original LTM theory using linear inequalities. The objective function handled this relaxation by penalizing vehicles' travel time in the network. During implementation of the LTM-NDP on the example network, the set of linear inequalities was found to be a suitable alternative to the strict minimum constraints, thus providing a valid LTM flow pattern. In contrast to the single-destination SODTA based on CTM, the proposed formulation required considerably fewer decision variables, thus potentially providing a more scalable approach. Furthermore, the model output of CTM-NDP showed that there is no incentive to allocate non-uniform budget (leading to non-uniform capacity improvement) to the cells of the same link, as the cell transfer flow is limited by the cell with the smallest capacity. This finding further advocates the use of LTM-NDP over CTM-NDP in terms of optimum budget allocation for a network design problem.

The proposed LP for the single-destination NDP could be applied for a range of network problems, including evaluating emergency evacuation strategies to a super-destination, and developing dynamic pricing policies. Future work will explore extensions of the proposed LTM-NDP approach to address networks with multiple destinations.

Author Contributions

The authors confirm contribution to the paper as follows: study conception and design: S.C., D.R., S.T.W.; data collection: S.C.; analysis and interpretation of results: S.C., D.R., S.T.W.; draft manuscript preparation: S.C., E.M., D.R. All authors reviewed the results and approved the final version of the manuscript.

References

1. Wardrop, J. G. Road Paper. Some Theoretical Aspects of Road Traffic Research. *Proceedings of the Institution of Civil Engineers*, Vol. 1, No. 3, 1952, pp. 325–362.
2. Beckmann, M., C. McGuire, and C. B. Winsten. *Studies in the Economics of Transportation*. Yale University Press, New Haven, CT, 1956.
3. Papageorgiou, M. Dynamic Modeling, Assignment, and Route Guidance in Traffic Networks. *Transportation Research Part B: Methodological*, Vol. 24, No. 6, 1990, pp. 471–495.
4. Gazis, D. C. Modeling and Optimal Control of Congested Transportation Systems. *Networks*, Vol. 4, No. 2, 1974, pp. 113–124.
5. D'Ans, G. C., and D. C. Gazis. Optimal Control of Oversaturated Store-and-Forward Transportation Networks. *Transportation Science*, Vol. 10, No. 1, 1976, pp. 1–19.
6. Vickrey, W. Congestion Theory and Transport Investment. *American Economic Review*, Vol. 59, No. 2, 1969, pp. 251–60.
7. Hendrickson, C., and G. Kocur. Schedule Delay and Departure Time Decisions in a Deterministic Model. *Transportation Science*, Vol. 15, No. 1, 1981, pp. 62–77.
8. Arnott, R., A. de Palma, and R. Lindsey. Departure Time and Route Choice for the Morning Commute. *Transportation Research Part B: Methodological*, Vol. 24, No. 3, 1990, pp. 209–228.
9. Ghali, M. O., and M. J. Smith. A Model for the Dynamic System Optimum Traffic Assignment Problem. *Transportation Research Part B: Methodological*, Vol. 29, No. 3, 1995, pp. 155–170.
10. Shen, W., Y. Nie, and H. M. Zhang. On Path Marginal Cost Analysis and Its Relation to Dynamic System-optimal Traffic Assignment. *Proceedings of the 17th International Symposium on Transportation and Traffic Theory*, Elsevier, 2007, pp. 319–352.
11. Qian, Z. S., W. Shen, and H. Zhang. System-optimal Dynamic Traffic Assignment with and Without Queue Spillback: Its Path-based Formulation and Solution via Approximate Path Marginal Cost. *Transportation Research Part B: Methodological*, Vol. 46, No. 7, 2012, pp. 874–893.
12. Merchant, D. K., and G. L. Nemhauser. A Model and an Algorithm for the Dynamic Traffic Assignment Problems. *Transportation Science*, Vol. 12, No. 3, 1978, pp. 183–199.
13. Carey, M. Optimal Time-varying Flows on Congested Networks. *Operations Research*, Vol. 35, No. 1, 1987, pp. 58–69.
14. Wie, B.-W. A Convex Control Model of Dynamic System-optimal Traffic Assignment. *Control Engineering Practice*, Vol. 6, No. 6, 1998, pp. 745–753.
15. Ziliaskopoulos, A. K. A Linear Programming Model for the Single Destination System Optimum Dynamic Traffic Assignment Problem. *Transportation Science*, Vol. 34, No. 1, 2000, pp. 37–49.
16. Carey, M., Nonconvexity of the Dynamic Traffic Assignment Problem. *Transportation Research Part B: Methodological*, Vol. 26, No. 2, 1992, pp. 127–133.
17. Shen, W., and H. Zhang. System Optimal Dynamic Traffic Assignment: Properties and Solution Procedures in the Case of a Many-to-one Network. *Transportation Research Part B: Methodological*, Vol. 65, 2014, pp. 1–17.
18. Yang, H., and Q. Meng. Departure Time, Route Choice and Congestion Toll in a Queuing Network with Elastic Demand. *Transportation Research Part B: Methodological*, Vol. 32, No. 4, 1998, pp. 247–260.
19. Chang, G.-L., H. S. Mahmassani, and M. L. Engquist. System-optimal Trip Scheduling and Routing in Commuting Networks. *Transportation Research Record: Journal of the Transportation Research Board*, 1989. 1251: 54–65.
20. Daganzo, C. F. The Cell Transmission Model: A Dynamic Representation of Highway Traffic Consistent with the Hydrodynamic Theory. *Transportation Research Part B: Methodological*, Vol. 28, No. 4, 1994, pp. 269–287.
21. Daganzo, C. F. The Cell Transmission Model, Part II: Network Traffic. *Transportation Research Part B: Methodological*, Vol. 29, No. 2, 1995, pp. 79–93.
22. Waller, S. T., K. C. Mouskos, D. Kamaryiannis, and A. K. Ziliaskopoulos. A Linear Model for the Continuous

Network Design Problem. *Computer-Aided Civil and Infrastructure Engineering*, Vol. 21, No. 5, 2006, pp. 334–345.
23. Ukkusuri, S. V. *Linear Programs for the User Optimal Dynamic Traffic Assignment Problem*. PhD thesis. University of Illinois, Urbana-Champaign, 2002.
24. Yperman, I., S. Logghe, and B. Immers, The Link Transmission Model: An Efficient Implementation of the Kinematic Wave Theory in Traffic Networks. *Proceedings of 10th EURO Working Group on Transportation Meeting, Poznan, Poland*. Publishing House of Poznan University of Technology, 2005, pp. 122–127.
25. Long, J., J. Chen, W. Szeto, and Q. Shi. Link-based System Optimum Dynamic Traffic Assignment Problems with Environmental Objectives. *Transportation Research Part D: Transport and Environment*, 2016, Vol. 60, pp. 56–75.
26. Levin, M. W. Congestion-aware System Optimal Route Choice for Shared Autonomous Vehicles. *Transportation Research Part C: Emerging Technologies*, Vol. 82, 2017, pp. 229–247.
27. Lighthill, M. J., and G. B. Whitham. On Kinematic Waves. II. A Theory of Traffic Flow on Long Crowded Roads. *Proceedings of the Royal Society of London A: Mathematical, Physical and Engineering Sciences*, Vol. 229, 1955, pp. 317–345.
28. Richards, P. I. Shock Waves on the Highway. *Operations Research*, Vol. 4, No. 1, 1956, pp. 42–51.
29. Newell, G. A Simplified Theory of Kinematic Waves in Highway Traffic, Part II: Queueing at Freeway Bottlenecks. *Transportation Research Part B: Methodological*, Vol. 27, No. 4, 1993, pp. 289–303.
30. Courant, R., K. Friedrichs, and H. Lewy. Über die partiellen Differenzengleichungen der mathematischen Physik. *Mathematische Annalen*, Vol. 100, No. 1, 1928, pp. 32–74.

The Standing Committee on Transportation Network Modeling (ADB30) peer-reviewed this paper (18-06694).

Sage Publications USA
sagepub.com

ISBN 978-1-5108-8254-6

CURRAN ASSOCIATES INC.
proceedings
.com

Highway Capacity and Quality of Service 2018

Transportation Research Record: Journal of the Transportation Research Board

Volume 2672, Issue 15

SAGE Publishing